Annual Student Conference: Competitions & Events 2023

AA009198

Held at the 2023 AIChE Annual Meeting

Orlando, Florida, USA
5-10 November 2023

ISBN: 978-1-7138-9322-6

Printed from e-media with permission by:

Curran Associates, Inc.
57 Morehouse Lane
Red Hook, NY 12571

Some format issues inherent in the e-media version may also appear in this print version.

AIChE
120 Wall Street, FL 23
New York, NY 10005-4020

Phone: (800) 242-4363
Fax: (203) 775-5177

www.aiche.org

Additional copies of this publication are available from:

Curran Associates, Inc.
57 Morehouse Lane
Red Hook, NY 12571 USA
Phone: 845-758-0400
Fax: 845-758-2633
Email: curran@proceedings.com
Web: www.proceedings.com

TABLE OF CONTENTS

UNDERGRADUATE STUDENT POSTER SESSION: CATALYSIS AND REACTION ENGINEERING

UNDERGRADUATE STUDENT POSTER SESSION: COMPUTING AND PROCESS CONTROL

UNDERGRADUATE STUDENT POSTER SESSION: EDUCATION AND GENERAL PAPERS

UNDERGRADUATE STUDENT POSTER SESSION: ENVIRONMENTAL

UNDERGRADUATE STUDENT POSTER SESSION: FOOD, PHARMACEUTICAL, AND BIOTECHNOLOGY

UNDERGRADUATE STUDENT POSTER SESSION: FUELS, PETROCHEMICALS, AND ENERGY

UNDERGRADUATE STUDENT POSTER SESSION: MATERIALS ENGINEERING AND SCIENCES

UNDERGRADUATE STUDENT POSTER SESSION: SEPARATIONS

Preparing a Silica-Supported Catalyst for Methane Pyrolysis with Hydrogen and Carbon Nanofiber Products

Samantha Harshberger, Jessica Hauck, Alan Weimer

Hydrogen is a key component for many essential chemicals, including ammonia. Hydrogen demand is expected to grow rapidly, up to 2.3 Gt annually by 2050[1] , yet industrially hydrogen is primary produced through steam methane reforming (SMR), generating approximately 9 kg CO_2/ kg of H_2.[2] Alternative sustainable and economically viable pathways for hydrogen production are needed. This poster will communicate progress from a collaborative project which aims to generate hydrogen and carbon nanoproduct (CNP) via catalytic methane pyrolysis.

In this research, the catalyst is formed via particle atomic layer deposition (ALD) of a metal nanoparticle (NP) such as nickel or iron onto a silica support. To produce a silica supported Ni catalyst, ALD is performed in a fluidized bed reactor via two self-limiting half reactions which are alternated until the desired amount of Ni is deposited onto the support. The resulting Ni nanoparticles are then sintered to prepare the catalyst for carbon nanofiber growth during methane pyrolysis. The resulting CNP/silica product is then investigated as an additive in concrete by the Hubler group in the department of Civil and Environmental Engineering. Various characterization methods are implemented, to include carbon combustion analysis, inductively coupled plasma optimal emission spectroscopy (ICP-OES), and X-ray powder diffraction (XRD). These results will further aid in the development of a catalyst for methane pyrolysis resulting in hydrogen and carbon nanofiber products.

Influence of TiO_2 Structure on Metal-Support Interactions in Rh/TiO_2 Catalysts Probed By Propylene Hydrogenation and Other Techniques

Lauren Hullender, Hanqin Zhao, Li-Yin Hsiao, Helena Hagelin Weaver

Interactions between rhodium metal and varying TiO_2 nanoparticle supports of similar sizes were analyzed to determine the dependence of resulting electronic interactions and reaction activity on catalytic structure. Nanoparticles of anatase (ANP), rutile (RNP), and brookite (BNP) were included in the study. Following an initial reduction at 200 °C, rhodium supported on ANP had higher activity in the propylene hydrogenation test reaction when compared to rhodium catalysts supported on BNP and RNP. The higher activity is likely attributable to a more electron-rich metallic rhodium species being present on the ANP support compared to other supports as revealed through X-ray Photoelectron Spectroscopy (XPS). A decrease in CO uptake and propylene hydrogenation activity was recorded in all Rh/TiO_2 catalysts following a reduction at 500 °C after which the activity could be effectively recovered by reoxidation at 400 °C and re-reduction at 200 °C in the rhodium catalysts supported on ANP and BNP. This suggests that ANP and BNP both result in the migration of a TiO_x layer over the rhodium nanoparticles, while a rutile TiO_2 support does not, at least under the conditions that are commonly used to induce strong interactions between an active metal and support. Additionally, while high reduction temperatures of BNP and ANP both cause TiO_x migration, the electronic interactions between the active rhodium metal and the different TiO_2 structures vary greatly. This study has revealed the importance of carefully matching the particle sizes of the active metal and supports when examining metal support interactions as they can be critically dependent on the stability and structure of the TiO_2 support.

Elucidation of Ce/Zr Ratio Effects on the Physical Properties and Catalytic Performance of $Ce_xZr_{1-x}O_2$ Catalysts

Yunfan Lu, Michal Luchowski, Wenhui Jiang, Mohammed Sifat, Amol Pophali, Gihan Kwon, Tae Jin Kim

Cerium oxide (CeO_2) is widely used as a catalyst support because of its oxygen storage capacity (OSC) and Ce^{4+} and Ce^{3+} redox cycle, which helps avoid catalyst deactivation, such as coke formation. In addition to its structural attributes, cerium is one of the few rare earth metals that is affordable to purchase in bulk (about $6 USD/kg), making them appealing for many industrial applications such as steam reforming, preferential CO oxidation (PROX), and emission control. However, cerium oxide's limitations include its low thermal stability and low density of defect sites on its surface. To combat this problem many researchers have incorporated different foreign metals such as zirconium, tin, titanium, and lanthanum to modify parts of the cerium oxide structure. In previous studies, the addition of zirconium, $Ce_xZr_{1-x}O_y$, shows a relatively better performance of the catalyst support in improving thermal stability and forming defect sites as compared to the addition of other metals. These supports can then be incorporated with various transition metals surface species for use in a wide range of industrial processes.

During this study, we investigated the effect of Ce:Zr ratio on the physical properties of a series of $Ce_xZr_{1-x}O_2$ catalysts, where x=1, 0.9, 0.6, 0.5, and 0. The UV and visible Raman spectroscopy, X-ray diffraction (XRD) spectroscopy, and thermogravimetric analysis (TGA) were applied to evaluate the defect sites, crystallite size, lattice parameters, and thermal stability of catalysts. With the collaboration with catalyst provider, DAICHI, reproducibility of tested catalysts was confirmed in addition to the surface area measurement data. As a model chemical reaction, CO oxidation was used to test the activity of the different catalysts as well as to describe the relationship between the structure and catalytic performance. Based on the characterization results, it is concluded that $Ce_xZr_{1-x}O_y$ could be as a potential support for transition metals, including nickel, cobalt, copper, and so forth. The availability of commercially available $Ce_xZr_{1-x}O_y$ catalyst supports make them a feasible choice for varying industrial processes.

Understanding Cation and pH Effects for Oxygen Reduction Reaction Activity and Selectivity

Angel Valles, Jay Bender, Joaquin Resasco

The optimization of electrocatalytic water splitting reactions has the potential to enhance the storage of intermittent renewable electricity. Traditionally, the structure and composition of catalyst active sites are engineered to optimize catalyst activity. However, recent work indicates that interactions between supporting electrolyte ions and the catalyst surface can have a significant impact on water splitting rates. Here, we study the role of electrolyte pH and alkali metal cations on oxygen reduction reaction (ORR) activity and selectivity using rotating ring-disk electrode and Koutecký-Levich experiments. Over oxophilic metals (Pt, Ru, and Ir), ORR rates are not strongly influenced by electrolyte pH. On the other hand, ORR rates significantly increase with electrolyte pH over noble metals (Ag and Au). In alkaline electrolytes, ORR rates increase with alkali metal cation size over all metals except Ag. Interestingly, both electrolyte pH and cations do not significantly influence ORR selectivity except over Au. This study demonstrates how electrocatalyst activity can be engineered beyond the catalyst active site.

Design of Promoted NiO Catalysts for Energy-Efficient Ethylene Production

Kajetan Leitner, Shuqiao Wang, Alyssa Hensley

Design of Promoted NiO Catalysts for Energy-Efficient Ethylene Production

Kajetan Leitner[1], Shuqiao Wang[1], and Alyssa J.R. Hensley[1]*

1 Department of Chemical Engineering and Materials Science, Stevens Institute of Technology, Hoboken, NJ, U.S.A.

* Corresponding Author: ahensley@stevens.edu

Ethylene is an integral chemical within modern society. From its uses in plastics to its uses in food ripening, ethylene is present in some form in almost every industry and household. Because of this, the market is expected to reach a value of $247 billion by 2027. Despite this, little innovation has been made in the way of making production more efficient. Commercial production proceeds via industrial steam cracking, a very energy intensive and low efficiency process. However, catalysts such as nickel oxide have shown promise to increase the process efficiency and decrease energy consumption through Ethane Oxidative Dehydrogenation (EODH).

Here, we use density functional theory to analyze NiO catalysts promoted with various transition metals to determine their stability with and without defects in the surface and subsurface. We will use the NiO(110) facet as a representative of under-coordinated, highly active facets on the NiO catalysts. NiO(110) surfaces were promoted with 8 different transition metal promoters (Al, Nb, Mo, Sn, Ti, V, W, Zr). The effect of the promoter of the catalyst surface structure was determined by calculating the Gibbs free energies of formation as a function of temperature and gas phase composition. After determining the dominant surface structures for all systems, we analyzed the adsorption energies of several reaction critical EODH intermediates on the catalyst surface. Overall, we found that the addition of a metal promoter will increase the stability of the surface, and that a metal defect will be less stable than an oxygen defect. Furthermore, we have identified three critical surface sites that bind EODH reaction intermediates. Taken together, this work will be used in the future as a foundation for promoter specific testing and can be used to inform and accelerate experimental catalyst screening.

Catalytic Conversion of Biomass-Based Platform Chemical: Triacetic Acid Lactone to Potassium Sorbate

Kyuhyeok Choi, Min Soo Kim, George Huber

Platform chemicals, also known as bulk or commodity chemicals, are chemicals that serve as building blocks or intermediates for the production of a wide range of other chemicals, materials, and products. The production of chemicals from renewable feedstocks is becoming increasingly attractive, driven especially by concerns over climate change and the rapid depletion of non-renewable fossil fuel sources. 4-hydroxy-6-methyl-2-pyrone, commonly denoted as triacetic acid lactone (TAL), has been highlighted as a platform chemical that enables the diverse conversion of renewable carbohydrates into families of commodity chemicals. Notably, it can be obtained from natural sources or synthesized from acetic acid. Potassium sorbate (KS) is a versatile chemical additive, primarily used as a food preservative.

In this research, a conversion of TAL to KS through three major steps was presented. The conversion process consists of: hydrogenation from TAL to 4-hydroxy-6-methyltetrahydro-2-pyrone (HMP), dehydration from HMP to 6-methyl-5,6-dihydro-2-pyrone (denoted as parasorbic acid or PSA), and ring-opening and hydrolysis from PSA to KS. During the hydrogenation step, it was found that a Ni/SiO_2 exhibited over 99.9% conversion of TAL and 99.3% selectivity to HMP. In the dehydration step, a reaction kinetics model was developed to maximize the PSA yield over Amberlyst 70 catalyst. Through the obtained kinetic data, the maximum yield to PSA was estimated to be 86.5%, and a subsequent experiment confirmed a yield of 84.2% of PSA with respect to HMP. In the ring-opening and hydrolysis step, KOH was used as a co-reactant and achieved >99.9% yield of KS from PSA in an ethanol (EtOH) solvent system.

Furthermore, the difference in THF solubility between the TAL-derived KS and its impurities resulted in a high level of purity (95.5%), which was confirmed by ^1H nuclear magnetic resonance (NMR) spectra. The overall yield of TAL-derived KS with respect to TAL was calculated to be 77.3%. A different geometrical configuration of TAL-derived KS compared to commercial KS was observed through NMR, but it was confirmed to have similarity in their antimicrobial properties against bacteria and fungi.

In conclusion, this study showed a new approach for producing KS from TAL, without the need for an intermediate step to yield sorbic acid. Through this research, the ability of TAL to serve as a potential platform chemical, transforming into multiple chemical derivatives with various downstream applications, was also confirmed.

Exploration of Indium Alumina As a Heterogenous Lewis Acid Catalyst for Propylene Oligomerization and Hydrogenation

Emmanuel Ortiz, Matthew Christian Gerona, Ted Kim, Jeffrey T. Miller

Within the U.S., there is a push to transition the production of transportation fuels from crude oil to natural gas liquids. Natural gas liquids (NGLs) serve as a bridge fuel to produce diesel fuel since NGLs produce less CO_2 emissions compared to petroleum feedstocks. Shale deposits yield natural gas liquids that provide a long-lasting energy source and feedstocks to produce high-value commodity chemicals. Single-site indium(III) alumina-supported catalysts will catalyze light hydrocarbon reactions to produce heavy hydrocarbon fuels such as diesel and gasoline. Literature suggests that indium is not known to perform these reactions. We synthesized indium-alumina catalysts using incipient wetness impregnation and tested various catalyst metal loadings. We also tested varying reactor conditions including gas flow rates and temperatures. We analyzed gas chromatography data to compare the conversion and product selectivity of indium-alumina and pure alumina catalysts. X-ray absorption spectroscopy suggests that indium-alumina contains indium single-sites. This study compares the catalytic properties of indium-alumina and alumina catalysts to determine the viability of using indium catalysts for light hydrocarbon reactions. Indium-alumina has proven to be a more effective catalyst for propylene oligomerization and hydrogenation than pure alumina. Establishing indium-alumina as a catalyst capable of performing hydrocarbon chemistry will deepen our understanding of the mechanisms that govern these reactions. Thus, allowing scientists to produce more efficient catalysts capable of manufacturing transportation fuels and commodity chemicals.

Furfural Oxidation Paired with Hydrogen Evolution on Nife/Ni Foam and Mxene-Supported Catalysts

Michael Galvin

The use of renewable energy sources for water electrolysis to generate environmentally friendly hydrogen faces challenges due to energy-intensive oxygen evolution reaction (OER) and potential risks of H_2 and O_2 mixing. To address these issues, a novel technique was proposed, replacing OER with electrocatalytic oxidative dehydrogenation (EOD) of aldehydes, producing carboxylic acids and hydrogen gas. Furfural oxidation was paired with hydrogen evolution to co-produce 2-furoic acid and hydrogen gas, achieving 100% efficiency with a 90% conversion rate in a H-type cell.

A two-dimensional early transition metal carbide (MXene) supported metal catalyst $Co/Mo_2TiC_2T_x$-700 was used as the cathode for hydrogen generation. MXenes offer advantages in catalysis due to their conductivity, stability, and high surface area. $Mo_2TiC_2T_x$ as the substrate facilitated cobalt (Co) distribution, preserving active sites and optimizing metal supported interaction (MSI), influencing the electronic structure of Co for enhanced hydrogen evolution performance. The optimized $Co/Mo_2TiC_2T_x$-700 exhibited low overpotentials of 243.99 mV @ 100 mA cm^{-2} and 320.56 mV @ 400 mA cm^{-2} in H-type cell tests.

The NiFe/Ni foam prepared through an electrodeposition method served as an anode, which was paired with the $Co/Mo_2TiC_2T_x$-700 cathode (both to have an area of 2cm x 2cm) into a two-counterpart flow cell for furfural oxidation and hydrogen evolution. $Co/Mo_2TiC_2T_x$-700 outperformed commercial Pt/C electrodes with lower cell voltage (2.204V vs 2.249V) up to 100 mA cm^{-2}, while maintaining less than 90% H_2 faraday efficiency and having the same catalyst loading of 0.5 mg cm^{-2}. This substitution improved the efficiency of furfural oxidation, co-producing hydrogen gas and 2-furoic acid. The research successfully developed efficient non-noble metal catalysts, offering an economic and safe solution for hydrogen gas production along with production of 2-furoic acid from a top bio renewable feedstock.

Radical Involvement and Active Sites in the Isomerization of 1-Butene

Risha Goel, Karoline Hebisch, Pawel A. Chmielniak, Carsten Sievers

Understanding the identity and location of active sites during catalyzed reactions is critical to enhance catalyst performance and mitigate deactivation. In the isomerization of 1-butene to iso-butene over ferrierite (FER), general consensus in literature exists that the reaction proceeds over strong Brønsted acid sites (BAS) during the first 24 hours on-stream. The formation of carbonaceous deposits inside the zeolite pores during this start-up phase renders the catalyst pores inaccessible beyond 24 h on-stream. Yet, FER maintains high catalytic performance over another 200 h. The nature of active sites and mechanism by which the isomerization continues remains an ongoing debate.

Carbonaceous deposits themselves have been suggested to be catalytically active during 1-butene isomerization. In a proposed pseudo-monomolecular mechanism, a benzylic carbocation located on a deposit reacts with a butene molecule to yield iso-butene. However, the presence of benzylic carbocations has not been experimentally confirmed (van Donk, S.; *et al.*, *J. Catal* **2002**, *212*, 86). Others have demonstrated radical species derived from carbonaceous deposits to be active during ethanol to hydrocarbon conversion over HZSM-5 (Pinard, L.; *et al.*, *Catal. Today* **2013**, *218*, 57).

Nitrogen physisorption of spent FER after butene isomerization confirmed that the internal pores of the zeolite are inaccessible after carbonaceous deposits have formed. FER possesses a unique 2-dimensional channel structure of perpendicularly intersecting 8-membered and 10-membered ring channels that offers an optimal pore size to stabilize the active coke species. Therefore, the active sites are likely located at the FER pore mouths. Pyridine adsorption followed by infrared spectroscopy was used to locate and quantify the available BAS. The loss of Brønsted acidity during the initial hours of the reaction indicates that BAS inside the zeolite pores cannot be a catalytic site beyond 6 h on-stream. Electron Paramagnetic Resonance analysis of our spent FER showed highly stable carbonaceous radicals over several months. Radicals and Brønsted-acidic benzylic cations both might be active species in butene isomerization. To elucidate the effect of the radicals on the reaction, 2,2,6,6-tetramethylpiperidine was added to the feed as a radical scavenger. Since this compound is also a base, its radical scavenging effects were distinguished from its basicity effects by using triethyl amine as a secondary probe molecule.

Attaining insight into the character of active sites will facilitate further investigation on how active sites can be stabilized, consequently increasing catalyst lifetime. This understanding can enable the design of optimized catalysts for similar hydrocarbon chemistry.

Exploiting Heat Transfer in CO_2 Hydrogenation

Szymon Kuzniar, Marc Porosoff

We are investigating tandem catalysts for hydrogenation of CO_2 into fuels in a single reactor. During the CO-mediated pathway, endothermic reverse water-gas shift (RWGS) is coupled with exothermic Fischer-Tropsch Synthesis (FTS). The CO intermediate diffuses to a secondary active site that catalyzes the carbon coupling reaction, with the diffusion seen as a kinetically integral step in the reaction pathway. However, the effect of heat transfer of the coupled exo- and endothermic reactions is neglected. We hypothesize that exploiting localized heat transfer will decrease the quantity of external thermal energy needed to run the tandem reaction, increasing energy efficiency and raising economic viability of the approach. While we are ultimately interested in exploiting heat transfer limitations to enhance CO_2 hydrogenation kinetics, FTS results in a wide distribution of hydrocarbon products, adding undesired complexity into the system. To focus on tuning reaction kinetics, we substitute FTS with CO methanation as the heat source. We first attempt to identify appropriate catalysts for the exothermic and endothermic reactions: the RWGS catalyst needs to be nearly 100% selective toward CO without CH4 formation, and be inert for CO activation; the CO methanation catalyst should be almost 100% selective towards CH_4, and inert for CO_2 activation. The catalysts are synthesized using incipient wetness impregnation, and screened using temperature programmed reactions at atmospheric pressure to evaluate their effectiveness at producing either CO or CH_4. The candidates we are evaluating are Pt/Al_2O_3 and Ni/ZrO_2 for RWGS and CO methanation, respectively. The former appears to be a promising RWGS candidate, active and selective for RWGS and inactive for CO methanation. On the other hand, Ni/ZrO_2 performs well for CO methanation, but remains active for the undesired CO_2 methanation side-reaction.

Scalable Synthesis and Characterization of Active Iron Aluminate Material for Solar Thermal Water Splitting

Liam Taylor, Kent J. Warren, Alan Weimer

Solar thermochemical fuels have emerged as a competitive alternative to fossil fuels as they are carbon-neutral and functionally interchangeable with conventional fuels. These fuels are produced by splitting water and carbon dioxide at high temperature, producing hydrogen and carbon monoxide. The reaction normally requires an extraordinarily high temperature, but it can be lowered to a more manageable temperature with the help of an iron aluminate oxygen carrier.

This work focuses on developing a scalable synthesis method for the active material, as well as characterization of the material under harsh chemical looping conditions. Spray drying is an attractive method for producing spherical, robust microparticles as it is cheap, scalable, and widely adopted in industry. A literature review indicated that, while similar particles had been produced in previous studies, this particular formulation was novel and would require some experimentation. In these experiments, factors such as solids vol%, use of additives, and spray drying conditions were varied, and the resulting slurry was sprayed in a BUCHI B-290 Mini Spray Dryer. The spray dried particles were then calcined. A formulation has been developed and is being tested at pilot scale. In a separate experimental campaign, the robustness of the active material under rapid cycling was investigated. Iron aluminate material was synthesized via the Modified Pechini method, and a small amount was subjected to repeated redox cycles at 1400 °C in a stagnation flow reactor. The Modified Pechini method is a sol-gel method that allows for a more homogeneous microstructure in the final product. The material has thus far undergone over 1000 redox cycles in this campaign, with 2000 total cycles planned, to demonstrate consistent reactivity over long periods of time.

Utilization of Metal-Organic Frameworks in the Degradation of Chemical Warfare Agents

Eamon Bartlett, Bradley Gibbons, Amanda J. Morris

Metal-organic frameworks (MOFs) have been an area of interest for filter applications because of their high surface area, porosity, and catalytic capabilities. MOFs are comprised of metal nodes connected via organic linkers to form complex geometries which can be finely tuned during synthesis. A zirconium-based MOF called UiO-66 has demonstrated particular promise in promoting the hydrolysis of nerve agents. Various modulators were added during synthesis to control particle size (100-1900 nm) and defect level (0-24%). Hydrolysis of a chemical warfare agent simulant, dimethyl 4-nitrophenyl phosphate (DMNP), was used to determine differences in reactivity between samples. The rate of hydrolysis was improved with decreased particle size, suggesting the reaction is limited to the external surface. Heightened defect levels allowed for some diffusion into the MOF resulting in quicker hydrolysis rates; however, the increase in rate was not substantial compared to the effect of particle size on rate.

Iridium Based Metal Oxides: Catalysts for Advancing the Oxygen Evolution Reaction in Acid

Zachary Reihl

In the face of a changing climate, it is clear that technologies that meet the world's growing energy demand in a sustainable way are needed. Green hydrogen, produced from the electrochemical splitting of water, could play an important role in decarbonizing electricity production, allowing for the storage of energy from renewables on the timescale of months. But wide scale implementation of hydrogen production requires advances in performance and cost of water electrolyzers. Proton exchange membrane electrolyzers for hydrogen production achieve high performance, but require the use of iridium, one of the world's rarest metals, as a catalyst at the anode. Mixed metal oxide structures can reduce the use of expensive iridium without compromising activity or stability, but the behavior of this broad class of chemicals in realistic oxygen evolution reaction (OER) conditions is poorly understood. Here, we investigate the performance of iridium containing double perovskites and strontium iridates and study their change in intrinsic activity and surface composition over time. All catalysts were synthesized via solid state reaction in air and characterized using X-ray diffraction. The activity was measured using chronoamperometry in 0.1M $HClO_4$ and normalized by the electrochemical surface area. In general, we find that both the double perovskites and strontium iridates converge to the same activity and the same surface composition, as confirmed by post-reaction X-ray photoelectron spectroscopy. Although these metal oxide surfaces can be made with an unlimited range of element compositions and stoichiometries, our results suggest that the post reaction activity and surface composition in acid do not depend on the pristine lattice.

Synthesis of CeO_2 Supported Period 4 Transition Metal Catalysts Via One Pot Chemical Vapor Deposition (OP-CVD) Method

Mary Calandra, Hannah Karkout, William Chen, Nicole Spencer, Amol Pophali, Ryuichi Shimogawa, Gihan Kwon, Anatoly I. Frenkel, Tae Jin Kim

Platinum group metals (PGMs) are recognized to be the most efficient, yet expensive surface species for heterogeneous catalysts. To mitigate this expense, extensive research is being done on alternative cost-effective transition metals as surface species for these catalysts. Period 4 transition metals such as Cr, Mn, Fe, Co, Ni, and Cu have been studied to determine which one(s) produces the most desirable characteristics for a catalyst, such as high dispersion, small particle size, and high reactivity. Cerium oxide (CeO2) was used as the support material because it's been proven effective due to its lattice configuration, which enables it to perform redox easily ($Ce3+ \leftrightarrow Ce4+$) and maintain a high oxygen capacity. Various synthesis methods to produce CeO2 supported catalysts, including dry or wet impregnation, precipitation, and spray pyrolysis, have exhibited limitations related to low surface species dispersion and metal agglomerations on the support material. In this study, we developed and applied a new synthesis method known as one pot chemical vapor deposition (OP-CVD) to combat these drawbacks. OP-CVD uses gas-solid interactions to produce MOx/CeO2 catalysts, where M is a period 4 transition metal, aiming to achieve full dispersion of these surface species onto the CeO2 support. These synthesized catalysts were then applied to CO oxidation as a model reaction. To validate the feasibility of OP-CVD in achieving high surface species dispersion and gain insights into the molecular structure, we employed multiple characterization techniques including thermogravimetric analysis (TGA), X-ray diffraction (XRD) spectroscopy, Raman spectroscopy, inductively coupled plasma optical emission spectrometry (ICP-OES), and scanning electron microscopy with energy dispersive X-ray spectroscopy (SEM/EDX). The characterization results confirm that surface transition metal oxides were effectively dispersed on the CeO2 support. The simplicity of the OP-CVD method holds great potential for synthesizing supported metal oxide (or metal) catalysts.

Understanding of Alcohol and Aldehyde Oxidation Behavior of Gold Nanoparticles with Carbonaceous Layers in HMF Oxidation

Kylie Park, Bohyeon Kim, Steven McIntosh

Biomass-derived feedstock provides a sustainable solution for biodegradable plastics production amid growing sustainability concerns. A key building block for such plastics is 2,5-furandicarboxylic acid (FDCA). FDCA is generally produced through the full oxidation of 5-hydroxymethylfurfural (HMF), which has two possible pathways due to its aldehyde and alcohol functional groups. Recently, gold-based electrocatalysts have emerged as effective catalysts for facilitating the HMF oxidation reaction (HMFOR) due to its remarkably low onset potential and great stability. However, existing literature lacks a definitive understanding of the favored reaction pathway and mechanistic details of HMFOR using gold-based catalysts. Herein, we examine the reduction-oxidation reaction of the gold surface to investigate its oxidation activity toward aldehyde groups and potential active sites of HMFOR. This gold surface under examination consists of gold nanoparticles covered with carbon film, synthesized through a facile electrochemical method and denoted as Au/C. Au/C exhibits a surface-enhanced Raman scattering (SERS) feature, enabling the use of in-situ SERS measurements as well as Au/C rotating disk electrode (RDE) experiments at varying rotating speeds. The RDE measurements provide valuable insights into the kinetics of alcohol and aldehyde oxidation, thereby revealing the favored pathway for HMF oxidation. The SERS measurements suggest that the oxidation (aldehyde or alcohol) in HMFOR depends on the coverage of hydroxyl ions on the Au surface. These findings further implicate the potential active site on this catalyst for HMF oxidation.

Effect of Zeolite Acidity and Solvent on the Conversion of Fructose to High-Value Compounds

Fabiola Y. Rodríguez-Rodríguez

The world is facing an imminent depletion of non-renewable fossil fuels, and the associated environmental concerns make it necessary to find sustainable alternatives. Biorefineries, using renewable biomass, are emerging as promising alternatives to traditional refineries and offer eco-friendly solutions to meet society's chemical demands.[1] However, a significant challenge in this transition is the optimization of the conversion of biomass-derived compounds, such as fructose into valuable chemicals.

This study examines the role of heterogeneous catalysts, specifically Lewis and Brønsted acid Beta zeolites, and solvents in controlling the conversion of fructose to potential platform molecules such as alpha-hydroxy acids (AHAs) and 5-hydroxymethyl furfural (HMF), or the commercially valuable difructose anhydride (DFA).

Using Sn-Beta Lewis acid catalyst to convert fructose in water results in low catalytic activity and generates undesirable by-products like furans and humins, leading predominantly to lactic acid with unsatisfactory selectivity. However, a noticeable enhancement in catalytic activity is observed with the introduction of the solvent gamma-valerolactone (GVL) in a ratio exceeding 2:8.[2] To achieve 61% conversion, it takes 3 hours in water compared to 15 minutes in GVL:water 9:1. In contrast, using pure GVL as a solvent the reaction shifts to the production of DFA, which is an intermediate to HMF production.[3]

Al-Beta, as a Brønsted acid catalyst can catalyze the dehydration reactions from carbohydrates to furans, including HMF in water, although with low catalytic activity. Again, the addition of GVL to water increased it.[4] At 3 h only 5% of conversion is obtained in water versus 8% of conversion in GVL:water 9:1. Using pure GVL as solvent also propelled the reaction towards producing the intermediate DFA.

Dependence of the Oxygen Reduction Selectivity to Hydrogen Peroxide on Hydrophilic Carbon Fiber Paper

Yiwen Sun, Teona Taseska, Connor P. Cox, Madeleine K. Wilsey, Kendra R. Watson, Lydia Schultz, Astrid M. Müller

Hydrogen peroxide is a vital product in industry due to its versatile applications, from environmental protection and manufacturing to healthcare and food safety. Hydrogen peroxide is used to sanitize water, food preparation surfaces, and medical equipment, as well as in the production of chemicals. The most common method for hydrogen peroxide production is the anthraquinone process, which necessitates large-scale plants and is energy intensive. Therefore, the accessibility of hydrogen peroxide production is constrained in some regions, and the utilization of fossil fuel derived energy is not environmentally friendly. To improve hydrogen peroxide production, several studies have been reported on hydrogen peroxide electrosynthesis by oxygen reduction in aqueous electrolytes. The main advantage of electrocatalytic hydrogen peroxide production is that it can be employed decentralized. Electrocatalytic hydrogen peroxide production is an emerging technology and faces challenges related to production efficiency. In this research, the pH dependence of oxygen reduction selectivity towards hydrogen peroxide was analyzed. Hydrophilic carbon fiber paper, developed in the Müller group,[1] served as catalysts. We systematically varied electrolytes and their pH values to study oxygen reduction selectivity to hydrogen peroxide, with the goal to increase the hydrogen peroxide production efficiency, towards broad applicability of hydrogen peroxide electrosynthesis as an environmentally sustainable technology to produce hydrogen peroxide.

1) M.K. Wilsey, K.R. Watson, O.C. Fasusi, B.P. Yegela, C.P. Cox, P.R. Raffaelle, L. Cai, A.M. Müller, Selective Hydroxylation of Carbon Fiber Paper for Long-Lasting Hydrophilicity by a Green Chemistry Process. *Adv. Mater. Interfaces* **2023,** *10*, 2201684.

A Versatile Approach Toward the Assembly of Diazirine Functionality Utilizing Bmida

Abby Mock

Our research focuses on a cost-effective approach which minimizes synthetic overhead when assembling compounds with diazirine functionality. The goals are to prevent the surplus use of ammonia in the formation of the diaziridines and use an IKA Electrosyn 2.0 in the oxidation of diaziridine. Success with either specific aim will significantly reduce the synthetic overhead currently observed. Diazirine functionality plays a vital role in a vast number of applications. Two of interest to us include diazirine functionality as a next generation propellant and as a photoaffinity label. With the latter, this strategy plays a vital role in drug discovery and biomedical research as it allows researchers to study, understand, manipulate, and control complex biological systems. Moreover, previous research has shown that diazirine functionality has the potential to be used as propellants, so this project also aims to explore the energetic properties of this functionality and develop a deeper understanding of their physical properties. Preliminary data has shown limitations with scope and low to moderate levels of conversion. Our current approach has as a focus, a cross-coupling reaction where scope is only limited by the pairing of N-methyliminodiacetic acid boronate ester (BMIDA) and substrate.

Cobalt-Based Mixed Metal Oxide Catalysts for the Oxygen Evolution Reaction

David Abgrab, Zhu Chen

Current methods of hydrogen production release an average of 12-13 kg CO2 per kg H2 produced; with fossil fuel-based production generating up to 27 kg CO2 per kg H2. An alternative process known as water electrolysis, has the potential to reduce CO2 emission per kg of H2 produced. In this process, the oxygen evolution reaction (OER) is the bottleneck limiting the rate of hydrogen production and the overall process efficiency. The sluggish reaction kinetics of OER can be improved through the use of mixed metal oxide catalysts which can improve device efficiency and reduce capital costs. Cobalt oxide has been found to be an effective catalyst for OER in alkali conditions, however, it has low activity and stability in lower pH conditions.

Herein, we aim to improve the activity and stability of cobalt oxide thin-film electrocatalysts for OER by changing the chemical compositions and solid state structures. We explored different thermal treatment conditions and heteroatoms doping strategies (e.g., manganese) to stabilize the cobalt oxide thin-film. We evaluated the activity and stability of the resultant catalysts under different pH conditions and found that activity is negligible in any electrolyte less than pH 13. The production of an active, affordable, and stable catalyst will incentivize greater penetration of the water electrolysis technology.

Stability and Activity Enhancement Effects of Mn-Doped Ni-Al Catalysts for Thermocatalytic Methane Decomposition

Oamfah Suwannapong, Fatou Baka Diop, Ceren Yilmaz Akkaya, Andrew R Teixeira, Pratap M. Rao

The market demand for hydrogen is increasing steadily as hydrogen has emerged as a clean alternative to fossil fuels with low greenhouse gas emissions. More than 50% of global hydrogen is produced from steam reforming of methane, a high temperature, pressure, and carbon footprint process that requires large energy inputs, mostly from fossil fuels [1]. Thermocatalytic decomposition of methane or methane pyrolysis is an alternative process to produce COx free hydrogen with solid carbon as a by-product. Various catalysts (including metal, metal and metal oxide supported, and ceramic and carbon based) have been utilized in this process to decrease the high activation energy required to break the C-H bond. Among those, Ni catalysts over metal oxide support gained significant attention due to their low cost and high performance at low temperature. Recently, metal additives including Mn were shown to improve the stability and activity of Ni based catalysts prepared by impregnation method [2]. In this work, we explore the effect of Mn doping on the methane conversion yield and the catalyst stability prepared through Ni-Al hydrotalcite precursor using a differential fixed bed reactor. The catalyst is evaluated for methane decomposition at 550 °C until their complete deactivation. The deactivated catalysts and carbon by-products are examined by scanning electron microscopy, elemental analysis, and Raman spectroscopy. These analyzes contribute to understanding of material preparation-structure-property relationships towards methane decomposition reaction which will be discussed in this poster.

[1] Kaiwen, L.; Bin Y. &Tao Z. Economic analysis of hydrogen production from steam reforming process: A literature review. *Energy sources, Part B: Economics, Planning, and Policy*. **2018**, *13:2*, 109-115

[2] Wang, D.; Zhang, J.; Sun, J.; Gao, W.; Cui, Y. Effect of metal additives on the catalytic performance of Ni/Al2O3 catalyst in thermocatalytic decomposition of methane. *International Journal of Hydrogen Energy*. **2019**, *44*, 7205-7215

Predicting CO_2 Adsorption Capacity in Amine-Functionalized Zeolites Using Bayesian Optimization with Language-Interfaced Fine-Tuning (BO-LIFT)

Phyllis Ong, Shane Michtavy, Marc Porosoff

Amine-functionalized materials are promising adsorbents for the direct air capture of carbon dioxide (CO_2). These materials consist of porous, inorganic supports, such as silica or zeolites, that are functionalized with organic amines. The amine functional group acts as a base to provide binding sites during interactions with the acidic CO_2 molecule, while the inorganic scaffolds upon which they are immobilized provide high surface area and stability. There exists a gap in the literature in understanding the importance of synthesis variables such as loading, solvent, and support porosity on CO_2 adsorption capacity. To explore these relationships, we are leveraging large-language models (LLMs) to develop a method of representing the amine-based materials by the text of the associated synthesis procedure. By using Bayesian optimization with language-interfaced fine-tuning ("BO-LIFT"), we will make predictions on adsorbent performance with text. Synthesis and adsorption experiments are performed ad-hoc upon request from the LLM, and predictions are made via Bayesian optimization of the black-box function mapping synthesis procedures to adsorption capacity. Our approach to representing adsorbents with the text of synthesis procedures is verifiable and actionable; it simultaneously eliminates the need for structural data of complex adsorbents when making predictions. Through eight iterative testing loops from two initial points, we achieve a maximum prediction accuracy of 96.9%, revealing the potential for the BO-LIFT approach in improving adsorbent performance through the interface of natural language.

Theoretical Investigation on Degradation Mechanism of Sarin over Graphene-Supported Single Atom Catalysts

Walter J. Cesarski, Sean T. Murray, Caspar Yi, Enoch Nagelli, Simuck Yuk

The direct exposure of chemical warfare agents (CWAs) has always been detrimental to Soldiers on the battlefield, causing an estimated 1.3 million casualties during World War I alone. Of particular concern are nerve agents, including sarin, soman, and VX. Novel materials with specific functionalities are necessary to provide advanced protection through layered coatings or respirators to ensure the safety of Soldiers and untrained citizens alike. Thus, gaining a mechanistic understanding on the degradation pathways of CWAs over chemical catalysts is a crucial area of active research. In this work, we propose to study some of the notable chemical warfare agents frequently employed in the field, specifically gas-phase hydrolysis of sarin over the supported single atom catalysts (SACs). We seek to investigate both thermodynamic and kinetic aspects of degradation hydrolysis mechanisms over graphene-supported SACs and identify the key rate-determining steps, which can further be manipulated through the choice of metals and, potentially, supports. We found that a linear relationship can be drawn between the energy difference and the activation barrier of rate-determining step for degradation process.

Investigating the Influence of Catalyst Structure on the Electrochemical Hydrogenation of *Cis,Cis*-Muconic Acid to Yield Adipic Acid

Huy Nguyen, Prathamesh Prabhu, Jean-Philippe Tessonnier

The conversion of plant biomass to synthesize sustainable and innovative chemicals is the focus of intense research efforts in the midwestern United States due to the abundance of crops and agricultural waste products (e.g., corn stover). In this context, *cis,cis*-muconic acid (ccMA) is emerging as a versatile compound derived from biomass through fermentation that can be transformed into large volume drop-in chemicals like adipic acid and caprolactam, as well as novel species like *trans*-3-hexenedioic acid and *trans*-2-hexenedioic acid. There is particular interest in the production of adipic acid (AA), as it serves as a precursor for nylon-6,6, a plastic that finds applications in apparel, food packaging, house goods, and transportation. Early studies showed that ccMA can be hydrogenated to AA in the presence of a catalyst using H_2 gas. However, H_2 is typically derived from fossil hydrocarbons using processes that are energy intensive and emit large amounts of greenhouse gases. To overcome this limitation, we hypothesized that ccMA could be electrochemically hydrogenated to AA using water (H_2O) as a source of hydrogen. Exploratory work performed in our group demonstrated the feasibility of this novel approach. This project built on these early findings and focused on understanding the pathway of this reaction in order to guide the design of high-performance catalysts. Understanding the fundamental reaction mechanism for converting ccMA to AA will also help us design a system for possible scale-up. Utilizing an electrolysis flow cell, two noble metals acting as the reaction catalyst, palladium, and platinum, were used to convert ccMA to AA. Samples of the electrolyte were collected at 15-minute intervals and analyzed through quantitative ^{1}H nuclear magnetic resonance. Nanostructured metal catalysts (Pd/C and Pt/C) were active for the desired conversion. In contrast, the corresponding bulk metals (Pd and Pt foils) only transformed *trans*-3-hexenedioic acid. These results are supported by theoretical calculations and demonstrate that the electrochemical hydrogenation of ccMA to AA is structure sensitive. This research has provided valuable insights into determining the ideal conditions involved in the electrochemical hydrogenation of *cis,cis*-muconic acid into novel and promising bio-derived compounds.

Exploring Catalyst Deactivation Routes on H-MFI and CHA Zeolites during Methanol-to-Hydrocarbons Chemistry

Stephanie Rodríguez Méndez

This study explored cyclization pathways toward forming polyaromatic species in MFI and CHA framework zeolites, which are relevant during the conversion of methanol-to-hydrocarbons (MTH) during zeolite-catalyzed chemistries. Polyaromatic species have been linked to catalyst deactivation because their size prevents them from diffusing out of zeolite pores. We specifically explored cyclization pathways to form naphthalene and diphenylmethane products, which are model compounds for "sheeted" and "branched" polycyclic species. The formation of these products was modeled with density functional theory (DFT) calculations as implemented in the Computational Catalysis Interface. For the case of the MFI framework, we developed calculations at both the straight channel (~0.55 nm) and channel intersections (~0.7 nm), and the preliminary data suggest that cyclization barriers for both naphthalene and diphenylmethane occur with higher activations at tighter environments. This hypothesis was further examined at CHA framework pores (~0.42 nm). While barriers in CHA are indeed higher than those occurring at MFI intersections, they are not higher than those occurring in the straight channels of MFI. This suggests that pore size alone is not a suitable descriptor for polyaromatic formation barriers. This, in turn, indicates that intrinsic differences also influence barriers in MFI and CHA frameworks, which stabilize the relevant cyclization transition states to different extents that cannot be decoupled into differences in pore size only. Future work will resolve these discrepancies by expanding the zeolite models subset to TON, BEA, and MEL.

First-Principles Investigation on the Effect of Al and Cu Density on the Mobility of Cu^{I} Ions in Cu-CHA Zeolites for NH_3-SCR

JohnJoel Sacris, William Schneider, Raghav Saxena

Selective Catalytic Reduction of NO_x by NH_3, also known as NH_3-SCR, is a critical technology for mitigating NO_x emissions from heavy-duty diesel engines. Cu-CHA zeolites demonstrate superior performance in NH_3-SCR, catalyzing the reaction through a redox mechanism involving both reduction and oxidation half cycles. The oxidation half cycle entails the pairing of two $Cu^{I}(NH_3)_2$ ions, activating an oxygen molecule to form Cu dimer species—the key active sites for low-temperature NH_3-SCR. The kinetics of the oxidation half cycle are governed by the mobility and pairability of $Cu^{I}(NH_3)_2$ ions through 8R windows of CHA zeolites, influenced by the aluminum distribution. In this study, we utilize ab-initio molecular dynamics and metadynamics sampling to compare the mobility of local and remote diffusion of $Cu^{I}(NH_3)_2$ ions from their respective charge-compensating aluminum atoms. Recent experimental observations have revealed that higher aluminum and copper densities lead to higher rate constants for the oxidation half cycle. To rationalize these experimental findings, we also model the local diffusion of $Cu^{I}(NH_3)_2$ ions in the presence of additional $NH4^{+}$ and $Cu^{I}(NH_3)_2$ ions. Our investigation shows that the presence of ammonium or copper cations does not significantly affect the local diffusion barriers. This lack of effect is attributed to the choice of the Al T-site within the chabazite cage, which governs the distance between the diffusing copper ion and the co-occupying cation. Additionally, we observe that local diffusion exhibits significantly lower energy barriers than remote diffusion, consistent with prior studies. These calculations provide evidence that the T-site location for the co-occupying ion in the CHA cage plays a pivotal role in the diffusion process, as outlying ions do not exhibit meaningful electrostatic interactions with the copper ion. Future work will investigate other T-sites closer to the T-site of the diffusing $Cu^{I}(NH_3)_2$ ion.

A Computational Investigation of the Urea Oxidation Reaction Mechanism Using Density Functional Theory: Promoting the NiOOH Active Phase By Introducing Effective Dopants

Matteo Garcia-Ortiz, Qiu Jin, Liney Arnadottir

Urea electrochemical oxidation is of general interest for water treatment in agriculture, but also a possible fuel for urea fuel cells. Urea oxidation is also of fundamental interest for amide chemistry. The urea oxidation reaction (UOR) in alkaline medium neutralizes urea into relatively benign products including H_2, N_2, and CO_2, rather than harmful NO_x compounds. Previous studies have suggested direct and indirect reaction routes for this full oxidation. The presence of metallic ions such as Fe, Co, and Cu in the nickel hydroxide surface may improve selectivity towards the indirect pathway where the $Ni(OH)_2$/NiOOH redox couple plays an important role. Since NiOOH is the active phase for UOR, the first step of transferring $Ni(OH)_2$ to NiOOH is crucial. In this work, we investigate the effect of Fe dopant on the Ni hydroxide transformation. Performing DFT calculations, as implemented in the Vienna Ab initio Simulation Package, the thermodynamics of the nickel dihydroxide to nickel oxyhydroxide mechanism was calculated. We found that doping the β-$Ni(OH)_2$ surface with Fe significantly reduces the free energy change for sequential surface dehydrogenation to β-NiOOH, from 2.27 eV to 0.93 eV, favoring the creation of UOR active sites. This indicates promising potential for the improvement of urea oxidation performance with Fe doping.

The Oxidative Coupling of Methane: A Mesoscale Kinetic Model of the Surface and Gas Phase Reaction Network

Lauryn Holgado, Srinivas Rangarajan

An open-source program called Mesoflow was used to model the intricate reaction network of the oxidative coupling of methane (OCM). This mesoscale model incorporates both the catalytic and gas phase reactions in addition to the interactions between these two phases, including 39 gas phase reactions, 14 surface reactions, and 32 species. Initially, a Python script was developed to model this reaction network and verify the kinetics. This was compared against Mesoflow's zero-dimensional dynamic model which checked the accuracy of the Mesoflow program. This preliminary model is kinetically driven. The Python script and the zero-dimensional model matched analytic solutions of simple reactions. Mesoflow's block catalyst case was then modified to reflect the OCM chemistry. This modified case modeled a batch reactor. The chemistry of this reaction network is very stiff, therefore the SUNDIALS integrator is required. Parameters, such as temperature, volume, and initial concentrations were held constant. Concentrations of key species, such as methane, methyl radicals, and ethylene were calculated as a function of time. With the surface reactions, the yield of valuable products, such as ethylene, was much higher than without. Therefore, a catalyst is required for high yield and selectivity. Over oxidation is also possible. An investigation into the effects of temperature on product selectivity is needed.

Characterization of Highly Dispersed Metal Catalysts Via Infrared Spectroscopy and X-Ray Diffraction: From Isolated Atoms to Clusters

Kolton Dean, Dr. Christopher T. Williams, John Regalbuto

Heterogeneous catalysis is a booming industry, driven by its ease of use and scalability. Recent focus on isolated atom metal-supported catalysts has led to various synthesis techniques being studied for industrial-scale production, aiming to enhance selectivity, activity, and uniformity of desired reactions. This project aims to study the effect of various synthesis techniques as they pertain to the size of the supported metal particles via several characterization methods. Various synthesis techniques were assessed, such as Dry Impregnation (DI) and Switched Solvent Synthesis (SwiSS). Several analytical techniques were used to study the effect on particle size, including Temperature Programmed Reduction (TPR), Fourier Transform Infrared Spectroscopy (FTIR), and X-Ray Diffraction (XRD). For catalyst preparation, each batch was made using 50 mg of Aerosil 300 SiO_2 support (300 m^2/g) at 2% weight loading, corresponding to a nominal areal loading of 0.377 Pd atoms/nm^2. Aqueous dry impregnation of the $Pd(NH_3)_4(NO_3)_2$ precursor was performed for each sample. For the SwiSS samples, the appropriate solvent was added afterward in a 3:1 molar ratio (i.e., switched solvent to stock solution ratio.) The solvents used were acetone and diethylamine, which were chosen due to previous preliminary investigations, their similar boiling points, but different properties such as dielectric constants and dipole moments. Samples were reduced at 170° C in 10% H_2/N_2. Particle sizes were estimated with the Scherrer equation from the XRD data, and a quantitative analysis of the FTIR data was performed. It was found that 170° C is a sufficient reduction temperature, and that the SwiSS method resulted in smaller particles compared to dry impregnation. With the present data and analysis from the FTIR, there was not found to be a meaningful difference between the ratios of terminally bound (Pt-CO) to bridge-bound CO (Pt_2-CO, Pt-$_3$CO), though there is a noticeable difference in the shape of the peaks, so more information may be revealed with more thorough analysis.

Unraveling the Stability of Trimetallic Nanoparticles with Machine Learning

Klaertje Hesselink, Maya Salem, Giannis Mpourmpakis

Metal nanoparticles have gained immense interest due to their wide application in various fields spanning from catalysis to nanoelectronics and drug delivery. With the vast configurational space of multimetallic nanoparticles, it is difficult to implement Density Functional Theory calculations to find the most thermodynamically stable chemical ordering (arrangement of different metals) for a specific nanoparticle. Here, we apply the Bond Centric Model (BCM)[1,2] coupled with an in-house developed Genetic Algorithm (GA)[2], which can accurately and efficiently capture the stability of nanoparticles of any size, shape and metal composition. We use a proof of concept on how the BCM + GA captures trimetallic nanoparticle stability of PdPtAu[3], to extend its usage to four more trimetallic nanoparticles of AgPdPt, AgAuPd, AgAuPt and AgCuPd made up of 2869 atoms, of cuboctahedral symmetry, with varying metal compositions. Our study reveals the total chemical ordering and, importantly, the surface composition of the nanoparticles resulting in the lowest cohesive energy (most thermodynamically preferred). We observe that for AuPdPt, Au tends to segregate to the surface, whereas Pd and Pt tend to reside in the bulk and subsurface. These observations for AuPdPt agree with other computational studies. We observed similar trends for AgPdPt; at high concentrations of Ag, there was more Ag on the surface, whereas Pd and Pt were in the bulk and subsurface, in agreement again with other computational studies. AgAuPt and AgAuPd showed trends that Ag and Au segregate to the surface while Pt and Pd occupy the bulk and subsurface layers. Our AgAuPt findings had similar surface composition trends compared to other computational studies and the AgAuPd findings agreed with experiments. We further extended the study to AgCuPd and we found that Ag resided on the surface, whereas the other two metals were in the bulk and subsurface, in agreement with experiments. Our work demonstrates the feasibility of using the BCM and the developed GA to reveal the exact chemical ordering of highly thermodynamically stable trimetallic nanoparticles and reveals the importance of metal cohesion and bimetallic bond strength in the overall chemical ordering trends.

References:

1. Z. Yan, M. G. Taylor, A. Mascareno, G. Mpourmpakis., Nano Lett. 2018, 18, 4, 2696-2704.
2. J. Dean, M. J. Cowan, J. Estes, M. Ramadan, G. Mpourmpakis, ACS Nano. 2020, 14, 8171-8180.
3. D. J. Loevlie, B. Ferreira, and G. Mpourmpakis, Acc. Chem. Res. 2023, 56, 3, 248-257.

Influence of Confinements and Solvents on the Dehydration of Tertiary Alcohols

Aurora Occhino

The dehydration of alcohols to olefins is an important reaction to many chemical transformations, including biomass conversion. Zeolites, which are crystalline aluminosilicates, provide confined environments where this reactivity may be enhanced by multiple orders of magnitude through a careful selection of solvents and confinement size. This effect can be enumerated through the measurement of activation enthalpies that enumerate the enthalpic barrier that a reactant (a tertiary alcohol) has to overcome, and activation entropy that enumerates disorder and the excess entropy that the reactant has to overcome at the transition state.

In this work, this information will be used for exploring the impact of solvent structure and properties that include solvent polarity and dielectric constants, and zeolite confinements, or pore size, on the dehydration kinetics of tertiary alcohols. A polar solvent, water and a nonpolar solvent, toluene will be used to explore the dehydration kinetics of 3-ethyl-3-pentanol and explore how changes in these properties will affect the reaction kinetics, including reaction rates, orders, activation enthalpies, and activation entropies.

Oxidative Dehydrogenation of Ethane: Molten Salt Promoted Catalyst with Passive Carbon Capture

Ryan Sedlacek

Ethylene is a valuable chemical that serves as an important component in many compounds and products such as polyethylene, polyvinyl chloride (PVC) and has various other applications. Ethylene is industrially produced from the dehydrogenation of ethane. Ethane is a significant product from petroleum refinement and other natural gas refinements. The current implemented method of converting ethane to ethylene, steam cracking, requires separate fuel to maintain high operating temperatures resulting in high COx (CO and CO2) yields, periodic shutdowns and the ethylene yield is limited to thermodynamic equilibrium. Our goal was to produce high ethylene and low COx conversion through a chemical looping, catalyst-promoted process while capturing CO2 with a molten salt promoter. The composite catalysts used were various mol combinations of $La_{O.9}$-$SrO_{0.2}$-FeO_3 (LSF) which were synthesized through reactive grinding. The catalysts were impregnated with various mol combinations of a molten carbonate salt mixture consisting of $LiCO_3$-Na_2CO_3-K_2CO_3. The carbonate-perovskite catalysts were run at 800°C with a total gas hourly space velocity of 150 h^{-1}. The chemical looping process consisted of a reduction phase (ethane and argon with ~20% ethane), an argon purge, a reoxidation using a flue gas stimulant, followed by another argon purge. Our initial results indicated moderate ethylene yields with low COx selectivity. However, more catalysts still need to be tested to see if chemical looping dehydrogenation is efficient enough to overcome steam cracking.

Nanomaterials for Photocatalytic Degradation of Pharmaceuticals Present in Wastewater

Amna Al Ajmi

Due to continuous global energy demand and industrial development, considerable interest has been attended to the health impacts of organic and inorganic pollutants in the environment. Such pollutants can severely damage our aquatic environment. Recently some research output has shown a measurable level of organic contaminants in the outlet of processed water. Therefore, it is necessary to design and synthesize artificial photocatalysts with high activity. Implementation of photocatalytic properties of semiconductor materials has shown the potential method to purify wastewater. This includes materials like metal oxides, metal sulfides, and oxygen nitrides. Recently, carbon-based materials like graphitic carbon nitride are found to show promising photocatalytic properties due to the conduction of higher electrical charges, aligned reduction and oxidation potential, band value in the visible light region, physical and chemical structure stability, and the easy production steps that allow the need for large-scale production from low-cost precursors. However, the photocatalytic efficiency of the pure sample is limited by the high recombination rate of the photo-induced electron-hole pairs. This project aims to design and fabricate pure and modified carbon nitride photocatalytic materials by direct thermal heating and microwave-assisted technique and to study the photocatalytic degradation of pharmaceuticals under visible light irradiation. The strategic objective of this investigation is to develop novel photocatalytic material from nonmetal-based material by chemical and thermal modification approach for water treatment from the organic pharmaceutical compound application. This will be achieved through fundamental experimental research focusing on the characterization of produced products and modifying it to use in different applications mainly for water treatment.

Copper-Catalyzed CO_2 Electroreduction Utilizing Zeolite-Templated Carbon Supports

Nelanne Bolima, Rajamani Gounder, Brian M. Tackett, Hwiyoon Noh, Justin Rosa-Rojas, Lauren Olesky

Global reliance on fossil fuels and the expansion of industry has led to a continuous rise in CO2 emissions and directly contributed to climate change since the 19th century. Consequently, global research efforts are now focusing on fuel-production technologies that mitigate CO2 emissions. One promising alternative is CO2 electroreduction, which utilizes renewable energy to convert CO2 to other fuels and chemicals. Previous studies of CO2 electroreduction report the production of light hydrocarbons when using copper-based electrocatalysts. However, data suggests that using microporous carbons may favor the formation of heavier hydrocarbons, which serve as a more efficient fuel source. Zeolite-templated carbons (ZTCs) are microporous carbons that can be synthesized using zeolite templates. The standard practice of ZTC synthesis used includes furfuryl alcohol impregnation, propylene chemical vapor deposition, heat treatment, and hydrofluoric acid etching to obtain the carbon product. Copper nanoparticles were then deposited on the surface of the ZTC via cetrimonium bromide (CTAB) reflux, followed by electrochemical deposition. Electrochemical tests were performed to assess the catalytic activity of the material. Gas diffusion electrodes were then used to reduce CO2 into alkenes with the copper containing ZTC serving as the catalytic site. Current, hydrocarbon weight, and gas/liquid product was subsequently measured and analyzed. This study aims to interpret the structural integrity of ZTC formation at each step of the synthesis. Moreover, this study looks to clarify how the synthesis and structure of ZTCs relate to their electrochemical performance and reduction of CO2, in comparison to commercial carbons. The results have provided an understanding of how adjustments to the synthesis protocol correspond to variability in porosity, structural order, and catalytic activity of the ZTCs when used for CO2 electroreduction.

Production of Carboxylic Acids from Waste Plastics: A Sustainable Pathway Via Pyrolysis and Hydroformylation

Enner Mendoza, Houqian Li, George Huber, Clark Landis

The current life cycle of plastic packaging highlights the potential to implement processes to increase recycling rates and reduce mismanaged plastic. Commercial production of carboxylic acids is primarily sourced from fossil feedstocks through a series of energy demanding processes. Carboxylic acids can be alternatively produced from waste plastics, offering a sustainable alternative for the production of mono- and di-carboxylic acids. Thermal decomposition of waste plastics yields an olefin-rich oil which produces a wide range of aldehydes after hydroformylation. To demonstrate a sustainable route for carboxylic acid production, two Pinnick oxidation procedures with different scavengers were employed to determine which process had a higher selectivity for the acid products. Gas chromatography-mass spectrometry was used to characterize the oxidation of the aldehyde model compound. The Pinnick oxidation which employed a hydrogen peroxide scavenger had a higher selectivity for the octanoic acid product. Gas chromatography was used to determine the retention times for C6-C12 acids to determine the concentrations of products after oxidation of the hydroformylated oil. Employing the Pinnick oxidation with a hydrogen peroxide scavenger may yield a product with C6-C21 carboxylic acids. Oxidation of the hydroformylated oil which produces carboxylic acids may prove a sustainable route for acid production from waste polyolefins.

Role of Phosphorus in Transition Metal Phosphides for Cleaving Hindered C-O Bonds in 2-Methyltetrahydrofuran

Angela Aguirre, Conor Waldt, David Hibbitts

Biomass can be upgraded into higher value chemicals, such as α,ω-diols, via C-O hydrogenolysis. However, this process requires the selective activation of sterically hindered C-O bonds. Here, we observe a model biomass-derived molecule, 2-methyltetrahydrofuran (MTHF). The molecule has two C-O cleavage pathways, the unhindered (^{2}C-O) cleavage and hindered (^{3}C-O) cleavage. Previous work examining C-O hydrogenolysis in MTHF demonstrated that an increase in P content to Ni shifts selectivity towards ^{3}C-O cleavage. Ni_2P (P62m spacegroup) showed the greatest selectivity towards ^{3}C-O cleavage over ^{2}C-O cleavage in comparison to $Ni_{15}P_5$ and pure Ni. Alongside this, other transition metals were also compared to their pure metal counterparts to investigate selectivity trends. A shift in selectivity towards hindered ^{3}C-O cleavage was seen for almost all transition metals investigated when P was added. Here we examine hindered ^{3}C-O cleavage using simulated models and calculating free energy barriers (ΔG) for Ni_2P using Density Functional Theory (DFT). Here we now more closely investigate structural effects on selectivity by modeling Ni_2P in both the Pnma and Fm3m spacegroups while comparing our results to previous work to determine if structural effects play a role in selectivity. It was found that Ni_2P(010) was the optimal surface for Pnma and Ni_2P(001) was the optimal surface for Fm3m among various Miller Indices. Probe molecules when then adsorbed to the catalyst surface to determine the likely preferred binding site for MTHF C-O cleavages. During our adsorbate calculations, it was observed that the Fm3m surface restructured, rendering it unsuitable for conducting chemical reactions due to its instability. Comparing our reaction coordinate diagram to previous work, it was found that the ^{3}C–O activation barrier was lower than ^{2}C–O activation in Ni_2P, making ^{3}C-O cleavage more favorable, as was seen in the prior work. Additionally, the previous relative energy difference (ΔΔE) calculated for Ni_2P P62m was 44 kJ mol-1, and our Ni_2P Pnma ΔΔE was 17 kJ mol-1. This suggests that structure does play a role in selectivity. Future work will include completing free energy difference (ΔΔG) calculations to further prove this suggestion true.

Reducing the Environmental Impact of Hydrogen with Chemical Looping Reactors

Anna Broerman, James R. McKone, Goetz Veser

After common hydrocarbon fuels are used and oxidized, they are often emitted into the atmosphere rather than recycled. Our laboratory is developing a new type of chemical reaction process we term "chemical looping hydrogenation" that enables efficient and selective hydrogen transfer to molecular hydrogen acceptors. Whereas our work to date on this reactive process focused on tungsten oxides (WO_x), this presentation will focus on a summer research project I undertook to explore mixed-metal oxides based on Mo and W as viable catalysts. My work focused on the following research question: to what extent does mixing Mo and W change the extent of hydrogen uptake in the oxide and thus hydrogenation of molecular acceptors? Hence, the objective of this research was to assess the use of $W_xMo_{1-x}O_3$ compounds over a range of metal ratios as catalysts in both electrochemical and thermochemical looping hydrogenation.

Five samples containing increasing ratios of W to Mo were synthesized from ammonium precursors and analyzed with X-ray diffraction (XRD) and a transmission electron microscopy(TEM). To examine the potential of these samples for hydrogen looping electrochemically, a series of cyclic voltammetry measurements were conducted. Oxide samples were further prepared for thermochemical analysis through wet impregnation with platinum as a catalyst for hydrogen insertion via spillover. The samples were then reacted sequentially with $H_2(g)$ for hydrogen uptake and C_2H_2 as a model hydrogen acceptor to complete full chemical looping hydrogenation. These results provided a promising basis for continuing work on mixed metal oxides in the context of chemical looping hydrogenation.

Efforts to Measure Reduction Potentials for Hydrogen Insertion in Transition Metal Oxides

Isabella Hughes, Evan Miu, James R. McKone

Hydrogen is a common industrial reactant that is almost exclusively sourced from nonrenewable resources. Electrochemical hydrogen insertion into transition metal oxides to form metal oxide hydrogen bronzes (H_xMO_y where M is a transition metal like W, Mo, or Ti) offers a potential alternative to conventional hydrogen generation. After electrochemically inserting hydrogen atoms from an abundant source like water into a transition metal oxide, hydrogen atoms can be stored inside the metal oxide lattice until oxidizing potentials are applied and the hydrogen atoms are de-inserted to make $H_2(g)$ or react with industrial feedstocks directly.

Ongoing work in our lab to understand the basic thermodynamics of hydrogen insertion/de-insertion processes depends on the ability to precisely measure the equilibrium reduction potentials ($E^{0\prime}$) for these reactions. My research project aimed to develop a standardized procedure to make these measurements across a wide range of metal oxide hydrogen bronze formers.

The procedure I developed involves depositing inks composed of powdered metal oxide, polyvinylidene fluoride (PVDF) as a binder, and variable amounts of acetylene black for enhanced conductivity onto glassy carbon working electrodes. Using a three-electrode cell setup and aqueous electrolytes at various pH values, I ran chronopotentiometry, cyclic voltammetry, and galvanostatic intermittent titration technique (GITT) experiments to measure the potentials at which hydrogen insertion/de-insertion occurs. Despite successive efforts to increase the reversibility of the reactions—adjusting the proportion of acetylene black; decreasing the amount of metal oxide on the working electrode; passing smaller amounts of current; and purging the electrolyte with inert gas—I was unable to identify a consistent set of conditions that approached microscopic reversibility to the extent required for precise measurements. We speculate that the reversibility is limited in part by side reactions with atmospheric oxygen that we have not yet been able to adequately exclude from the cell; we hope to improve upon measurements in the future by implementing a design to minimize oxygen exposure during experimentation. We further conjecture that it may be impossible to reversibly de-insert all hydrogen from most metal oxides due to inherent changes in conductivity as a metal oxide incorporates hydrogen into its crystal structure. If such a limitation exists, it may be appropriate to pursue non-electrochemical approaches to these measurements of equilibrium dynamics in future studies.

Pt-M Thin Film Catalyst Synthesis for Olefin Epoxidation Using Water As O Atom Source

Yamilet M. Rivera Cintrón, Chenyu Jiang, Karthish Manthiram

Production of propylene oxide with hydrogen peroxide requires the direct burning of fossil fuels to reach elevated reaction temperatures. Electricity can serve as a greener alternative to fossil fuels, sustainably functionalizing olefins with the aid of efficient and selective electrocatalysts. However, understanding mechanisms of electrochemical reactions remains untrivial due to dynamic changes of the catalysts under applied potential. Olefin epoxidation using water as the oxygen atom source has been previously demonstrated in ambient conditions with PtM bimetallic nanoparticles as the catalyst drop-cast onto carbon paper (CP) electrodes (M. Chung, PhD Thesis, MIT), achieving a Faradaic efficiency (FE) higher than 90% for cyclooctene oxide. Due to its porous nature, CP is regarded as an unsuitable substrate for *in situ* spectroscopies that require an ultra-high vacuum environment, such as soft X-ray absorption spectroscopy. By changing the substrate to a nonporous material and depositing the catalyst without sacrificing its high catalytic performance, the mechanism of this oxidation reaction can be investigated to reveal the reason why PtM excels as a novel catalyst for olefin epoxidation. This study provides insight on how deposition methods of PtM affect its structure and catalytic properties by analyzing the FE obtained from the use of CP, indium tin oxide, and glassy carbon as substrates. PtM thin films were prepared by drop-casting, sputtering, and spin coating, and chronopotentiometry was carried out at 40 mA/cm^2 until a total charge of 20C was passed. Scanning electrode microscopy characterization was performed to obtain information on the surface morphology of sputtered PtM thin films.

Analysis of Molybdenum and Iron Catalyzed Non-Oxidative Methane Coupling

Ethan Robey, Mihir Kulkarni, Bryce Clutter, Madelyn R. Ball

Methane conversion reactions are of high value to the chemical and energy industries, as a wide range of products such as acetylene, ethane, ethylene, or methanol can be produced from the abundant feedstock supplied by natural gas. Non-oxidative methane coupling in particular is a clean way to produce these chemicals with minimal coking, greater production of useful byproducts such as hydrogen, and no production of undesirable byproducts such as carbon monoxide. Due to the high stability of methane, however, this class of reactions is considered one of the holy grails of catalysis research. If the low activity and selectivity of this reaction can be overcome, it could lead to significant advancements in the production of high value and widely used chemicals such as ethylene. In this investigation, molybdenum and iron supported monometallic and bimetallic catalysts were synthesized via incipient wetness impregnation and tested in the non-oxidative methane coupling reaction. Various characterization techniques including H_2 temperature programmed reduction, CO chemisorption, N_2 physisorption, and X-ray diffraction were employed on the different catalysts. Reaction performance was measured in a fixed bed flow reactor with a feed of methane and nitrogen at 700°C. Pre carburization was conducted by flowing pure methane for a brief period of time once the catalyst reached the target temperature and then beginning the designated reaction feed gas mixture. The outlet gas concentration was analyzed by gas chromatography. Through this method we have determined appropriate reaction conditions, i.e. temperature, gas flow, catalyst mass, and time, for investigation of these catalysts. Ongoing work is investigating the performance and properties of these catalysts to inform future catalyst modifications. The improvement of catalysts for non-oxidative methane coupling could eventually lead to a higher selectivity of desirable products for industrial methane conversion processes, increasing the economic gain and decreasing the waste generated by this reaction.

Re-Tooling Recycling: Liquid Metal Catalysis of Polycarbonate Pyrolysis

Nicholas Burnette, Goetz Veser

Polycarbonates (PC) are widely used polymeric materials with many desirable physical properties including high-impact resistance and a transparent optical grade. However, PC also account for a significant portion of the growing plastic waste crisis that is disrupting the livelihood and environmental well-being of communities around the globe. The vast majority of waste polymers are released into the environment as a result of booming demand for plastics and the inability of current recycling methods to keep up with consumption rates. The most prominent of such recycling methods are mechanical processes that degrade the physical qualities of recycled material, while more robust chemical recycling methods have yet to become viable alternatives. Pyrolysis, or thermal decomposition in anoxic conditions, is one chemical recycling process that converts polymer waste into a spectrum of products; the most valuable of which being the liquid products (oil) that can be regenerated into fresh polymer or used for other industrial applications. However, the main disadvantage of pyrolysis is the vast energy input required to drive the highly endothermic reactions. This study investigates the pyrolysis of PC in a liquid metal (LM) catalytic media using thermogravimetric analysis (TGA) and gas chromatography-mass spectroscopy (GC-MS) to assess the performance of LM catalysts compared to the non-catalytic reaction. LM's were evaluated as catalysts due to key advantages over conventional, solid catalytic materials. Bismuth and indium were chosen for their compatibility with pyrolysis conditions. The low-melting nature of these metals (relative to PC pyrolysis reaction temperature) allows for an analysis of their catalytic activities in the liquid phase. Both LM catalysts showed promise in reducing the energy penalty of pyrolysis by significantly reducing the reaction onset temperature when compared to the non-catalytic reaction. Each catalyst influenced the selectivity of products in the oil collected from the experiment; illustrating the different possible mechanistic pathways documented in the literature. Bismuth promoted the formation of heterocyclic compounds, and the indium catalyzed reaction favored the formation of phenol and large-branched phenol derivatives. The further development of this process stands to improve the efficiency of PC pyrolysis and the circular lifecycle of plastics.

Design and Construction of a Vibrating Fluidized Bed Reactor for Particle Atomic and Molecular Layer Deposition

Bergen Evans, Hailey Loehde-Woolard, Kent Warren, Alan Weimer, Mark Wallace

Atomic and Molecular Layer Deposition (ALD/MLD) is a series of self-limiting gas phase reactions that are used to deposit films uniformly. ALD is used to deposit inorganic films, whereas MLD is used to deposit organic films. These processes use the concept of self-limiting surface reactions in alternating dosing and purging phases to achieve a dense, uniform coating [1]. The amount of time with each reactant dose and purge can be altered, as well as how many cycles are used, to optimize the maximum amount of uniform coverage on selected support materials. Both ALD and MLD have a wide range of applications including catalysts and barrier coatings. Particle ALD was introduced in the early 2000s at the University of Colorado Boulder [1] and involves a specialized vibrating fluidized bed to achieve conformal coatings on primary particles, which will be constructed herein.

The build of this reactor involved the knowledge of the desired chemistry, fluidization, mechanical mixing, material choices and their restrictions, heat transfer, instrumentation and control, accessibility, economics, and safety. Proper fluidization and mixing ensure that the concentration of the reactants are dispersed equally, prevents agglomeration, and reduces mass transfer limitations. This reactor was designed to operate and perform ALD and MLD to create CO2 capture materials which will require multiple precursors and a range of temperature conditions. An extensive Standard Operating Procedure (SOP) outlines the entire system and its individual components. Safety being the utmost important aspect in any reactor, the SOP is meant to detail the overall process and guide and identify any internal failures that might be occurring.

After the construction of this reactor, numerous tests were performed including temperature monitoring, fluidization testing, and performance of hardware and software checks. The baseline of ALD in this reactor was demonstrated through alumina ALD with trimethyl aluminum and water. Developing these skill sets has provided a deeper, thorough understanding and hands on experience of chemical engineering fundamentals. The design and construction of this reactor was performed by a graduate student and an undergraduate student with support from a research associate and the machine shop.

References

[1] Alan W. Weimer. J. Nanopart Res 21:9 (2019)

Development of Nickel Catalysts for CO_2 Methanation – Investigating the Effect of Support and Synthesis Method

Rachel Phillips

One approach to mitigate climate change is the reduction of atmospheric CO_2 via CO_2 conversion into useful chemicals. CO_2 conversion by hydrogenation to methane is catalyzed by heterogeneous Ni-based catalysts. For these catalysts, the support has been reported to have an impact on catalytic performance; we seek to understand these support effects in order to design better catalysts. In this work, we have synthesized a series of Ni nanoparticle catalysts using several synthesis methods to investigate how the support choice and synthesis parameters influence the structure and the effectiveness of the catalyst. Ni catalysts were prepared on SiO_2, mesoporous silica (SBA-15), Al_2O_3, TiO_2, and CeO_2 supports. These catalysts were made by incipient wetness impregnation (IWI), strong electrostatic adsorption (SEA), or colloidal synthesis of nickel nanoparticles. The resulting catalysts were characterized by H_2 or CO chemisorption and TEM to determine the impact of support and method on Ni particle size. Ni/SBA-15 prepared by SEA had the highest dispersion of the studied materials. Furthermore, during the methanation reaction, the Ni/Al_2O_3 prepared by SEA had the highest CO_2 conversion rate. With the goal of preparing uniform sized Ni particles across the series of supports, we prepared colloidal Ni particles which were found to be 3.16±0.5 nm as prepared. Once deposited onto the supports, however, we determined that the Ni/SiO_2 particles were much larger than the Ni/TiO_2 particles. Ongoing work is investigating alternative approaches to maintain uniform Ni particle size across the series of supports. From this study, we have made advancements in our ability to utilize support materials to improve catalysts for CO_2 conversion and other reactions.

Parameters Optimization for Improved Catalytic Activity and Current Densities on Iron-Kudzu Based ORR Catalyst

Valentina Milano-Benitez

The development of a novel kudzu biomass-derived catalyst for enhancing the oxygen reduction reaction (ORR) is presented in this research. Leveraging kudzu, an invasive species and abundant carbon source, our approach offers an effective, sustainable and cost-effective alternative to precious platinum catalysts commonly employed in portable vehicles and stationary applications. The catalyst synthesis involves an optimized kudzu carbon precursor, iron salt, and nitrogen source. The carbon precursor incorporates an improved percentage of KetJenBlack, a unique electro-conductive carbon black material. Structural insights into the catalyst are obtained through scanning electron microscopy (SEM), elucidating the overall material structure. Catalytic performance is quantified through a Rotating Ring Disk Electrode Experiment (RRDE), allowing for the visualization and data collection of the onset potential in the redox reaction. This experiment is conducted in a 3-electrode cell using a 0.1M KOH electrolyte. The catalyst ink is deposited on a polytetrafluoroethylene-treated carbon working electrode until the desired loading is achieved. The study contributes to the advancement of sustainable catalysts with potential applications in energy storage devices.

Addressing Mechanistic Studies on Electrochemical CO2 Reduction Based on ATR-Seiras

Tatiana Giraldo, Adriano Braga, Brian M. Tackett

Strategies to reduce net CO2 emissions have had a significant progress in recent years. In particular, electrochemical reduction of CO2 presents a sustainable path to produce valuable fuels and industrially relevant chemicals while contributing to the process of global decarbonization. High activity and selectivity toward carbon products, and avoidance of CO2 crossover in alkaline electrolytes are crucial to make this process viable. To this end, recent studies have shown that selective electrocatalysis of this reaction on acidic conditions is possible only in the presence of metal cations, however It´s still uncertain the underlying mechanism by which the cation enhancement occurs or the impact that the identity of the cations have. This study aims to advance our understanding of this mechanism by exploring the molecular interactions on the catalyst-electrolyte interface with the use of in-situ/operando techniques, for this we conduct surface-sensitive ATR-SEIRAS experiments supported by grazing incidence XRD characterizations and electrochemical performance testing, which are also critical to shed light into key intermediates during CO2 reduction reaction. The key to achieve strong SEIRAS activity relies on appropriate reflection, conduction and catalytic activity of the metal film, giving that it serves as both the source of surface enhancement and the working electrode, so consequently, we also develop a detailed methodology for tailoring the catalyst surface, particularly building on Au deposition procedures in ITO-coated silicon chips in the way of improving the IR, taking advantage on surface plasmon resonance These efforts in catalyst design will translate into a hybrid ITO-Gold film electrode, that should reveal cation-dependent fundamental electrochemical phenomena that promote CO2 reduction in acid electrolytes, given that gold is active for CO2 reduction and ATR-SEIRAS.

Oxidation of Cyclohexane over Metal Nitrogen Doped Carbons and Zeolite-Supported Metal Phthalocyanines

Miles G. Miller, Ethan Iaia, Martin G. Bakker, James W. Harris

Oxidation of cyclohexane is of industrial importance for production of cyclohexanol and cyclohexanone, together termed "KA oil". KA oil is a precursor used for synthesis of Nylon-6 and Nylon-6,6. Nylon has wide applications across multiple industries; thus, KA oil synthesis is critical for modern life. Heterogeneous catalysts can be used for cyclohexane oxidation in the liquid phase at mild conditions with high selectivity to the alcohol and ketone products. Metal phthalocyanine catalysts (MPCs) and metal-nitrogen-doped carbon catalysts (MNCs), have similar molecular structures; most notably isolated single metal atom active sites coordinated to four nitrogen atoms which are themselves within carbon rings. MNCs and MPCs deposited onto and encapsulated within faujasite catalysts, MPC/FAU and MPC@FAU, respectively, were prepared and characterized for use in comparative studies of the reaction rate, apparent reaction orders, and product selectivity with respect to the liquid phase batch oxidation of cyclohexane utilizing *tert*-butyl hydrogen peroxide (TBHP) as the oxidant. Additionally, experiments were conducted to elucidate the effects of the zeolite support. Reactions were performed in a glass batch reactor sealed with a rubber septum and placed within an oil bath held at 303 K and stirred at 350 RPM. Products were identified using a gas chromatograph (GC) equipped with a mass spectrometer and quantified using a GC equipped with a flame ionization detector, using n-propanol as an external standard. Rates measured over sodium faujasite and nitrogen doped carbon without transition metal active sites were measured and subtracted from the measured rates over metal-containing samples. Comparisons are drawn regarding the relatively reactivity of each material by calculating the per metal atom turnover frequencies (TOFs). Our ongoing work indicates that MNC, MPC/FAU, and MPC@FAU catalysts with Fe active sites outperformed all other tested transition metals within these supports. TOFs for each of the sample types followed the following reactivity trends: Fe > Co > Ni for MNCs, Fe > Mn > Co > Cu > Cr > Ni for MPC/FAU, and Fe > Mn > Cr ≈ Cu for MPC@FAU samples. With these results and further study, we hope to further probe the effects of transition metal identities on catalytic performance in cyclohexane oxidation and to determine the effectiveness of studying well-defined MPC@FAU materials as models for the less tunable, less uniform metal active sites of MNCs.

Catalytic Lignin Depolymerization of Corn Stover

Maria Camila Castro Sr., Rajdeep Deka

Abstract

The aim of this work was to find from mass balances the extent of the catalytic depolymerization reaction of Corn Stover lignin. A reaction extent of 77.2% by weight of the final product was found with respect to a total lignin content in the biomass sample of 14.5%. For this, a 20 L Parr BSTR reactor, nickel catalyst on carbon support and methanol as solvent were used. The input biomass was previously sieved (0.15 mm to 50 mm) and free of dust. After the reaction, pressure filtration was used to eliminate solids and for liquid-liquid separation a rotary evaporator was used where methanol was removed, resulting in a solution rich in lignin. Then, the lignin-rich samples were subjected to a gas chromatography flame ionization detector (GC-FID) where 11 compounds were identified, including: vanillin 20.8%, methoxyisoeugenol 22.6%, coniferyl alcohol 25.6% and dimethylphenylpiperazinium 29.6%. This depolymerized lignin is then used as the main component of a resin that represents a safe alternative to formaldehyde in the production of composite boards and wood adhesives. Additionally, give added value to the lignin that remains as waste in the paper industry and biorefineries.

Characterization of an Electrocatalyst $Fe_4n(CO)_{12}^-$ for CO_2 Reduction and H_2 Evolution

Franklin Guevara, Kevin Lee, Louise Berben

Energy from renewable, periodic sources like solar or wind must be stored during times of less than peak production. If we could store the electrons produced by renewable means in an energetically dense fuel, it would aid in the economical scale-up of renewable energy. Additionally, fuels that are both energetically dense and can be sustainably produced without the use of fossil fuels or without intensive heat or pressure are ideal due to global sustainability goals. Thus, the chemical reduction of atmospheric carbon dioxide or carbon dioxide present in flue gas to value added products such as formic acid, which can be used in fuel cells, is an important factor in reducing global emissions and transitioning to net zero.

It has been shown that certain metal carbonyl clusters have electrochemical redox properties that might prove useful for the catalytic conversion of carbon dioxide into formic acid. Through the use of cyclic voltammetry techniques, the order and rate of the reaction of these clusters can be determined, providing further insight into the reaction mechanism and thus allowing for manipulation of the catalyst to deliver the desired product. Under certain parameters determined by pH, Et4N[Fe4N(CO)12] can selectively catalyze the production of either hydrogen or formate.

Magnetic Field Enhancement of the Electrocatalytic Oxygen Evolution Reactivity in Amorphous Cobalt Oxide Thin Films

Hermann Klein-Hessling Barrientos, Ruben Mirzoyan, Ryan G. Hadt

Photoelectrochemical water-splitting presents a promising avenue to generate sustainable molecular hydrogen as a fuel source and molecular oxygen as a biproduct on the commercial scale. The kinetic and thermodynamic barrier behind this process is the oxygen evolution reaction (OER), which involves the transfer of 4 protons and 4 electrons to form the oxygen-oxygen bond. This study investigates the effects of a directional, uniform, and tunable externally applied magnetic field on the oxygen evolution reaction (OER) activity of the phosphate-based and borate-based amorphous cobalt oxide thin films. These films are of particular interest since they consist of abundant earth materials, are low-cost, operate at neutral pH, and have self-healing properties. A magneto-electrochemical setup was designed and 3D printed to investigate how varying strengths of a magnetic field influence the OER performance in these catalysts. Increases in OER catalytic activity of up to an order of magnitude, as defined by Tafel Slope analysis, were observed at the highest attainable field strength of 1.4 T. The observed magnetic field enhancement was found to be reproducible across various films of the same thickness and was found to be most significant in CoPi films of 5mC thickness. The results offer unique insights into the capabilities of magnetic field enhancing OER catalysis in amorphous cobalt oxide films, and further supports the yet-inconclusive intramolecular oxygen radical (O*) coupling mechanism for the oxygen evolution rate determining step between adjacent Co oxo units.

Development of Nanostructured NiMo Oxides for Oxygen Evolution Reaction.

Jillian Richter, Jean Pascal Fandré, Harun Tüysüz

Water electrolysis is a method of splitting water molecules into their O_2 and H_2 components and is an imperative part of green hydrogen production. The sluggish oxygen evolution reaction (OER) is a half reaction that requires an electrocatalyst to improve the overall process. Green hydrogen has many applications including hydrogen fuel-cells, hydrogen feedstocks, and alternative fuels. Recurring electrocatalyst metals used for this half reaction include Ni, Fe, Co, and Mn. These metals have been combined with numerous other transition metals for overall improvement of the reaction. We aim to expand the collection of knowledge on promoters possible for the NiO electrocatalyst for increased stability, selectivity, and activity. In this study, the synthesis of a Ni-Mo metal oxide electrocatalysts with varying molar ratios and synthesis methods was studied to observe the differences in OER activities. All catalysts were synthesized through hard templating methods, using SBA-15 as a support. The electrocatalysts synthesized included a Ni_4MoO_4, Ni_8MoO_4, and Mo-doped NiO catalyst. Reflection X-ray diffraction (XRD), Brunauer-Emmett-Teller (BET) surface area analysis, Fourier-Transform Raman Spectroscopy, and energy-dispersive X-ray analysis (EDX) are used to characterize the prepared materials. Tafel slope, Nyquist plots, and cyclic voltammetry measurements are considered for the OER evaluation. The experimental results show the modifications necessary to utilize hard templating methods with Mo, as well as the resulting material if these modifications are neglected. This is especially highlighted through the XRD and EDX data. The OER screening showed that the Ni_8MoO_4 electrocatalyst was overall the most efficient based on an evaluation of the overpotential, the current density, the resistance, and the Tafel slope. Our approach to synthesize Ni-Mo-O electrocatalysts will provide an evaluation of the addition of molybdenum to the NiO electrocatalyst for the progression of efficient OER reactions.

Oxygen Reduction to Hydrogen Peroxide Electrocatalyzed By Hydrophilic Carbon Fiber Paper

Teona Taseska, Yiwen Sun, Connor P. Cox, Madeleine K. Wilsey, Kendra R. Watson, Lydia Schultz, Astrid M. Müller

Hydrogen peroxide is one of the most important chemicals in the world because of its many applications in energy storage, chemical synthesis, and water treatment. Currently, hydrogen peroxide is synthesized using the very energy-intensive anthraquinone process that generates large amounts of chemical waste and requires a complex and large-scale infrastructure. Therefore, multiple studies have been conducted to develop a more cost-effective and environmentally friendly synthesis method. A process that satisfies those requirements is electrocatalytic oxygen reduction in aqueous electrolytes. Electrocatalytic processes are powered by electricity, can utilize renewable sources, and do not need large, complex infrastructures, thereby allowing decentralized use. In this research, we investigated the production rate of hydrogen peroxide as a function of electrolyte pH, using oxygen reduction electrocatalyzed by hydrophilic carbon fiber paper in aqueous media. We employed a green chemistry process that was recently developed by the Müller group to imbue carbon fiber paper with long-lasting hydrophilicity,[1] which served as cathodic catalysts for hydrogen peroxide electrosynthesis. We systematically varied applied potential and the composition and pH of the aqueous electrolytes to gain mechanistic insights and optimize the production rate of hydrogen peroxide. Overall, we have developed a clean, cost-effective, energy efficient, and environmentally friendly approach to produce hydrogen peroxide.

1) M.K. Wilsey, K.R. Watson, O.C. Fasusi, B.P. Yegela, C.P. Cox, P.R. Raffaelle, L. Cai, A.M. Müller, Selective Hydroxylation of Carbon Fiber Paper for Long-Lasting Hydrophilicity by a Green Chemistry Process. *Adv. Mater. Interfaces* **2023,** *10*, 2201684.

Diffusion and Sedimentation in Suspensions of Shape-Anisotropic Nanoparticles

Penelope Kovakas, Yashraj M. Wani, Arash Nikoubashman, Michael Howard

Nanoparticles are a crucial emerging technology in the fields of material science and chemical engineering, for example, to deliver medicines, to improve industrial processes such as oil recovery, and to enhance or modulate properties of other materials by making composites. Nanoparticle dynamics, such as diffusion or sedimentation, play an important role in determining how nanoparticles are processed and ultimately behave in these applications. The dynamics of spherical nanoparticles are well understood but far less is known about the dynamics of the many different shapes of nanoparticles that can be synthesized in the laboratory. To address this knowledge gap, we performed computer simulations of tetrahedral, cubic, octahedral, and spherocylinder nanoparticles suspended in a Newtonian solvent. We used the multiparticle collision dynamics (MPCD) method to model important solvent-mediated hydrodynamic interactions between nanoparticles. Diffusion and sedimentation coefficients were measured as a function of nanoparticle volume fraction using equilibrium and nonequilibrium simulations, respectively. The simulation results were compared to experimental data and theoretical predictions where available. This work provides useful reference data on the transport properties of nanoparticles with anisotropic shapes, and the methodology can be readily extended to new shapes.

Enhancing Sabatier Reaction Efficiency with Membrane Reactors

Owen Gerdes, Vitor Renan Vitor Gama, Oishi Sanyal, Fernando Lima

With growing concerns about rising CO_2 levels and their environmental effects, the relevance and opportunity for impact of the Sabatier reaction are increasing. This reaction converts CO_2 and hydrogen into methane, which makes it crucial for carbon utilization and generating sustainable energy. The Sabatier reaction has a unique application for spacecraft fuel generation using cabin air rich in CO_2. Intensified unit operations, like membrane reactors, may improve efficiency by increasing CO_2 conversion at equivalent temperatures, overcoming energy consumption challenges. Considering the membrane reactor as a shell and tube reactor, cabin air is passed through the tube side, H_2 is passed through the shell side, and CO_2 passes through the membrane to the shell side, where the catalyst is packed, and the reaction takes place. A membrane reactor and a traditional packed bed reactor were simulated using AVEVA Process Simulation (APS) in which both carried out the Sabatier reaction and operated with equivalent space time and space velocity. The two methods' heat duties, capital costs, and operating costs were assessed to compare the viability of the membrane reactor. The respective conversions of the two reactors were obtained with the reactors operating at similar temperatures, H_2/CO_2 feed ratios, and catalyst amounts. Sensitivity analysis was then performed to find the ideal temperature range for identical conversion for the two reactors.

Modeling Blood Flow in Repaired and Healthy Aortic Arches Using Computational Fluid Dynamics

Dominic Lippa

The aortic arch is connected directly to the heart. It is responsible for delivering blood to three arteries, known as head vessels, and the lower extremities. Birth-defects in the aortic arch, known as coarctations (CoA's), hinder the flow of blood in the body, which can manifest into many health problems. Research was conducted this summer using Computational Fluid Dynamics (CFD) software to analyze flow patterns in healthy and surgically repaired aortic arches, as well as some unrepaired arches to ultimately improve patient outcomes.

Patient MRI and CT scans were provided by the University of Rochester Medical Center, and modeling the flow of blood through the aortic arch was done using the CFD software ANSYS Fluent. The three models used to collect data were the SST k-omega, k-epsilon, and laminar. Patient blood pressure (BP) data was implemented at each of the four major openings in the arch. Some trials used the diastolic BP at the ascending aortic arch, and the systolic BP at the descending aortic arch. The three head vessels were left unassigned so the simulation could calculate pressures at those outlets. Simulations were run to obtain velocity streamlines, pressure contours, and wall shear contours for nine patients.

The SST k-omega model was the most representative. The best residuals were produced using the SST k-omega model and simulation times were efficient (2-3 hours), which corresponds to a smooth model and properly designed mesh. In contrast to the k-epsilon model, the SST k-omega model consistently converged in trials when the k-epsilon model did not. Figures produced by both the SST k-omega model and the laminar model were similar, but additional parameters in the SST k-omega model resulted in more accurate near-wall simulations.

In all, patients with CoA's displayed more complex velocity streamlines, and greater pressure gradients near the third head vessel on the arch. However, the wall shear in healthy arches compared very closely to those with CoA's.

Exploring Machine Learning Models for Predicting Vapor Pressure

Eli Kaplowitz, Charles McGill

The current way that vapor pressure is predicted is through fitted empirical equations, such as Antoine's equation. The problem with the current methods is that the coefficients of these equations need to be found experimentally for each different molecule and temperature range. Fitting to these empirical forms do not extrapolate well outside the fitted temperature range. Also, these equations cannot be used to estimate vapor pressure for species where experimental data are not already available. To address these problems, the goal of this experiment is to create a machine learning model to predict vapor pressure of molecules. The machine learning models are built by using Chemprop, a software package for creating machine learning models for molecular properties.

In this study, we explore several variations of model structure. The first is the most basic in that it is a standard model predicting the pressure from the training data directly, with the temperature as an added model feature. A more complex version of the model involves predicting the Antoine coefficients as an intermediate model result, instead of predicting the pressure directly. Then, using those coefficients, along with temperature, to get the vapor pressure. A third option is to use a separate model to predict extra data, like the critical temperature and pressure, and adding those as extra features, along with temperature. A fourth option is to introduce a level of noise to the temperature included as an extra feature to the model in order to avoid overfitting the Antoine's coefficients. The method for this experiment involves taking data gathered from different sources (DIPR, NIST, Yaws Handbook) and using it with Chemprop to create models with the different structures outlined above. We discuss the effect of different model structure variations on the accuracy of the resulting models, such as which model variations perform the best when used in conjunction with each other. In evaluating model performance, we address both potential use-cases where data is not available for a molecule and cases in which the tested data is outside the fitted range of a model.

Development and Optimization of Robotic Nanodroplet Dispenser with Hydrophobic Needles and Computer Vision

Yuntian (Areopl) Bai

- The goal is to develop a robotic printer that rapidly dispenses nanoliter-scale droplets of polymer solution for evaporative deposition of thin polymer films onto spiral waveguides of biosensing chips.

- Advantages of the custom instrument include an increase of printing speed, lower cost, and no nozzle clogging issues compared to previous piezoelectric printing processes.

- A process of surface chemical treatment to RN needles was developed of gauge 28-32. They were tested with air tight syringes to eliminate dead volume.

- First, cleaned needle surface was roughed through acid etching using 3:1 hydrochloric and nitric acid mixture. After etching, the needle was dip-coated with 1H,1H,2H,2H-Perfuorodecanethiol to create a fluorinated surface.

- The purpose of hydrophobic treatment is to increase the contact angle between polymer solution droplet and nozzle wall to facilitate consistent droplet detachment.

- In the semi-contact printing process, the liquid droplet forms a bridge between the nozzle and printing surface, with the nozzle not making direct physical contact with the printing surface.

- The use of computer vision was implemented through the mounted microscope enabling more precise and automated printing processes.The onboard microscope captures the photonic chip inside the calibration pattern. Perspective warping correction and edge detection algorithms are applied to identify the shape of the chip, X/Y coordinates of the chip's edges are calculated.

- Using the known design of the chip, approximate X/Y coordinates of individual spiral waveguides are calculated, and these are superimposed on the image for precise alignment and automated printing process.

- Further research will focus on quantitatively measuring droplet sizes and optimizing the distribution of polymer solution deposition for enhanced control.

- Ongoing efforts will optimize droplet volume, solvent properties, and polymer concentration to achieve desired polymer thin film thickness and uniformity. Ways to reduce the Coffee Ring effect and improve contact pinning will be further investigated.

CFD Analysis of Mixing in the Transition Regime: User-Defined Functions and Viscosity Variations

Julia Baker, David G. Foster

Mixing is a common process used in the manufacturing and processing industry. To optimize the mixing efficiency, it is critical to understand what is going on inside the tank. In a physical mixing tank, several flow regimes exist simultaneously. At the impeller, turbulent flow is likely to exist. In the bulk and the surface of the tank, the fluid can be in laminar flow. Between the two extremes, there will be a wide range of Reynolds Numbers, all falling into the transitional flow regime. The process for industrial mixing is typically modeled in the laminar or turbulent regimes, as the equations governing these regimes are well understood[1-2]. Due to the complexity of transition mixing, there is minimal modeling done in the regime despite it being a common working regime for some industries. The lack of accurate transition mixing models hinders these industries in fully understanding their processes[1]. Computational Fluid Dynamics (CFD) is used in modeling various physical systems to analyze velocity profiles, streamlines, volume fraction distribution, and numerous other fluid and flow properties. In this case, ANSYS Fluent, a commercial CFD software package, was used to model the volume fraction distribution of dye that is released in the tank while being mixed. Inaccuracies of transition mixing models may arise from the inability of CFD to predict large scale instabilities in the flow domain. To circumvent this issue, multiphase transient simulations of the mixing of Carbopol were conducted to study the transition mixing region with higher accuracy than before. Carbopol is a non-Newtonian, shear-thinning fluid whose viscosity profile can be altered with change in pH and/or concentration. Two different viscosities, 7500 cP and 3350 cP, were used to investigate their dependence on flow patterns. Different mixing regimes were achieved by changing the speed of the impeller. Two turbulent models, the Transition Shear-Stress Turbulence Model and the Spalart-Allmaras Model, were used to model the mixing in the tank and compared for analysis. Both models accurately depicted the behavior of the Carbopol at high Reynolds numbers, as expected from the two turbulent RANS models. A user-defined function (UDF) was then created from the Carreau-Yasuda shear-thinning viscosity model to better represent the dispersion of dye inside the mixing tank in the transition regime. The UDF successfully modeled the shear-thinning of the Carbopol throughout the entire mixing tank.

Designing Automated Systems to Monitor the Effect of Humidity on Membrane Permeability

Nathaniel Rudman

Polymer membranes have promise to reduce the cost of separating carbon dioxide after the burning hydrocarbon based fuels. Humidity often changes membrane properties due to a phenomenon called plasticization that often reduces the selectivity and performance of a membrane. In the Moon lab, we are interested in understanding how humidity impacts gas transport mechanisms in facilitated transport membranes. This requires equipment capable of measuring gas permeabilities of newly synthesized materials under controlled humidity conditions.

I am designing and constructing a custom, automated system to monitor the effect of humidity on the gas permeability of membranes synthesized in our lab. I am designing a hardware controller with a graphical user interface (GUI) with the goals of ease, safety of operation and automated data collection that requires minimal user input. Pre-scripted routines will allow for batch measurements of membrane gas permeability as a function of pressure, temperature, and relative humidity. The real time data the system generates will be interpreted, displayed, and logged for further analysis. The operator of the system needs to be capable of both manually controlling pneumatic valves within the system and running pre-set procedures that control the valves. Additionally, valve operation needs to have software interlocks implemented to ensure the safety of the operator cannot be compromised. Finally, the GUI needs to be capable of displaying real time data and ensuring ease of access to any operator of the system.

Cooking with Chem-E: Analysis of Sensitivities of Assumptions in Multilayered Beef Wellington Comsol Model

Hannah Smith, James E. Maneval, Margot Vigeant

The analysis of what makes a complex simulation "good," or accurate, is highly relevant in chemical engineering. Assumptions are constantly being made with engineering design aspects, and the hope is to illustrate this significant chemical engineering concept with a model that people can resonate with: the example of food. A model of the cook time of beef wellington, the popular British dish, takes around 40-45 minutes to cook at 425°F. The goal of the research was to create a transport-based model that would predict cooking time based on different assumptions and compare it to actuality.

Principles of heat and mass transport coupled with the equations of energy and appropriate boundary conditions were used to provide the mathematical model regarding the beef wellington. Like the traditional dish, the steak fillet was at the center, followed by a cream-based mushroom sauce, and ending with a puff pastry wrapped around the entire piece of steak. This dish is normally served rare, or about 48°C at the center. The computational model was developed via COMSOL Multiphysics software, version 6.1. The material and temperature properties of these layers were included in the COMSOL Multiphysics model. For the purposes of the model, the beef wellington was considered a multi-layered cylinder with constant volume. One of the many assumptions that were implemented in the model was that there was that moisture transport by diffusion was neglected, so the equations of continuity were not used in the first case of the model.

This "base case" model with its relevant assumptions, differential equations, and boundary conditions, predicted cook time of 40.2 minutes. When determining the sensitivity of the model to thermal conductivity of the meat, the cook time increased to 45.8 minutes when thermal conductivity decreased by forty percent and decreased to 34.6 minutes when thermal conductivity increased by forty percent. In accord with physical expectations, cooking time decreased as the thermal conductivity increased, further validating the model. The next phase of research will address the impacts of the addition of moisture transport will truly affect the predictions by measuring cook time at different levels of doneness: rare, medium, and well-done.

Computer-Verified BET Analysis Using the Lean Theorem Prover.

John Velkey, Parivash Feyzishendi, Maxwell P. Bobbin, Tomáš Skřivan, Tyler R. Josephson

Chemical engineering frequently requires calculations for solving problems. Because of the wide range of computational tools and methods available for performing calculations of this nature, inconsistencies between methodologies are common. For example, when adsorption datasets were sent to sixty-one different labs, there was wide variation in the corresponding BET surface area reported, despite the fact that determination of BET surface area is a well-known calculation [1]. Consider the utility then, of programs that perform calculations using functions that have provable mathematics, that are "bug free" as a result of their mathematical verification. To illustrate this concept, we use Lean 4 to write a computer-checked derivation of BET theory, and then link this derivation to functions that regress the BET parameters for a data set.

Lean 4 is a novel, multi-purpose programming language that also functions as an interactive theorem prover. We first use Lean as a theorem prover, and formally verify the derivation of the BET adsorption theory. This proof is first presented in [2], originally implemented in an earlier version of the Lean interactive theorem prover. Proofs in Lean 4 are not identical to executable computing programs, which can be separately defined as functions. Not all computing functions can have proofs written on them, if they are strictly typed to function with floating-point numbers. This is due to the disjoint relationship between Real numbers and floats. Reals can be the subject of proofs, but because they potentially have infinite decimal digits, they cannot be computed or displayed by computer systems. Instead, scientific computing uses floats to perform actual calculations, but proofs about floats are not the same as proofs about reals; even $a + b - b = a$ is not true for all floats. To make verified scientific computing possible in Lean, we introduce a method using polymorphic functions to connect proofs about real-valued functions to computations using the same functions with float-valued inputs. We use the developed code to analyze datasets from [1] performing BET regression, and compare our formally-verified results to those previously reported.

[1] Osterrieth, Johannes W. M., et al. "How Reproducible Are Surface Areas Calculated from the BET Equation?" Advanced Materials, vol. 34, no. 27, 23 May 2022, p. 2201502, https://doi.org/10.1002/adma.202201502.

[2] Bobbin, Maxwell P, et al. "Formalizing Chemical Physics Using the Lean Theorem Prover." ArXiv (Cornell University), 21 Oct. 2022, https://doi.org/10.48550/arxiv.2210.12150.

Enabling the Structure-Property Predictions of New Organic Photovoltaic Chemistries with Quantum Chemical Parameterizations

Gwen White

We develop computational tools to fill in force field parameter gaps for molecular dynamics (MD) simulations. Specifically, we focus on high-efficiency organic photovoltaic (OPV) materials Y6 and BTO, which have recently achieved power conversion efficiencies exceeding 17%. We implement the open-source QUBEKit software stack on high-performance computing infrastructure to parameterize molecular fragments, defined using SMILES strings. We examine the structural characteristics of these materials using real and frequency space techniques through MD simulations varying in density and temperature. Our findings reveal the absence of discernible crystal structures or long-range periodicities within these morphologies. Despite their lack of order, these factors suggest that the studied molecules have charge transport capabilities that do not rely on short-range order, setting them apart from many other OPV materials. Additionally, we conduct a comparative assessment of the QUBEKit force field against the established OPLS amber force field, providing insights into their relative suitability for OPV materials simulations.

Machine Learning for Predicting Cradle-to-Gate Life Cycle Inventory (LCI) Data

John Pazik, Austin Lehr, Emmanuel Aboagye, Jared Longo, Ethan Blanda, Robert Hesketh, Kirti Yenkie

Improving the sustainability of chemical processes is crucial for mitigating the effects of climate change. One way this can be accomplished is through incorporating considerations for environmental impacts during the early stages of process design while modifications can still be easily made. Unfortunately, there are currently no simple methods for determining environmental impacts this early on, with most of them involving predictions based on molecular structure via large-scale simulations involving Density Functional Theory or Molecular Dynamics. These simulations are time-consuming, and therefore not very practical for use by design firms that often have limited time. By using Machine Learning (ML), it is possible to predict the environmental impacts of the chemicals used within these processes much faster and with much less computational expense, allowing for this data to be available to design engineers in a much more practical manner.

In this work, a ML algorithm was developed to predict cradle-to-gate life cycle inventory (LCI) data for both pre-existing and novel chemicals. The LCI data predicted by the algorithm focused on four different environmental metrics: human health (HH), ecosystem quality (EQ), global warming potential (GWP), and recourse utilization (RU). Two variants of the algorithm were developed: one employed Artificial Neural Networks (ANN), and the other employed eXtreme Gradient Boosting (XGBoost). Both variants used over 350 data points, split between a training, testing, and validation set. Data was sourced from EcoInvent, a chemical information database, and the feature set included 200 molecular descriptors and 16 thermodynamic properties for each datapoint. A stepwise feature selection process was used to reduce the number of features from 216 to 10. Following hyperparameter tuning, the performance of both variants was assessed on the test set using the R-squared and Root-Mean-Squared-Error (RMSE) values. Following this analysis, it was determined that the XGBoost variant was more effective for the HH, GWP, and EQ metrics, producing RMSE values equaling 2.73, 1.18, and 0.574, respectively. In addition to the algorithm itself, a case study representing the extraction of polyphenols from wine pomace using acetone solvent was also implemented to demonstrate the utility of the LCI prediction algorithm in enabling the creation of a detailed cradle-to-grave Life Cycle Analysis of the entire process.

Automating Self-Driving Laboratory Growth ASSAY Protocol to Efficiently Measure Cell Antibiotic Resistance

Alp Demirtas

Antibiotics have saved millions of lives over the past century, but their efficacy is decreasing because bacterial strains, especially in hospital settings, are developing resistance to antibiotics faster than new antibiotics can be produced.[1] Finding new antibiotic treatments that inhibit bacterial growth requires testing multiple combinations of antibiotics, which becomes strenuous and time-consuming for human researchers. Through the combination of Artificial Intelligence (AI) and robotics in self-driving laboratories, these treatments can be discovered more efficiently. AI can more intelligently select which treatments to test while the robotic experimentation platform can perform the necessary wet lab validations on a faster and larger scale than human researchers. With self-driving laboratories, these treatments can be discovered more efficiently due to a higher number of experiments being run continuously and machine learning algorithms to predict combinations. This paper seeks to create a fully automated biological assay with automated data processing to determine a bacterial cell's susceptibility to an antibiotic. The robotic experimentation platform for the experiment includes a liquid handling robot, optical density reader, microplate manipulator, peeler, incubator, and sealer. All robotic actions are initiated from a Python script that invokes a Workflow Execution Interface (WEI) server to run workflows. Each workflow serves as a protocol of automated actions that are executed when the WEI server sends a request to the associated robot node. The automated experiment, intended to generate dose-response curves after a 12 hour incubation, is tested with *Escherichia coli* (*E. coli*) and tetracycline antibiotic as a proof-of-concept, and could later be used to test other antibiotic resistant bacterial strains. The results indicate that *E. coli* is susceptible to tetracycline antibiotic and there is negligible contamination because of the robot actions. These results were consistent with previous literature on the interactions between *E. coli* and tetracycline, demonstrating the promise of self-driving laboratories as a high throughput experimentation service for Argonne researchers.

[1] (Ventola, *The antibiotic resistance crisis: part 1: causes and threats* 2015)

Development of Automated Active Learning Loop for the Prediction of Zeolite-OSDA Binding Energies

Nga Vu, Rafael Gomez-Bombarelli, Mingrou Xie

The production and consumption of fossil fuels contributes significantly to environmental pollution and climate change. A recent effective alternative is to utilize zeolites, nanoporous materials, to act as a sustainable catalyst in innovative separations processes of transforming biomass into chemicals and fuels because it generates considerably less pollution than traditional methods. When synthesizing a suitable zeolite, an OSDA (Organic Structure Directing Agents) and its binding energy with zeolite guide the formation of the zeolite's porous structure. To guide zeolite synthesis experimental work, computational techniques in calculating binding energy between zeolites and OSDAs have been utilized. However, this is an extremely expensive and resource-intensive process, especially in the scope of 6 hundred billion possible pairings between over 3 million OSDAs and 85 thousand zeolites. Therefore, there is a need to build a machine-learning model to predict the binding energy between zeolites and OSDAs to save computational time and resources. Our project aims to develop a streamlined codebase that employs active learning to continuously modify and improve the existing machine learning model. The code will enable the system to automatically screen the chemical space for data points of interest and set up calculations of actual binding energies. Then it will extract these computed energies from the database and retrain the model with those new data. This process iterates until the model is capable of robust inference in the desired chemical space of zeolites and OSDAs. This exploration could lead to the identification of suitable syntheses of new, desired zeolites that have desirable properties for catalyzing these sustainable chemical processes.

A Machine-Learning Approach to Simulate Confined Liquid Lithium Polysulfides for Battery Applications

Joshua Zhou, Brandon Bukowski

Lithium-sulfur batteries have great potential to be a high-capacity and environmentally friendly alternative to existing lithium-ion batteries but are heavily limited by their inability to retain charge after cycling. There is a polysulfide shuttle effect where the battery loses effective mass during cycling due to transport of liquid polysulfides resulting in non-conducting Li_2S gradually coating the cathode during use. It has also been shown that embedding a metal-organic framework (MOF), a porous material, is able to dramatically improve the battery's cycling capabilities by reducing liquid polysulfide diffusion into the electrolyte. However, the underlying mechanisms are not yet well understood. Here, we investigate how lithium and sulfur interact as a confined liquid. Our simulations use ORCA, a quantum chemistry software, and the extended Tight Binding (xTB) semi-empirical algorithm. We also use Parinello meta-dynamics to ensure sampling of a wider area in chemical configuration space. The energy, force, and coordinate data generated from these configuration searches are then used to train a machine learning model using DeePMD. This model is then used to simulate a lithium-sulfur-MOF system with an order of magnitude more atoms than can be conventionally done using strictly ab-initio techniques. We also intend to integrate other compounds, such as solvents, and battery anodes and cathodes, into our simulations. We hope to understand why lithium and sulfur atoms interact differently in the presence of a MOF, what features of these MOFs best suit our needs, and how may we best utilize them.

Development of Machine-Learning Models for the Prediction of Membrane Active Peptides with Cell-Penetrating Capabilities Based on Amino Acid Periodicities

William DeFreese, Atefe Alimirzaei, Christopher Kieslich

Development of Machine-Learning Models for the Prediction of Membrane Active Peptides with Cell-Penetrating Capabilities Based on Amino Acid Periodicities

Recently, membrane active peptides have demonstrated promise as alternatives to conventional small molecule-based therapeutics. Upon interacting with the cell membrane, these peptides exert their activity either by translocating through it to deliver cargo or to disrupt it resulting in cell lysis. Specifically, cell penetrating peptides (CPPs) pass through the cell membrane, transporting bioactive molecules such as plasmid DNA, nanoparticles, imaging agents for disease diagnostics, other peptides, and proteins. CPPs are composed of less than 30 amino acids and are cationic, amphipathic, and hydrophobic in nature. All these properties make them ideal for both carrying cargo and interacting with cell environments without disruption. However, experimental design of CPPs is time consuming and expensive in the wet lab. Therefore, computational prediction of their activity is highly desirable for most efficient experimental design. To this aim, we have developed and trained support vector machine models (SVMs) to discriminate between active and nonactive CPPs. MCLPP 2.0 datasets have been used which contain both training and independent test sets. We utilized the Blocks Substitution Matrix (BLOSUM100) to first vectorize the sequences. It has been previously shown that oscillations of amino acid periodicities can categorize structure and function of peptides. Therefore, we applied the Fourier Transform to each of the generated BLOSUM100-based representative vectors to measure the amplitude of the amino acid periodicities. This resulted in 160 frequency descriptions (Fourier transform coefficients) for each peptide which served as input features to our SVM models. We performed 10-fold cross validation to train and tune the models based on 10 training testing sets. After assigning best hyperparameters, we carried out a feature selection algorithm based on certain criteria to rank the features. This subsequently allowed us to determine the most predictive features contributing to cell penetrating activity of peptides. We then exposed our models to the blind test sets utilizing the best features found and evaluated the models' performance using various metrics. Furthermore, we have incorporated other sets of physicochemical features and properties of amino acids into our models to try our prediction accuracy. Upon comparison to state-of-the-art studies, our novel featurization and feature selection methods produced promising results.

Keywords: Cell Penetrating Peptides, Machine Learning, Support Vector Machines, Feature Selection, BLOSUM100

A Computational Tool for Taxadiene Synthase Mutation Site Screening

Liv Toft, Eleanor Dong, Codruta Ignea

Taxol is a high value pharmaceutical who's potent antitumoral properties have been used to treat over one million patients since its discovery. Efforts have been made to optimize the Taxol biosynthesis pathway in engineered yeast strains by identifying a mutated variant of the bottleneck enzyme, taxadiene synthase (TXS), that exhibits improved product specificity and catalytic activity. This project focused on creating a model of TXS activity that can be used as a computational tool for screening mutation site candidates. The 3D structure of the active enzyme was constructed using homology modeling. It includes a previously unresolved N-terminal region and "JK-loop" which serve the purpose of ensuring that the substrate remains in the active site during catalysis. Following this, quantum mechanics/molecular mechanics molecular dynamics (QM/MM MD) simulations were performed for the active TXS model docked with the initial substrate, the five intermediates, and the two taxadiene isomer products. The resulting models reveal that different sets of residues interact with each of the different substrates. Furthermore, two mutations sites that were previously shown to inactivate TXS were found to play a key role in stabilizing one of the intermediates. These findings demonstrate how these models can be used as a tool for screening mutation site candidates before wet-lab testing, greatly accelerating the discovery pipeline.

Autonomous Image Segmentation for Single Nanoparticle Tracking in Liquid Phase Transmission Electron Microscopy

Naisargi Goyal

Solutions of nanoparticles are extensively used in catalysis, energy storage, drug delivery, and inkjet printing. To understand the structure-property relationship of such solutions, it is necessary to analyze the interactions and motions of nanoparticles in the liquid environment at the single particle level. Liquid-phase transmission electron microscopy (Liquid Phase TEM) is a promising technique for studying the dynamics of these solutions. Liquid Phase TEM is an emerging in-situ microscopy technique that allows us to visualize the dynamics of nanoparticles in their native liquid environment with an unprecedented spatial (nanometer) and temporal (millisecond) resolution. However, the analysis of liquid-phase TEM images is challenging due to the presence of noise caused by the thick liquid layer and the interaction of the liquid cell window with the incident imaging electrons. This lowers the contrast between the nanoparticles and the liquid background, making it challenging to identify and track the nanoparticles accurately. Moreover, with the rapid advancement of microscopy technologies, vast amounts of data are generated. To address this data deluge resulting from multimodal in-situ microscopy data, automated algorithms for efficient data processing are needed. Traditional image segmentation methods, such as ones based on thresholding, are time-consuming and labor-intensive. In recent years, machine learning-based approaches such as deep learning architectures have shown remarkable performance in image segmentation. These models can be trained on diverse datasets and hence can implement image segmentation on wide variety of samples, including those with low contrast or high noise. Furthermore, these models eliminate the need for manual parameter adjustments, streamlining the image segmentation pipeline and removing any human bias from the image analysis. Here, we developed an unsupervised autoencoder deep learning model for image denoising and segmentation in liquid phase TEM. The algorithm identifies a single nanoparticle by isolating the region of interest, removing background noise, and outputting the boundary of nanoparticles, position, and orientation. To train and test the model, tens of thousands of real, raw TEM images were used as input and the corresponding segmented images generated by the traditional image segmentation thresholding algorithm developed previously were used as the output training dataset for this model. This algorithm can be used to study the dynamics of nanoparticles of various shapes and sizes in the liquid phase. The results from this project will be an essential part of an artificial intelligence-based workflow for automating electron microscopes to analyze samples faster, more efficiently, and without any human intervention.

Zero-Point Energies from Bond Orders and Populations Relationship

Colin Rude

Our group recently reported a bond energy from bond orders and populations (BEBOP) model that can quantitatively predict atomization energies and bond energies using well-conditioned orbital populations that are obtained from *ab initio* single point energy calculations. We now show that a similar procedure can be used to predict zero-point vibrational energies (ZPEs) of a molecule using just one *ab initio* single point energy calculation. Like the BEBOP model, our ZPE-BOP model is based on Mulliken orbital populations obtained from a minimum basis set projection of Hartree-Fock orbital populations using a well-conditioned basis set. Seven atom-pairwise parameters were fit to reproduce B3LYP-scaled ZPEs for molecules containing H, Li, Be, B, C, N, O, and F atoms using expressions that involve an extended Hückel-type bond energy term, a short-range anharmonic energy term, and a coupled three-body oscillator energy term. Notably, this first-generation model has better accuracies (as benchmarked to 109 species from the G2/97 test set and an additional 83 reference species) than other highly efficient and approximate quantum chemistry methods (e.g., AM1, PM6, and PM7 semiempirical methods). Moreover, this model appears to be able to extrapolate ZPEs for molecules that are not part of our training set, indicating that ZPE-BOP would be transferable across chemical space.

Melting Point Predictions of Thermally Stable Ionic Liquids Using a Machine Learning Approach.

Jarod Morris, Brooks D. Rabideau

Thermally robust ionic liquids hold promise as high temperature solvents, heat transfer fluids, and high-performance lubricants. These ionic liquids, inspired by thermoplastics like PEEK and PES, incorporate perearyl functional groups that lend them high thermal stability, however they often have melting points greater than 100 °C. A broad liquid range is usually desirable in most applications, so finding thermally robust ILs with lower melting points would significantly enhance their applicability. In this study, machine learning models are developed to make predictions of melting points for these thermally stable ionic liquids using a process called quantitative structure-property relationships (QSPR). The methodology used to model these ionic liquids was adapted from a previous study [Paduszyński et al., Journal of Molecular Liquids, 344, 2021] that examined multiple different machine learning techniques in depth. Modeling was performed on the large and diverse set of 929 ionic liquids from this study, then on a set of 53 thermally stable ionic liquids, then on the combined data set. The performance of each model was evaluated using 10-fold cross validation and the calculation of Q^2, which measures the predictive capacity of the model. This cross validation only considered the predictions of thermally stable ionic liquids since that is the overall intent of the modelling. From this cross validation, it was found that the model trained on both sets of data and had no outliers removed showed the greatest predictive capability (with a Q^2 of .6764) for thermally stable ionic liquids. This model is then used to predict ionic liquids from the thermally stable family with low melting points.

CFD Analysis of Repaired Coarctation of the Aorta

Olivia Dillenbeck, Dominic Lippa, David Foster, Jason Mandell

This project focuses on modeling aortic arches of those who had procedures due to Coarctation of the aorta (CoA) which is defined by a stenosis of the descending thoracic aorta (DAo). CoA is a congenital heart defect that occurs in newborns and results in the narrowing of the aortic arch. This defect impacts approximately 6-8% of the population and if left untreated, other serious heart conditions can arise later in life. Current clinical guidelines recommend intervention when there is significant aortic narrowing, and when a significant pressure gradient is measured during cardiac catheterization. Patients with CoA remain at high risk of long-term morbidity due to complications including increased mortality, hypertension, stroke, and decreased exercise capacity.

Computation Fluid Dynamics (CFD) is a branch of mechanics that creates simulations for a variety of applications. CFD can be used to understand the flow and pressure patterns in geometries of patients with CoA by accurately modeling the patients aortic arch for velocity gradients, pressure gradients, and flow fields to fully understand how the geometry and repairs to the geometry effect patients. CFD is a useful tool in medical applications and can be used to study the flow patterns within the human body. In our project specifically, we use CFD to predict the flow behavior within aortic arches that had surgery due to CoA. We compare the flow within these repaired arches to those with normal geometries to enable better analysis. Under an IRB (Internal Review Board) approved study, clinical data was supplied by the University of Rochester Medical Center. This data included aortic arch anatomy, flow data, and heart rates from cardiac magnetic resonance imaging studies, as well as blood pressures from outpatient visits. ANSYS fluent, a commercial CFD software package, was used to analyze velocity streamlines, cross sections, pressure, and wall shear contours within each arch. All simulations were conducted with normal blood parameters for density and viscosity and governed using the Continuity and Navier-Stokes equations.

Through the examination of velocity, pressure, and wall shear contours, we can analyze blood flow simulations through repaired arches. Our main goal with this project is to accurately depict a model of the aortic arch with CFD and eventually use this information to contribute to better surgical procedures in the future.

Characterizing Dialysis Efficacy, Kidney Failure, and Renal Recovery Using Blood Urea Nitrogen (BUN) Modeling

Kaniyah Purcell, Annabelle Lint, Gilles Clermont, Robert Parker

Characterizing Dialysis Efficacy, Kidney Failure, and Renal Recovery using Blood Urea Nitrogen (BUN) Modeling

Kaniyah Purcell, Annabelle Lint, Gilles Clermont, Robert S. Parker

Observation:

When patients experience renal failure, they require treatments like intermittent hemodialysis. To describe a patient's renal health, dialysis adequacy (Kt/V) and renal recovery times can be analyzed. However, because Kt/V and renal recovery are not routinely collected, blood urea nitrogen (BUN) concentrations can be used to estimate Kt/V and renal recovery instead. Blood urea nitrogen (BUN) concentration levels can be used to estimate a patient's Kt/V and renal recovery time. A model was created to estimate dynamic BUN trajectories before, during, and after dialysis using clinical medical data.

Method:

A 2-compartment ordinary differential (ODE) model was created to predict BUN levels before and after dialysis. BUN generation and dialyzer clearance model parameters were fit using a differential evolution algorithm to minimize normalized root mean squared error (RSME) between the predicted and actual trajectories within clinically reasonable parameter bounds. Parameters for BUN levels were updated every dialysis session. Bounds for parameters were established using high-density data simulation results. The model was tested on 4 patients who had no renal function and 2 patients with renal recovery. For patients who experienced renal recovery, additional parameters were included: K, β, and . K represents how much the patient recovers, β represents the time that renal recovery begins, and represents the time it takes to reach 63.2% of max Kr.

Results:

For the 4 patients who experienced no renal function, the model fit relatively well. The RMSE for those patients was under 2. Additionally, Patients who experienced renal recovery fit the model relatively well. The RMSE for those patients was under 5, and the parity line between our mathematical model and medical data demonstrated a good fit.

Conclusion:

The developed model predicted BUN trajectories over time and captured variance in BUN clearance and generation in patients with and without renal recovery. The model calculated estimates of dialysis adequacy that could support clinicians through intermittent hemodialysis support systems. Simulations produced using the renal recovery mathematical model demonstrated a direct correlation between levels of BUN decreasing as renal function increases. Simulations produced by the model for patients with no renal function demonstrate that high-adequacy dialysis sessions removed high BUN concentrations and excess fluid volume.

Deciphering the Unfolding Pathway of Bovine Serum Albumin Under Varying External Stressors: Insights from Molecular Dynamics Simulation

Clare Cocker, Yinhao Jia, Janani Sampath

The therapeutic potential of proteins has revolutionized the treatment of a wide range of diseases, from diabetes to cancer. However, the effectiveness of these therapeutics is often compromised by the sensitivity of proteins to environmental factors during manufacturing, delivery, and administration. Environmental variables such as pH, temperature, and shear force can cause proteins to destabilize and unfold, leading to a loss of functionality and therapeutic efficacy. To address this challenge, it is crucial to engineer proteins with enhanced stability. A key aspect of this engineering process is understanding how proteins respond to external stressors and identifying the specific regions within the protein that need to be targeted to enhance stability. Thus, deciphering the mechanisms by which a stressor destabilizes a protein is pivotal for the targeted design of more stable therapeutic proteins.

In this study, we employed all-atom molecular dynamics simulations to elucidate the unfolding pathway of Bovine Serum Albumin (BSA) under varying external stressors, including temperature, pH, and shear stress. We characterized global changes in the protein during denaturation using a range of structural metrics, including root-mean-square deviation (RMSD), hydrogen bond number, and hydrophobic solvent accessible surface area (SASA), and validated these results through secondary structure assessment and representative simulation trajectory snapshots. Additionally, we divided BSA into three domains and tracked the sequential order of their destabilization to further explore the unfolding pathway. Our results revealed distinct denaturation patterns across different stressors, providing insights into the underlying mechanisms that can be leveraged to design novel strategies for enhancing the stability of therapeutic proteins.

Leveraging Large Language Models for Molecular Property Prediction in Scientific Research

Shatha Al Adawi, Anthony J. Crespo, Usman Abbas, Qing Shao

Recent advancements in large language models (LLMs) have seen their implementation across various domains, providing assistance in numerous tasks from text classification and translation to complex document searches. The adaptable nature of general-purpose Large Language Models (LLMs) prompts a pivotal question: Can these models be effectively fine-tuned to serve as artificial intelligence assistants in scientific research, particularly in predicting molecular properties? In this work, we conducted two experiments focusing on fine-tuning OpenAI's Ada model to predict both the antimicrobial potential of peptides and the binding affinity of ligands to target proteins. The results show that a well-designed prompt-completion can successfully empower the Ada model to predict molecular functions of peptides or small molecules.

Predicting Vapor-Liquid Equilibria Using Physic-Inspired Machine Learning Models

Abdullah Alqallaf, Charles McGill

The prediction of Vapor-Liquid Equilibrium (VLE) properties has been one of the ongoing issues in modeling complex systems to produce desired products. The Traditional methods of predicting VLE data often need significant amount of data to capture irregular molecular interactions, which constrains the predictive scope of said methods. In our method, we utilize Chemprop which is a state-of-the-art software package that implements D-MPNN (Directed Message-Passing Neural Networks) architecture to offer fast and versatile open-source software solutions that predict complex chemical properties. We apply this D-MPNN architecture to predict VLE binary mixture properties of complex systems. Our models take in the Simplified Molecular-Input Line-Entry System (SMILES) representation of molecules along with temperature, liquid composition, and pure component vapor pressures (T, , x_i) in order to predict the gas phase composition and pressure y_1, y_2, and P.

Three experimental models have been developed to demonstrate the utilization of this approach, all models take in individual SMILES for both components 1 and 2 in a mixture and then process them into separate D-MPNN's resulting in individual molecule vectors while having their temperatures and compositions (T, x_1, and x_2) embedded to this molecule vector before being fed to the FFN to get the target properties. The base model predicts the target properties directly as outputs of the models. The second and third models predict intermediate values used in traditional activity coefficient models to calculate target properties. In detail, the second model predicts activity coefficients (γ_1 and γ_2) which combined with the input properties (T, , and x_i) can be used to calculate the target properties. Instead of predicting activity coefficients, the third model predicts parameters of the Wohl activity coefficient model (A_1, A_2, and A_3), which is used to subsequently calculate activity coefficients to predict more accurate VLE data. The three models developed were trained on approximately 28,000 data-points that includes isobaric and isothermal systems sourced from the Dortmund Databank explorer version and a high-quality benchmark paper by Jean-Noël Jaubert et al. With our models, we show how the inclusion of traditional activity coefficient structures leads to improved representation of equilibrium behavior. These layered methodologies set the stage for more robust understanding of VLE predictions via machine learning.

Eco-COMP: Towards Responsible Computing in Materials Science

Sai S. Lingampalli, El-Tayeb Bentria, Fadwa El-Mellouhi

Bridging the time and length scales and the use of large molecular dynamics (MD) simulations in material science is expected to surge in the next few years, partially due to the development of highly accurate machine learning inter-atomic potentials that enable the simulation of multi-million atomic systems. The emergence of easy-to-use machine learning potentials (MLP), such as the Machine-Learned Spectral Neighbor Analysis Potential (ML-SNAP) is indicative of the growing accessibility and versatility of this tool in the scientific community. It allows scientists and engineers to understand key chemical reactions at the atomic level relevant to the development of green energy and environment solutions such as catalysts, batteries, and solar cell. We also expect a high demand for material science simulations using multiple nodes within high-performance computing facilities (HPCs) due to their computational intensity. Through the analysis of catalysis simulation setups consisting of bulk metallic systems with adsorbed molecular species on the surface, we identified various factors that affect parallel computing efficiency. To foster sustainable and ethical computing practices, this study employs the Large-scale Atomic/Molecular Massively Parallel Simulator (LAMMPS) to find the optimal allocation of computing resources based on the simulation input. We thus propose guidelines to promote responsible computing within HPC architecture: Eco-Comp is a user-friendly automated Python tool that allows material scientists to optimize the power consumption of their simulations using one command, by recognizing, evaluating, and recommending choices on resource usage to a user based on their simulation data, saving time and energy. This work serves towards ensuring that the developing technology of our future will be build upon a foundation that prioritizes sustainability and the ethics of responsible computing.

Molecular Dynamics Discovers Ligand Binding Regions on Molybdate Transport Protein, Moda

Angelina Moncrieffe, Jannette Carey

In the pharmaceutical industry, many protein crystal structures are labeled as "undruggable" due their lack of suitable crevices for binding of ligands, when viewed under static frames, through static docking. Our hypothesis is that the majority of proteins do have ligand binding sites when assessed dynamically. The focal point of this research was to utilize computational molecular dynamics in order to study the binding behavioral patterns of the Molybdate Binding Protein, ModA. This protein was tested on all 20 amino acids and bound to additional common functional groups, which revealed unknown binding sites on protein. The binding energies of the compounds on the protein increased by double, resembling energies considered promising for Pharmaceutical lead compounds. This molecular docking strategy mimics the ligand fragment approach used in the Pharmaceutical Industry and reveals that readily available computational tools for dynamic evaluation allow identification and improvement of ligand binding.

Mathematically Modeling Chemokine Ligand-Receptor Kinetics in the Context of Pancreatic Cancer

Joseph Mennicucci, Ashlee N. Ford Versypt

Pancreatic cancer is an aggressive form of cancer that has poor prognosis and survival rate. Although this cancer is difficult to detect and treat, experimentation has illustrated that chemokines largely impact the tumor microenvironment as they have specific roles in developing and restricting the immune interactions of the tumor. Chemokines include ligands that bind to specific receptors and give signals for their cell development and movement. In a complex biological environment, there can likely be many competitors to ligand binding to the receptor, affecting the movement and development of a particular cell. To model chemokine ligand and receptor interactions, we developed an ordinary differential equations (ODEs) model to quantify the ligand-receptor binding kinetics for a radioligand competing with either an agonist or antagonist to bind with a heterodimeric receptor, described using uncompetitive inhibition. To test our model, we chose published data that reported binding dynamics of two ligands CXCL10 and CXCL12 to the heterodimeric receptor CXCR3-CXCR4, competing with a series of agonists and antagonists to either the singular receptor or the heterodimer receptor. We estimated the kinetic parameters in the ODEs using the reported PIC50 values in the article. Our model showed similar dynamics of the ligand and receptor competition model compared to the experimental data. The model is compared to data from three published experimental cases. The first case involves binding of the radioligand CXCL10 or CXCL12 with increasing concentrations of an unlabeled CXC10, unlabeled CXCL12, and VUF10661 as agonists to the receptors. The second case details the competition of the radioligand CXCL10 and CXCL12 with increasing concentrations of VUF10085, TAK-779, and AMD3100 as antagonists to the receptors. The third case details the full system dynamics with the radioligand CXCL12 competing with an unlabeled CXC10, unlabeled CXCL12, and VUF10661 as agonists, binding to the CXCR4 singular receptor and the CXCR3-CXCR4 heterodimeric receptor, and considers ligand-receptor dissociation half-lives. By creating an in-depth mathematical model of this complex system, the time and resources of wet-lab researchers can be saved during their experimentations and a greater understanding can be gained about the physical nature of their experiments. We aim to apply this model to pancreatic cancer for designing ligands to enhance immune infiltration into these immunologically cold tumors.

Magma: A Robust, User-Friendly Kinetic Modeling Platform for Simulating and Optimizing Algal Bioprocesses

Vincent Xu, Yinjie Tang

Micro-algal processes have broad impacts on wastewater treatment, environmental remediation, and bio-production. For monitoring natural algal growths or optimizing photobioreactor (PBR) operations, it is important to predict and control algal cultivation parameters including light, bioreactor size and configurations, temperature, CO2 and nutrient conditions. Kinetic models of micro-algae systems are widely used for elucidating photoautotrophic growth phenomena and performance. However, programming numerical methods to evaluate systems of ordinary differential equations (ODE) representing micro-algae systems (e.g., Monod Models) can be a time-consuming and error-prone process. The Micro-Algae Growth Modeling Application (MAGMA) is a user-friendly, simple, yet powerful tool designed to streamline the development of kinetic growth models based on empirical data for predicting growth phenomena. MAGMA offers the following functionality: 1. Specification of system environmental conditions, biological and chemical species in the system, and the governing differential equations that describe the rate of change of each species in the system, 2. Specification of mass transfer flows in and out of the system, 3. Parameter regression to fit models to empirical data, 4. Optimization for PBR cell and chemical productivity, 5. Kinetic visualization tools, 6. Cloud database integration to enable users to access model templates and models created by other users, and 7. Integration of kinetic model with PBR controls (e.g., model predictive control).

MAGMA is free for academic use and is designed so that a user's lack of prior programming experience will not be a hinderance to their usage of the tool. For validating MAGMA functionality, a case study on micro-algae growth and toxin production have been performed. Based on experimental data, MAGMA was used to evaluate harmful algal growth and toxin production by cyanobacterium *Microcystis aeruginosa*. Specifically, the model evaluated the toxin reduction effectiveness of nutrient reduction, herbicide application, and competitive co-culture with a fast-growing, non-toxic species of cyanobacteria (Synechoccocus elongatus UTEX 2973). The models generated with MAGMA based on the data from this study included mono-culture and co-culture batch growth kinetics, batch nutrient consumption kinetics, batch microcystin toxin production kinetics, batch herbicide consumption kinetics, final cell concentration in mono-culture and co-culture across varying initial nutrient concentration conditions, and final microcystin toxin concentration across varying initial nutrient concentration conditions. MAGMA is a robust kinetic modeling framework designed to enhance knowledge of micro-algae growth phenomena by connecting microbial growth theory to experimental observations.

Keywords: Kinetic modeling, micro-algae, photo-bioreactor (PBR), MAGMA, ordinary differential equations, mass transfer, regression, optimization, model visualization, cloud database

Interfacial Competition between Surfactant and Polymer Excipients on a Drug Nanocrystal Surface

Dien Nguyen, Lucas Attia, Devashish Gokhale, Talia Zheng, Patrick Doyle

Oral administration is preferred among all drug administration routes due to its simplicity, low cost, and convenience. However, nearly 90% of drug candidates in the pharmaceutical development pipeline and 40% of commercially available drugs are hydrophobic, which significantly limits their dissolution kinetics in the gastrointestinal (GI) tract and consequently decreases oral bioavailability. The Doyle Group has developed a suite of 'bottom-up' methods to template drug nanocrystals by embedding active pharmaceutical ingredients (API) in polymeric matrices as an approach to reduce drug aggregate sizes for enhanced bioavailability. However, a mechanistic understanding of excipient-drug interactions and their influence on crystallinity during processing remains a significant hindrance in developing new formulations.

In this work, we present the use of all-atom molecular dynamics (MD) simulations to advance our understanding of interfacial competition between surfactants (Tween 80/Span 20) and polymers (methylcellulose) on the surface of a fenofibrate nanocrystal, a model hydrophobic API. To simulate this system, we establish a novel workflow for simulating two excipients on a triclinic crystal surface, including modification of the crystals to fit the desired simulation box dimensions, structural and topology file preparation for multi-component crystal simulation in GROMACS. Our setup allows convenient access to layer-by-layer structural analysis of the drug crystal, as well as any pairs of interactions among the excipients and the crystal layers. The molecular interactions between drug, polymer, and surfactant on the drug surface control the self-assembled nanostructure, leading to varying degrees of crystallinity based on excipient composition. Furthermore, excipient Lennard-Jones potentials and density along z-axis suggest that strong bulk polymer-surfactant stabilization contributes to preferential polymer surface delocalization. The simulation results are compared against experimental crystallinity data with various polymer-to-surfactant ratios, supporting our hypothesis for interaction mechanism and characteristics on each face of the fenofibrate crystal. This mechanistic insight can inform the design of new drug formulations using our promising bottom-up approach.

Sequence Patterning, Morphology, and Dispersity in Single-Chain Nanoparticles: Insights from Simulation and Machine Learning

Sophia Colmenares, Roshan Patel, Michael Webb

Single-chain nanoparticles (SCNPs) are intriguing materials inspired by proteins that consist of a single precursor polymer chain that has collapsed into a stable structure. In many prospective applications, such as biocatalysis and nanomedicine, the utility of a single-chain nanoparticle will intricately depend on the formation of a mostly specific structure or morphology. However, it is not generally well understood how to reliably control the morphology of single-chain nanoparticles. In this study, we use molecular simulation in tandem with machine learning analyses to assess how the patterning of cross-linking moieties on precursor chains impacts the morphology and properties of SCNPs. We adopt a coarse-grained, phenomenological model to study the formation of SCNPs from precursor polymer chains. To comprehensively probe the sequence–structure space, we simulate the formation of 7,680 SCNPs from precursors chains that are predominantly distinguished by the fraction and distribution, or blockiness, of reactive cross-linking groups along the backbone. These simulations are analyzed to characterize resulting SCNPs via structural analysis and unsupervised manifold learning. Specifically, the uniform manifold approximation and projection (UMAP) algorithm is used to learn a low-dimensional numerical embedding of the SCNP morphologies and the manifold over which they are distributed. These analyses reveal not only the landscape of possible SCNP morphologies but also the dispersity of structures arising from given precursor specifications, which has not been previously well-characterized. We show how the overall fraction of functionalization and blockiness of cross-linking moieties biases the formation of certain local and global morphological characteristics. Specific archetypal morphologies, including rod-like structures, were found to be strongly associated with certain regimes of different fractions of cross-linking and blockiness. Importantly, we illustrate and quantify the dispersity of morphologies that arise due to the stochastic nature of collapse from a well-defined sequence as well as from the ensemble of sequences that correspond to a given specification of precursor parameters. Moreover, we show that specific sequences enable more precise control over morphology. We note that morphological dispersity depends on sequence patterning and that a lower fraction of cross-linking groups and a high blockiness illustrate the widest range of possible morphology and dispersity outcomes. Overall, this work critically assesses how precursor chains might be feasibly tailored to achieve given SCNP morphologies and provides a platform to pursue future sequence-based design.

Group Theory and Artificial Intelligence

Amani Midgette

This research delves into the innovative application of group theory, particularly symmetric groups, in the realm of artificial intelligence algorithms for image identification. It also explores the distinctive architectures of Convolutional Neural Networks (CNNs) in the context of computer vision. CNNs have brought about a paradigm shift in computer vision, playing a central role in image classification, object detection, and diverse visual recognition tasks. Among these architectures, AlexNet, VGG16, and VGG32 have garnered significant attention for their impactful contributions and architectural innovations. In this study, our objective is to explore and compare these three CNN architectures, elucidating their design principles, performance, and suitability for various computer vision tasks.

The core aim is to develop advanced techniques that enhance the robustness of image recognition by accounting for image rotations and sub-image transformations. Traditional image recognition systems often falter when confronted with rotated or transformed images, constraining their practical utility. To surmount this challenge, we harness the principles of symmetric group theory to devise innovative algorithms adept at handling these variations. This research systematically evaluates and compares the image recognition performance of the three networks (AlexNet, VGG16, and VGG32).

The research is bifurcated into two primary aspects: image rotations and sub-image transformations. To address rotations, we delve into the mathematical underpinnings of symmetric groups and their ability to model various rotation angles. We employ these group-based representations to devise AI algorithms that proficiently identify images, irrespective of their orientation. The study encompasses a diverse range of datasets, underlining the effectiveness of the proposed algorithms. Our results show substantial enhancements in image identification accuracy, especially in scenarios involving image rotations and sub-image transformations. The integration of symmetric group theory in artificial intelligence holds the promise of broadening the capabilities of image recognition systems, rendering them more adaptable to real-world applications in fields such as computer vision, robotics, and autonomous systems. This research serves as a bridge between theoretical group theory and pragmatic AI applications, showcasing the potential of interdisciplinary approaches in propelling image recognition technologies.

Long COVID: Elucidating the Impact of COVID-19 on Black Male Engineering Students- Highlighting Their Resilient Assets

Fatima Elmouden, Jared Davis, Tyron Slack, Le Shorn Benjamin, Jerrod Henderson, Erik Hines

Long COVID: Elucidating the Impact of COVID-19 on Black Male Engineering Students- Highlighting their Resilient Assets

Authors: Fatima Elmouden, Erik Hines, Jared Davis, Tyron Slack, Le Shorn Benjamin, Jerrod Henderson

In 2020, the COVID-19 pandemic had a widespread effect on university students, and accompanying it was the social unrest following the murder of George Floyd (e.g., dual pandemic). From the shutdown of university campuses nationwide to the switch from in-person to online teaching, the impacts of these sudden changes have been studied on a general and university-to-university basis. Despite the proliferation of studies highlighting the effects of the dual pandemic on students, fewer have foregrounded the experiences of Black men in engineering undergraduate programs and how they experienced these phenomena. Yet Black men constitute a missing segment of the population who could contribute to the global engineering knowledge economy (Burt et al., 2019). Therefore, in this study, we aimed to investigate how Black male engineering students experienced these two phenomena to understand better how to support Black men's success in engineering. We conducted semi-structured interviews with 20 Black male engineering students. The guiding framework for this study was alienation and engagement. Using reflexive thematic analysis, we developed three themes: (1) feelings of alienation, (2) student response to alienation, and (3) resilient reintegration. Though participants reported setbacks due to the effect of the pandemic and social unrest on their academic pursuits, they also reported an unwavering ability to adapt and reframe those setbacks in positive ways that allowed them to push forward.

Redesign of a First Year Undergraduate Chemical and Biomolecular Engineering Laboratory

Shane Haycock, Chiwon Yu, Joseph Menicucci, Cody Turner, Paul Bader

The purpose of our work was to enhance the first year undergraduate chemical and biomolecular engineering laboratory experience at Lehigh University. After extensive student feedback from years past, it was clear that the Introduction to Engineering Practice chemical and biomolecular engineering laboratory - taken by engineering students in the fall of their first year - needed a significant redesign to better engage students while still giving them valuable insight into chemical and biomolecular engineering practices. Three experiments were redesigned and/or re-implemented to achieve this goal. These new experiments introduce students to process control, leaching, and power generation. Each experiment gave students the opportunity to learn the basics of crucial aspects of their future careers, while creating an engaging, open-ended, and - most importantly - fun learning environment. This poster describes the newly redesigned experimental procedures with early qualitative student feedback from implementation.

A Power Series Approach for Heat of Vaporization Predicted By the SRK and PR Equations of State

Joseph Kuklewski, Michael Misovich

The Soave-Redlich-Kwong (SRK) and Peng-Robinson (PR) equations of state can be used to describe thermodynamic properties such as heat of vaporization. In this project, the adjusted heat of vaporization, a similarity variable defined as reduced heat of vaporization divided by reduced temperature , was studied. The adjusted heat of vaporization was expressed as +, where was the acentric factor function and and were two infinite series of the common adjusted temperature η. The first 29 coefficients of and were found for the SRK equation, while the first 25 were generated for PR. Results from exact SRK and PR algorithms were compared to the power series truncated to various numbers of terms. For the maximum number of terms used in the series, the relative truncation error was as low as 10^{-9}. Plots were produced to express the relationship between η, ω, and the number of terms used. Work is ongoing to find alternative methods to express the truncation error.

Experimental Design for Biomaterials Education: 3D Bioprinting

Zareena Al-Shehab, Janina Li, Mekhi Vazquez, Olivia Kim, Andrea Vernengo, Jennifer Weiser

Biomaterials as a field has become increasingly important for the development of medical applications and restoration of function for people post-injury and disease[1], raising the need to extend knowledge of biomaterials to students. To educate students about the importance and properties of biomaterials, educators designed experimental models to establish a foundation of knowledge about 3D bioprinting (3DBP). 3DBP is an additive manufacturing technology that utilizes biomaterials to synthesize living, functional tissues for regenerative medicine[2]. 3DBP is widely utilized because of its ability to create customized and precise objects efficiently and omnidirectionally. This was the motivation to develop an accessible, inquiry-based learning (IBL), hands-on educational module to teach the fundamentals of 3DBP. The focus of the educational module created was to understand how different factors affect the fidelity of a 3DBP print.

In the IBL experiment, the main variable explored by students was altering the viscosity of the support bath, which impacts print fidelity of a bioink. Two types of commonly used support bath polymers were utilized in this module: gelatin hydrogels and Carbopol® 691 polymer jammed microparticles. Both polymers are safe and accessible. The module was designed to highlight the difference in the type and degree of secondary entanglements gelatin and Carbopol® 691 polymer have regarding their effect on support bath viscosity and printing. Moreover, a secondary variable students explored was altering the concentration of the bioink (gelatin containing Coomassie blue for visualization). In doing so, students can learn how bioink composition can affect printing fidelity.

Learning gains for the module will be assessed using a pre-/post-test with questions pertaining to the factors affecting viscosity in 3DBP printing fidelity. Additionally, a validated pre-/post-survey based on those from the Scientific Literacy and Student Value in Inquiry-guided Lab Survey (SLIGS) was given[3]. The 3DBP IBL educational module was designed for undergraduate students to have a hands-on way to explore how factors affect print fidelity.

References

[1] Huebsch, N.; Mooney, D. J. Inspiration and application in the evolution of biomaterials. Nature 2009, 462, 426–432 DOI: 10.1038/nature08601.

[2] Vanaei, S.; Parizi, M. S.; Vanaei, S.; Salemizadehparizi, F.; Vanaei, H. R. An overview on materials and techniques in 3D bioprinting toward biomedical application. Engineered Regeneration 2021, 2, 1–18 DOI: 10.1016/j.engreg.2020.12.001.

[3] Ankeny C, Stabenfeldt S. Cost-effective, Inquiry-guided Introductory Biomaterials Laboratory for Undergraduates. 2015 ASEE Annual Conference and Exposition Proceedings, ASEE Conferences; 2015, p. 26.412.1-26.412.16. doi:10.18260/p.23751.

Black Male Engineering Pathways

Jared Davis, Jerrod Henderson

The underrepresentation of Black men in engineering is well-known and documented. Black Males account for only 3.9% of male engineering undergraduates compared to their White and Asian American counterparts, at 53.1% and 13.2%, respectively. (American Society of Engineering Education, 2021). The lack of representation of Black men in engineering is linked to isolation, marginalization, having a lower sense of belonging, and a disconnect from traditional STEM pedagogical practices and environments which are often positioned as objective, colorblind, and apolitical (Malone & Barabino, 2009; McGee, 2020; McGee & Martin, 2011; Baber, 2015; Henderson et al., 2021). Meanwhile, Black men constitute a missing segment of the population who could contribute to the global engineering knowledge economy (Burt et al., 2019). Black students who take "non-traditional" pathways to become engineering students might be a potential source for the future engineering workforce. For those reasons, in this study, we sought to understand how Black male college students navigated paths into engineering as internal transfers. We closely followed the transitions of three Black male undergraduate students who intend to major in engineering. During year one, each student participated in three semi-structured interviews. Using an Interpretive Phenomenological Analysis (IPA) approach we developed 9 themes: Engineering Transitions, Community Building, Personal Motivators, Leveraging Resources, Self Confidence, Escapes, Spirituality, Negative Peer Experiences, and Early Exposure. We found that these students also have similar experiences to those of current first-year engineering majors and gained insight into how students transition into engineering mindsets.

Montmorillonite-Catalyzed Dehydration of Terpene Alcohols: A Greener Alternative for Organic Chemistry Lab

Neha Nichakawade, Tausif Tamim

Montmorillonite was used to catalyze the dehydration of the terpene alcohols menthol, isopulegol, and terpineol. Montmorillonite is a natural clay that serves two roles- as a heterogeneous acid catalyst and as an absorbent of water formed in the reaction. These terpene alcohols are also derived from nature, as these can be extracted from mint and cardamom. These agents are commonly found in fragrances and produce many different isomers (and scents) when done reacting.

The dehydration reactions were carried out using conventional heating with reflux columns and distillation apparatus as well as microwave heating. The reaction proceeds to full conversion when heated in the MW synthesizer at 200 °C for 20 minutes. The identity of the products obtained was established by separation and analysis of the reaction mixture using GCMS. The formation of multiple products can be attributed to the reversibility of the hydration-dehydration reaction and carbocation rearrangements. A mechanism to explain the formation of thermodynamic and kinetic products can be proposed using the relative stability of products and intermediates. These energy values were calculated using the molecular modeling software Spartan.

Currently in the organic chemistry lab, this experiment is done with cyclohexanol and phosphoric acid to produce cyclohexene. This reaction requires the use of concentrated phosphoric acid and the cyclohexene formed has a pungent smell. Therefore, using montmorillonite clay and terpene alcohols, which are derived from nature, provides a greener approach to the laboratory experience. In addition, these reactions require a thorough analysis of the product mixtures using analytical techniques such as FTIR, GCMS, and ^{1}H-NMR. Students also have the opportunity to compare the relative stability of the carbocation intermediates and subsequent products.

Can Oscillations of a Simple Pendulum Explain the Oscillations of a Fluid Interface?

Sabrina Campos

Understanding the phenomena of resonance in fluid systems is important due to its relevance in heat removal applications in microgravity environments in space. In this work, we utilize the effect of resonance on the oscillations of a simple pendulum to elucidate the interface behavior of a fluid layer subjected to vertical oscillations. A simple pendulum suspended from a base plate will oscillate at its natural frequency when slightly disturbed. If the pendulum bob is heavy, the damping of these oscillations due to the air resistance will be negligible, and the natural frequency will depend only on the gravitational acceleration and the inverse of the pendulum length. When the base plate of the pendulum is vertically oscillated near its natural frequency, the pendulum bob will oscillate violently due to resonance. Analogously, fluid systems also have a natural frequency, and the interface of the fluid layer will exhibit large undulations or patterns when vertically oscillated near its natural frequency. The natural frequency of a fluid layer in the absence of viscous damping depends on the gravitational acceleration, surface tension and the wavelength of the pattern observed at the interface. To this end, we demonstrate the effect of resonance on different pendulum lengths and use it to explain the emergence of various patterns with different wavelengths at the interface of a fluid layer undergoing vertical oscillations.

Using Educational Comics to Promote Student Interest in the Breadth and Depth of Chemical Engineering

Ira Hysi, Luke Landherr

To encourage students to pursue chemical engineering at undergraduate and graduate level engineering programs, the motivations behind student involvement and success in this field must be examined and adapted to accordingly. In this study, an educational comic has been created and distributed to university students to investigate whether career prospects impact student motivation in pursuit of chemical engineering on an institutional level. This comic was then distributed to students beginning their first undergraduate year at Northeastern University and who had not yet definitively declared their major.

This comic dismantles two key misconceptions that may discourage students from pursuing chemical engineering: limited scope of industrial opportunities and restrictive skill sets relating exclusively to chemistry and math. To overcome these misconceptions, the comic demonstrates a student questioning their career interests and learning about the fluidity of chemical engineering work in various industrial contexts. This educational tool delves into industries such as oil and gas, material science, drug discovery, sustainable energy, environmental preservation, agriculture, and food science. Through the inclusion of these industries, the comic provides a clear but thorough introduction into the breadth of environmental, social, and scientific impact that chemical engineers can have. The comic also emphasizes the diversity of knowledge bases and skill sets chemical engineers use to combat the fallacy that their work is exclusively based on chemistry or math.

For a baseline, students were instructed to complete the Felder Soloman Learning Style survey to assess their preference for visual or verbal learning tools. Further survey questions assessed the impact of the comic to boost student engagement, comprehension, and retention of learning materials. Some students were also personally interviewed to gain a more holistic evaluation of their attitude and understanding towards pursuing chemical engineering as a career. Further analysis with the MUSIC model of student motivation will be completed to further understand student incentive in pursuing chemical engineering academically and professionally.

The results of this study indicate greater student interest in pursuing chemical engineering and showcase the power of expanding educational tools to better showcase the possibilities that await graduates.

Building a Problem-Solving Community of Scholars: Peer-Facilitators' Talk Moves

Eric Sims, Jeanette Jarvis, Jaiah Steele, Jerrod Henderson

Despite efforts (e.g., interventions and significant financial investments) to improve diversity, degree completion, and retention in engineering, many institutions continue to struggle to realize these goals. This qualitative study examined one student success intervention at a Hispanic Serving Institution (HSI) that seeks to improve retention rates and GPAs for rising sophomore chemical engineering students through workshops and a preview of gateway sophomore-level courses through a problem-solving workshop led by upper-level students who previously earned an A or B in the course. Our study aimed to understand the discourse in these sessions and its impact on student success. We used observations and video recordings of the class sessions to analyze the discourse. We observed 5 sessions during the five-week intervention, which included 4 video recorded sessions. These sessions consisted of peer facilitators and students working together to solve a problem as a large group. We analyzed the videos using a deductive approach. We focused on the specific words used by noting quotes and memoing the context in which the quote was said. For the analysis, we built on the framework by Grinath & Southerland (2019) and Miel et al. (2023) using the three broad categories of discourse moves: conservative, ambitious, and inclusive. We considered conservative to be an instance where a statement was meant to elicit an anticipated answer or deliver information, ambitious when a comment is meant to encourage reasoning, and inclusive when a statement is used to include all voices in the classroom. Throughout the program, peer facilitators primarily used conservative talk moves, followed by ambitious and inclusive moves at almost equal amounts. Although conservative talk moves occurred more frequently, as seen in other studies, our findings illustrate a greater emphasis on inclusive and ambitious moves. To improve outcomes for engineering students, it is essential to understand how they learn and what discourse patterns work best to succeed. Learning from how peer facilitators interact with students as opposed to typical educators could prove valuable in engineering education, improving student outcomes and future intervention design.

Creation, Compilation, and Analysis for Educational Chemical Engineering Comics

Nethra Iyer, Luke Landherr

There is a stigma surrounding chemical engineering that the field and major are difficult and complex. While this can be true as evidenced by nuanced topics such as fugacity and kinetics, one technique to make the theoretical challenges more approachable is to present the concepts more visually, given the general preference of students in STEM for visual learning. As such, this research study focuses on the creation, compilation, and analysis of chemical engineering comics geared towards K-12 students. Comics have become increasingly more prevalent as visual learning tools, with a long history of their educational potential dating back to the 1940's, so they could be used to help guide more students into chemical engineering without the stigma of the field being too complex.

A series of comics was created displaying how chemical engineering is found in everyday common applications, including chocolate and shampoo. These comics were shared with early engineering students via anonymous survey. All survey participants were evaluated in terms of learning preferences using the "Index of Learning Styles Questionnaire" to determine their general approach to learning, including visual versus verbal and sequential versus global. Students were then asked how comfortable they were with basic ideas of chemical engineering, establishing a baseline for their understanding. After reading the comics, students were further assessed through the MUSIC Model of Motivation, a method of teaching analysis that quantifies the strength of an educational activity through the five perceptions of empowerment, usefulness, success, interest, and caring. Analysis from the surveys focused on gauging the potential and validity of chemical engineering comics as a gateway from K-12 curricula. With the characters in the comics being depicted of various backgrounds, additional questions were asked to assess students' perception of representation in the curricula.

This paper will share the results from this investigation. Further studies will seek to determine how K-12 students feel when they have no knowledge of college-level engineering courses, how students from public and private schools understand these topics, and the effects of diversity, equity, and inclusion within the characters and storylines.

Blending in Baffled and Unbaffled Mechanically Agitated Vessels – Effect of Reynolds Number and Liquid Height

Austin Colon

The standard vessel for mixing in the process industry is a baffled tank. This design is very common especially for low viscosity fluids. Nonetheless, unbaffled vessels are found in both industry and the laboratory. The reasons cited for using unbaffled vessels are primarily concerns around cleaning the vessel between batches and the potential for stagnation in high viscosity fluids, particularly when properties change over the processing time. We will examine the effect of baffles on overall blend time using one vessel with removeable baffles. The same axial flow-down pumping impeller is used. The blend time was measured using the iodine color change method over a range of viscosities and Reynolds Numbers. The depth of the vortex was also measured. In this work we observed that the flow patterns are dramatically different between unbaffled and baffled tanks. At high Reynolds Numbers with the baffled configuration, we see a familiar turbulent diffusion throughout the vessel. Without the baffles the fluid segregates into two zones: a swirling inner zone near the shaft and a more turbulent outer zone. The overall blend time is longer in the unbaffled tank at the same rotational speed compared to the baffled tank. We identified two blend times in using unbaffled tanks, a short blend time for the outer zone and a much longer time for the inner vortex centered zone which defines the overall blend time for the vessel. At low Reynolds Numbers the baffled blending is in the transitional flow regime between turbulent and laminar with a combination of eddies and striations. The last region of the tank to become completely mixed is located behind the baffles. Unbaffled blend times are shorter than baffled and as the Reynolds Number decreases the difference increases. The vortex is reduced with decreasing Reynolds number. We consider this work just a beginning step in a long study and hope this work will be continued by other researchers.

Hygroscopicity Study of Aerosolized Amino Acids

Alexander Kremer

During the summer of 2023 I worked alongside Professor Tim Raymond of the Chemical Engineering department on a research project concerning aerosols. This specific experiment used both the Cloud Condensation Nuclei Counter (CCNC-100 by DMT) and the Scanning Mobility Particle Sizer (SMPS, TSI Inc) to count and characterize a series of aerosols which were made in the laboratory. There were six total compounds which were tested to determine their behavior as an aerosol. The six compounds were selected based on previously observed behavior and activity.

First, the individual compounds were made into a solution and aerosolized, which allowed for its particle size distribution and particle counts to be recorded for later comparison. Second, two compounds were selected and mixed into one solution. After this solution was made it would then be aerosolized and run through the CCNC and SMPS to record the particle size and counts of the mixture. This process was then repeated for three compounds mixed into a singular solution. Thirdly, two compounds were chosen and made into separate solutions. These separate solutions were aerosolized and then mixed in a chamber. This gaseous mixture was run though the CCNC and SMPS to collect the particle size and count data. After all of this data was collected, it was analyzed and compared using software written by the department for aerosol analysis. Each of these trials gave a value denoted as Kappa, which was a measure of how active a compound or mixture was. This Kappa value, the average particle size, and particle count was then compiled and compared. This set of values was the ultimate goal of the experiment this summer.

Heroma VV Sensor, Water Toxicity Sensor

ANA Vega, Valentina Mayorca Gonzalez, Maria Paula Heredia Fuentes

ABSTRACT

Life revolves around water, so a strict control on its quality is vital to guarantee human health and protection of biodiversity. However, nowadays, industrialization accelerated by human population growth and people's lack of information about possible environmental problems end up polluting water bodies with biological and artificial pollutants. Including, the presence of heavy metals in drinking water can have serious repercussions on health such as intestinal and kidneys diseases, accompanied even by severe brain damage. Therefore, this work aims to develop a biosensor that is at the service of detecting two harmful heavy metals, mercury, and lead. Among biosensors, the amperometric enzymatic biosensor was selected due to the advantages they offer over others, for instance, simplicity, short time responses and miniaturization potential are some of these features. At same time, an innovative design will allow to detect heavy metals in an easy and cost-effective way without neglecting the ease of transport and handling with a series of simple instructions that allow access not only to scientists but also to standard or unskilled users. The acting principle includes the analysis of urease enzyme behavior in the presence of specific heavy metals since it is inhibited at -0,03 V to 0,0 V reaching the formulation of calibration curves that show the concentration of an analyte by considering the variation in the current flow. The sensor has maximum dimensions of 76.2 mm long, 39.54 mm high and 62 mm wide with an approximate weight of 218 grams, having an initial cost for each sample measured of 12 USD.

Keywords*:* Biosensor, amperometric, mercury, lead and, water.

Soil Microbial CO2 Respiration Testing, Effect of Partial Pressure

Shaina Larsen

As the need for more sustainable and efficient agricultural operation develops, a more real time and finer grained mapping of the influence of soil quality and composition on crop production and properties is necessary. Few real time tools are available for farmers to profile their soil, particularly on a molecular level, where microbial decomposition synthesizes the nutrients required to grow crops. Thus, this overall research project focuses on the development of wirelessly powered subsurface soil sensors categorizing soil quality by profiling the gaseous byproducts of organic decomposition within the soil. As a subproject, the present report details the qualification of hydrophobic membrane compatibility for passive transport of CO_2 to the sensor chamber. A larger scale prototype has been constructed to test the membrane, which will later be scaled down. The device has been designed and assembled in its initial form in Spring 2023, utilizing an Isco high-pressure syringe pump(s), Matheson flow controllers, and high-pressure steel tubing. Flows of pre-pressurized water and CO_2 were fed from the syringe pump to a high-pressure acrylic vessel while N_2 serves as a countercurrent carrier gas. Early results from pure CO_2 feed in terms of permeability as a function of carrier gas flow will be described.

Characterizing the Activity of Electrochemically Active Bacteria in Microbial Fuel Cells in the Presence of Toxic Cyanobacteria

Nada Shetewi, Amanda Simson, Anthony Napolitano, Radmila Janjusevic

The frequent occurrence of cyanobacterial harmful algal blooms (cHABs) poses serious threats to freshwater habitats and drinking water resources due to the impairment of the ecosystem and the production of *Microcystis Aeruginosa*, a highly toxic strain of cyanobacteria. *Microcystis Aeruginosa* produces a cyanotoxin, MC-LR, that is hard to destruct and causes serious damage to the liver if ingested. This research aims to treat cHABs by utilizing the natural metabolic processes of electrochemically active bacteria (EAB), a group of bacteria that are naturally present in aquatic ecosystems. EABs are capable of oxidizing organic matter, including algal biomass, through biocatalytic reactions, releasing electrons in the process. A microbial fuel cell (MFC) can harvest the electricity produced from EAB in an anaerobic anode chamber while exchanging protons with an aerobic cathode chamber. By using cHAB infected lake water as a substrate, an MFC can treat lake water while generating electricity. In this work, we have set up the system to test the electrical activity of various EAB strains from Central Park Pond water by carrying out cyclic voltammetry tests in a dual-chamber MFC, with each isolated strain at a time. The most active strains were identified by the intensity of their reduction and oxidation peaks in a current vs. voltage plot (CV plot). Longer tests will be carried out with *Microcystis Aeruginosa* present in the media as substrate for EAB. The efficacy of EAB in treating cyanobacteria will be identified by the concentration of intracellular and extracellular MC-LR toxin. In this work, we also tested multiple methods to isolate intracellular toxin from cyanobacteria and to accurately identify MC-LR levels in samples pre and post treatment including plate essay, Western blot, and HPLC.

Blended Cover Treatability Study: Activated Carbon Performance

Juan Aguilar Lopez

Chemical contamination is a major environmental concern that deteriorates ecosystems and human health alike. Leakage and poor management of chemical process plants where pesticides are or were produced are a prime example, such as the United Heckathorn Superfund site located in the San Francisco Bay area whichcreate disturbance in marine ecosystems. The risk in the harbor rises primarily from dichlorodiphenyltrichloroethane (DDT), a highly stable, carcinogenic, organic compound whose production was banned decades ago. However, improper management and abandonment allowed DDT to contaminate the surrounding land and marine ecosystem. DDT's hydrophobic properties suggests it should concentrate in the sediment reducing bioavailability, however, DDT concentration has remained high in the water due partially to natural water flows. Furthermore, persistent boat activity such as from tugboats continuously uplift contaminated sediment allowing mixing with the water above, maintaining a hazardous environment for wildlife and humans alike. Previous efforts to clean up the site included dredging up the contaminated sediment, which failed to significantly reduce DDT contamination in either the water or sediment layer. Therefore, a new engineering solution is required. Research is focused on creating a blended cover which will further reduce DDT's bioavailability in the marine harbor ecosystem. The blended cover, a mixture of gravel and activated carbon, can be placed on top of the sediment layer. To test activated carbon's performance to attract and sequester DDT from the water, aquarium tank experiments simulating similar conditions to those in the bay area were constructed in which the activated carbon and gravel size were varied. After a month, a passive sampler was used to test DDT concretion at different layers. Utilizing a Gas Chromatography-Electron Capture Detector (GC-ECD), presence and concentration of DDT and its derivatives was analyzed in different blended cover conditions. Preliminary analysis suggests that the smaller the activated carbon and gravel size, the better performing the blended cover will be at removing DDT. Moreover, physical tests are ongoing to test the erodibility of different blended cover mixtures. Research into alternative clean-up techniques for contaminated sites is of great value for ecosystem remediation where much improvement is left to be made.

A Tracker Tool for Estimating Chemical Releases and Emissions in End-of-Life Plastic Management

Matthew Conway, John D. Chea, Austin Lehr, Gerardo Ruiz-Mercado, Kirti Yenkie

Plastic production and continuously increasing usage globally present considerable challenges with waste disposal, creating a linear economy in which most plastic is used once and then landfilled or incinerated. Many efforts have been made to improve recycling in the US, but chemical additives in plastics have impeded the recyclability of end-of-life (EoL) plastics. This work aims to address this globally pressing issue by analyzing the movement of additives in plastic, and tracking their migration across the EoL stage pathways, including recycling, incineration, and landfilling. The municipal solid waste (MSW) data from the 2018 Advancing Sustainable Materials Management report was used as a basis for these calculations, and a case study was completed to measure changes in additive releases, environmental impacts, and energy consumption resulting from the different recycling efficiencies [1]. A Python-based graphical user interface (GUI) was programmed to allow users to change various parameters in the EoL stream data to make predictions and assist informed choices regarding the current MSW end-of-life procedures. Based on user input, this tool estimates the chemical additive releases, greenhouse gas emissions, and energy consumption associated with these EoL procedures. This python-based GUI chemical release tracker tool is designed to support the transition towards a circular plastic economy by providing valuable insights into plastic EoL management, and work is ongoing to develop the program into a web-accessible application for free and public use.

[1] Advancing Sustainable Materials Management: 2018 Fact Sheet. U.S. Environmental Protection Agency, December 2020. https://www.epa.gov/sites/default/files/2021-01/documents/2018_ff_fact_sheet_dec_2020_fnl_508.pdf.

Verifying Aerosol Particle Size Distributions from a Drum Sampler Using Fluorimetry and ICP-MS for Elemental Analysis

John Piorkowski, Timothy Raymond

It is estimated that the average person spends 90% of their time indoors. The quality of indoor air can cause a variety of health effects from long and short term exposure. Therefore, monitoring and identifying harmful particles suspended in the air is critical, especially in environmentally hazardous occupations such as metal working. Instruments have been developed to monitor air quality over time, through use of a series of rotating drums that separate particles by size and with time resolution. This project investigates two main goals: (1) characterization of a rotating drum sampler by investigating particle size separation across drums through fluorimetry and (2) quantification and speciation of elements collected on the drum through inductively coupled plasma mass spectroscopy (ICP-MS). Fluorescent particles are aerosolized, size selected using a differential mobility analyzer (DMA) and collected by the drum sampler. Air samples were collected from an on campus makerspace that contains metalworking equipment such as CNC routers and welding equipment. Samples were analyzed in the ICP-MS for reading of harmful heavy metals, including zinc, manganese, and cobalt. Preliminary results indicate trace amounts of cobalt and copper present on drum samples.

Effect of Free Chlorine Exposure on Fouled Engineering-Scale Reverse Osmosis Membranes and Their Surface Characterization

Hannah McCollum, Bianca Miguel De Souza Chaves, Mohammed Alhussaini, Andrea Achilli

Reverse osmosis (RO) membrane technology has proven to be an essential approach to ensuring high quality water for potable reuse. The disposal of end-of-life (EoL) membranes continues to be a critical challenge. RO membranes are periodically disposed of, typically in landfills, creating a global environmental and economic concern. Chlorination has emerged as a potential solution for extending the lifetime of RO membranes. Engineering-scale results showed that chlorination does not compromise selectivity while restoring permeability. To further explore how chlorination changes EoL RO membrane characteristics and mass transfer properties, two EoL RO spiral-wound membranes were removed from the first (PV1) and last (PV6) pressure vessels of an engineering-scale system, to represent organic fouling (PV1) and inorganic scaling (PV6) propensity. Then, the membranes were unrolled, cut into coupons, oxidized with chlorine at different doses, and tested in a crossflow configuration bench-scale system with reclaimed water effluent. The fouled coupons were oxidized by progressive exposure to free-chlorine doses (up to 48,000 ppm-h, namely: 1k, 2k, 4k, 6k, 12k, 24k, and 48k ppm-h). The water samples were analyzed for organic matter content in relation to chlorine dose using UV254 absorbance, EEM fluorescence, and total organic carbon (TOC). Water was tested for inorganic material by measuring conductivity, total dissolved solids, and ion concentration. The membrane surface modification was assessed by measuring Zeta potential and contact angle. Organic results show a significant increase of UV254 absorbance and TOC with chlorine dose increase, suggesting that the membrane surface was progressively modified. At the 4k ppm-h dose, both permeate samples had similar organic and inorganic content to the 0k dose, indicating minimal degradation of the polyamide layer. EEM fluorescence also significantly increased with chlorine dosage, consistent with TOC and UV254 absorbance results. Inorganic results showed an increase of ion presence with chlorine dose, more pronounced after 6k. TOC results were correlated to sulfates and indicate a linear relationship between inorganic and organic matter. Finally, membrane analysis revealed an overall decrease in zeta potential and water contact angle with the 0k (no chlorine exposure) PV1 and PV6 membranes measuring charges of -4.88 and -24.59 mV, respectively, while the 4k membranes reached -30.42 and -33.57 mV. Contact angles for PV1 and PV6 were reduced by 23.8% and 29.8% from 0k to 6k. The outcomes of this study provide insights into mass transfer mechanisms of chlorinated membranes and the chlorine dosage at which a compromise between permeability and selectivity is achieved.

From Creek to Community: Unveiling Water Quality in Union County

Omuhle Ndhlovu

The state of Pennsylvania is surrounded by numerous water bodies that not only serve as habitat for aquatic life, but also provide drinking water and offer recreational opportunities for its residents. As a result, it is imperative that these water bodies be kept clean and safe. There are, however, several streams in Union County that are impaired, which means they cannot not be used for recreational purposes, let alone consumed.

Over this summer we conducted research in the Buffalo Creek with the goal of determining the water quality of the Creek through collaborating with a community partner, Buffalo Creek Watershed Alliance (BCWA) and raising awareness on the safety of the water. We collected water samples at 13 locations along the Buffalo Creek, five times over a period of 30 days. We measured standard water quality parameters such as temperature, pH, total suspended solids (TSS), turbidity, conductivity, and dissolved oxygen (DO). We compiled the data sets in order to compare them to the standard values required by the Department of Environmental Protection (DEP) for recreational waters.

In the lab, we filtered water samples and stored them in the -80°C freezer until DNA extraction. For this summer we extracted DNA from last year's water sample filters using Qiagen's DNeasy PowerSoil Kit and added Salmon Sperm as a positive control. Later on, the DNA will be used for PCR with various primers we would have screened for accuracy in identifying fecal coliforms from identified animal sources.

Moreover, we made use of Microbial Source Tracking (MST), a molecular tool used to determine the animal source of fecal coliform (i.e., bacteria) contamination. DNA sequences of particular bacteria from various animal sources are known, therefore we can use this information to screen water samples and identify the animal source. This is accomplished by isolating DNA from water samples and using Polymerase Chain Reaction (PCR) to make millions of copies of the DNA of interest, if it is present. For this research project we targeted bacteria present in the feces of pigs, cows, dogs, poultry, geese, horses, and humans.

Throughout the summer we held bi-weekly meetings with our community partners to inform them on our progress with the research project. These meetings were also an opportunity to receive suggestions from them (as community members) on what other sites may be of concern.

Exploring Sustainable Practices in Plastic Recycling and Manufacturing: A Case Study in Georgia

Riddhi Bhattacharya, Elisavet Anglou, Fani Boukouvala

The burgeoning environmental challenges posed by the accumulation of plastic waste underscore the imperative for sustainable recycling solutions. Polyethylene Terephthalate (PET), being among the most prevalent plastics, garners significant attention for its recycling potential. The ecological threats of plastic waste accumulation, coupled with PET's widespread use, necessitates innovations in recycling techniques. Thus, for the effective realization of closed-loop recycling technologies, a deep understanding and evaluation of the status of the supply chain is necessary.

In this work, a robust analysis of the supply chain of PET recycling and manufacturing within the state of Georgia is presented. Rigorous data collection from various state and national sources, a meticulous inventory of regional facilities is compiled. [1][2] This database offers a comprehensive depiction, charting the PET journey from collection, sorting, landfilling, recycling, to reuse [3][4]. With a focus on real-world data, this work extensively investigates transportation efficiencies and facility capabilities, emphasizing environmental preservation throughout.

A computational framework is developed to evaluate the supply-chain of waste PET in Georgia through an equation-based formulation. The potential of a selection of chemical recycling methodologies if integrated within the current supply chain network- especially those in recent limelight such as hydrolysis, glycolysis, methanolysis, mechanocatalysis and dissolution – are compared [5][6]. By leveraging real data from our inventory, this comparison serves to quantify the impact of each method on CO2 emissions throughout the supply chain. In summary, this work not only offers a strategic roadmap for fortifying PET recycling infrastructures in Georgia but also presents a modifiable blueprint for broader regional adaptation. The accrued insights prove invaluable for a spectrum of stakeholders, from policymakers and facility operators to researchers, united in their pursuit of a sustainable PET recycling landscape.

References:

[1] Georgia Department of Economic Development. 'Georgia Recycling Infrastructure Map.' (2023)

[2] U.S. Environmental Protection Agency. 'Recycling Infrastructure and Market Opportunities Map.' (2021).

[3] Thakker, V. and B.R. Bakshi, "Designing Value Chains of Plastic and Paper Carrier Bags for a Sustainable and Circular Economy". (2021).

[4] Chaudhari, U.S., et al. 'Systems analysis approach to polyethylene terephthalate and olefin plastics supply chains in the circular economy: a review of data sets and models.' (2021).

[5] Uekert et al. 'Technical, Economic, and Environmental Comparison of Closed-Loop Recycling Technologies for Common Plastics.' (2023).

[6] Schwarz et al. 'Plastic recycling in a circular economy; determining environmental performance through an LCA matrix model approach.' (2021).

Identifcation of Electrochemically Active Bacteria and Their Potential in Water Treatment

Anthony Napolitano, Nada Shetewi, Amanda Simson, Radmila Janjusevic

Electrochemically active bacteria (EAB) are microorganisms capable of releasing their intracellular electrons to the extracellularly redox active materials, like electrodes. EAB can thrive under low oxygen conditions by respiring and metabolizing anaerobically, the process through which they generate electrons that could potentially be exchanged in a microbial fuel cell (MFC). Therefore, EABs have the potential to treat lake water affected by cyanobacterial Harmful Algal Blooms (cHABs), which poses a serious threat to drinking water bodies. The main source of water contamination in cHABs comes from various cyanotoxins produced by cyanobacteria. In the water contaminated from algal blooms, EAB can be employed to consume cyanobacteria through bio-catalyzed reactions and, together with other bacteria capable of breaking down cyanotoxins, used for treatment of polluted water. Therefore, this study set out to identify species of EAB present in an aquatic ecosystem, Central Park Pond (CPP) in NYC, affected by cHABs. The goal of this study was to learn the types and properties of identified EABs to tailor the MFC in treating the polluted water.

We have employed plate-based kinetic fluorescence tests to screen CCP bacteria isolated on Luria-Bertani agar plates with meat peptone as a nitrogen source. Riboflavin was used to monitor the electrochemical activity of tested bacterial cultures as oxidized Riboflavin is fluorescent, but the reduced molecule becomes non-fluorescent. The fluorescence of a sample containing EAB was expected to decrease over time as the Riboflavin was reduced by the EAB in increasing anaerobic conditions in samples covered by mineral oil and tape. To be considered an EAB candidate, the sample should display rapid drop in fluorescence within the first two hours, with fast fluorescence recovery following aeration. We have identified 76 EAB potential candidates, isolated DNA from 38, and PCR-amplified variable parts of 16s rRNA gene. All PCR fragments were gel-purified and sent out for sequencing. The sequences of rRNAs were analyzed using BLAST. 33 isolates were identified as members of the *Aeromonas* family, with some members known to be EAB. One sample was identified to be in the *Pseudomonas* family, two as *Pseudomonas guineae*, one in the *Bacillus* family, and one as *Shewanella putrefaciens*. The identification of *Shewanella* is significant as it is a known EAB, proving the efficacy of the testing method. Future research directions include experimental characterization of identified strains for their electrochemical activity.

Growth and Decomposition Pathways for Ammonium Nitrate Clusters

Ubaidullah Hassan

The decomposition of ammonium nitrate has important applications in storing and transporting ammonium nitrate safely. Decomposition pathways observed in sputtered mass spectra from literature were explained using binding energies found from Monte Carlo geometry optimizations followed by ωB97X-D3 and ωB97M-V density functional theory calculations. It was found that decomposition of cluster ions due to the loss of ammonia or nitric acid are favored. Additionally, all peaks in the positive and negative mass spectra were identified with their corresponding structures. With the reasonable confidence gained from consistency with experiment, the neutral series $(NH_4NO_3)_n$ was studied for $n \leq 40$ in hopes to learn about the early stages of particle growth. This led to an interesting pattern being observed in the structures. Better understanding new particle formation can lead to a better ability to model haze formation in polluted urban environments and model the global climate.

Evaluating the Environmental Impacts of Fertilizer Production Using Photosynthetic Recovery of Nutrients in Livestock Waste

Celeste Mills, Leonardo Gonzalez, Aurora Del Carmen Munguia Lopez, Victor Zavala

Current waste management methods in the dairy industry yield significant environmental repercussions. Failure to implement proper management practices will result in the degradation of farming soil, air, and water resources. These impacts manifest as emissions of greenhouse gases and the leaching of phosphorus and nitrogen into the environment [1]. Two of the main practices that are employed are: i) solid-liquid separators (SLS) to make the manure easier to transport to long-term storage containers and ii) anaerobic digesters (AD) that can capture the methane emissions and clean the captured methane, which can be used as a renewable biofuel. Neither of these practices mitigates nutrient leaching. A promising avenue to reduce nutrient leaching and improve greenhouse emissions involves utilizing cyanobacteria (CB) [2]. By using CB, it is possible to create the Renewable Nutrients from Algae (ReNuAl) process.

The ReNuAl process presents three distinct versions: ReNuAl (0), which integrates CB into an AD system, harnessing the captured methane in an on-site generator; ReNuAl (1), which similarly employs an AD but utilizes all the captured methane as an additional revenue stream; and ReNuAl (2), which eliminates the AD and methane collection. CB functions as an optimized fertilizer, able to deliver the required nutrients to crops in the ideal ratio. Additionally, CB exhibits the potential to curb emissions by sequestering carbon dioxide.

Multiple integration pathways are feasible for incorporating CB into the waste management system for dairy farms. We performed a life cycle assessment to assess the environmental impacts of these diverse systems and compare them with prevailing practices employed by permitted farms (with more than 1000 Animal Units). We used the software openLCA with the Environmental footprint method [3]. The current practices for this analysis are assumed to be using i) an SLS and ii) an SLS and an AD. Our analysis shows that despite the larger energy requirements of ReNuAl, the emissions are substantially reduced for all proposed ReNuAl systems. However, other factors must be considered, such as the environmental effects of the current grid composition and the use of water, plastic, and synthetic nitrogen.

[1] Hu, Y., et al. Managing Conflicting Economic and Environmental Metrics in Livestock Manure Management. ACS ES&T Eng., 2022

[2] Ma, J., et al. Infrastructures for phosphorus recovery from livestock waste using cyanobacteria: Transportation, techno-economic, and policy implications. ACS Sustain. Chem. Eng., 2021

[3] Ciroth, A., et al. openLCA 1.10: Comprehensive user manual, GreenDelta, Berlin, Germany, 2020.

Impact of Morphology and Polymorph on Behavior of Succinic Acid Aerosols in Mixtures with Ammonium Sulfate

Will Feldscher, Ryan Snyder, Dabrina Dutcher, Timothy Raymond, Julia Reynolds

Particles on the scale of nanometers are incredibly important in atmospheric chemistry due to their complex role as aerosols in the atmosphere. Atmospheric aerosols and their morphologies directly impact cloud formation, which in turn affects global warming and climate change modeling. Many factors can affect the distribution and morphology of atmospheric aerosols, such as humidity, temperature, and the presence of other chemical species. In this project, a system of succinic acid, a dicarboxylic acid, and ammonium sulfate, an inorganic salt, was studied as both are commonly found in atmospheric aerosols.

There are a variety of ways to model and study atmospheric aerosols; this project utilized a syringe pump to feed aqueous solutions of varying concentrations and mixing ratios of succinic acid and ammonium sulfate into an atomizer to produce nanoscale particles. These particles were then dried and analyzed in a Scanning Mobility Particle Sizer (SMPS) system to obtain size distribution data. This method of analyzing particles typically results in a Gaussian particle size distribution. However, for specific mixtures of succinic acid and ammonium sulfate, a bimodal aerosol size distribution was observed. This phenomenon represents a gap in understanding of the atmospheric chemistry in the succinic acid/ammonium sulfate system.

It was initially hypothesized that the observed bimodal aerosol size distribution is due to the polymorphism of succinic acid, as a similar phenomenon is observed in aerosols of glutaric acid, another dicarboxylic acid. Succinic acid has two known polymorphs: the β polymorph which is stable at room temperature, and the α polymorph which is stable above 137°C. To test for the presence of these polymorphs in different solutions of succinic acid and ammonium sulfate, particles were collected using a cascade impactor to then be analyzed with powder x-ray diffraction (PXRD). To further investigate the morphology of the particles collected, scanning electron microscopy (SEM) was used to take characteristic images of collected samples. This combination of SMPS, PXRD, and SEM data was collected for aerosols formed from aqueous solutions of succinic acid and ammonium sulfate to better understand the system and its relevant chemistry.

Economic and Environmental Analysis a National-Scale, Thermo-Chemical Plastic Upcycling Infrastructure in the United States

Evan Erickson, Philip Tominac, Jiaze Ma, Horacio Aguirre-Villegas, Victor M. Zavala

Global rates of plastic production demand comprehensive and reliable plastic waste management systems to avoid the worst impacts of terrestrial and marine plastic pollution. However, overall plastic recycling rates remain insufficient, while the bulk of plastic wastes are incinerated, landfilled, or allowed to accumulate in the biosphere.

Traditional plastic recycling by mechanical methods is a mature technology but faces significant inadequacies as a plastic management strategy given that it has been developed only to accept limited types of plastics. This limitation reinforces the parallel need for environmentally damaging disposal methods, such as landfilling and incineration, to accept waste products not fit for mechanical recycling.

A complementary strategy for plastic waste management, known as plastic "upcycling," uses chemical methods rather than mechanical ones. Upcycling involves a chain of processing technologies: beginning with the sorting and purification of plastic fractions of common recyclable waste streams at municipal recovery facilities (MRFs) and plastic reprocessing facilities (PRFs), plastic wastes undergo a series of chemical conversions including pyrolysis and steam cracking to arrive at pure plastic monomers, which can finally be repolymerized into products (here, low-density polyethylene and polypropylene, or LDPE and PP) with the same properties as virgin plastics.

To date, plastic upcycling has not received sufficient analysis exploring how an upcycling circular economy may be realized. Our study will use a framework to examine the potential for an upcycling system to be instituted at a large scale (the continental US), allowing us to design an upcycling network of facilities that reaches all residential waste producing sources in the continental US. We determine that upcycling provides a vastly more favorable economic picture than current downcycling, in which the value of polymer end-products propagates through the value chain and even provides an opportunity for residential waste producers to be paid for their recycled plastic, and quantify its economic sensitivity to a number of key parameters. Finally, we comment on the potentially favorable environmental performance of upcycling relative to current US plastic disposal methods.

Investigating Effect of Thermal Gradient on Diffusion of Microbial Respiration Byproducts through PTFE for Soil Profiling

Marton Varga

The agricultural industry utilizes crop rotation to take advantage of plant species differing in nutrient requirements to lessen the adverse impacts of successive crop planting on soil quality. As the need for more sustainable and efficient agriculture develops, a deeper understanding of the influence of soil characteristics on crop production and quality is necessary. Few tools are currently available for farmers to profile their soil, particularly on a molecular level where microbial decomposition synthesizes the nutrients that crops require. The research group focuses on the investigation of gaseous diffusion through hydrophobic membranes protecting sensor components for use in an encompassing project designing a subsurface soil sensor profiling soil characteristics via the detection of gaseous byproducts of organic decomposition by microbes. Prior to data collection, the project required the development and construction of a membrane testing structure and standardized protocol to evaluate diffusive properties of hydrophobic membranes. The acrylic testing structure encloses a thin membrane between its two halves where gas flows on each side, contacting the membrane surface through fritted glass connected to gas flow. PTFE (polytetrafluoroethylene) is selected as a selectively permeable membrane to evaluate the effects on diffusion rate with an induced temperature gradient inside the membrane testing device to determine efficacy of a heated chamber in the soil sensor. Carbon dioxide, a primary byproduct of microbial respiration, is selected as the gas investigated while nitrogen is used as a carrier gas on the opposite side of the membrane and testing device. Computational methods (COMSOL, Python) are used to verify the increased diffusion detected and calculate mass balances.

What Is Contaminating Streams in the Buffalo Creek Watershed?

Demi Gonzalez

The state of Pennsylvania is surrounded by numerous streams and bodies of water, some as easily accessible as being in your backyard. However, the majority of these waterways are inadequate for recreational activities due to nutrient pollution, bacteria and pathogen contamination, and sediment pollution. Despite waterways being assessed for impairment, the general public remains to have access to these bodies of water with no warning of the water's quality. After collecting and evaluating water samples from 13 locations in the Buffalo Creek Watershed, five times over a period of 30 days for the summer of 2022 and 2023, we have determined that the majority of the streams demonstrate high bacteria concentrations from fecal pollution. It is hypothesized that this fecal pollution is derived from farming practices as Union County is rural.

From the water samples collected, filtration for DNA extraction was done using *Millipore Express PLUS 0.22 μm* filters. The addition of 100 μL of 1.2 mg/L Salmon Sperm DNA to filters was done before using *Qiagen's DNeasy PowerSoil Kit* for DNA extraction. The collection of animal fecal matter was also included in this research and involved samples of pig, cow, goose, chicken, dog, deer, human, and horse fecal samples collected within the area. From these samples, DNA extraction was performed using *Zymo Research's Quick-DNA Fecal/Soil Microbe Miniprep Kit*. After DNA extraction was done for water and fecal samples, a Polymerase Chain Reaction (PCR) was performed for each DNA extraction to make millions of copies of the DNA of interest. Gel electrophoresis was then performed with the PCR samples of animal extracted DNA to determine the specificity and sensitivity of our animal primers; CowBACB2, CowM3, Pig-2, Avian GFD, Goose CG1, Deer EF447, Dog DG3, Human HF183, and Horse HoF597F.

Once we have established the specificity and sensitivity of our animal primers to the correct animal DNA, we can move towards determining what animal fecal matter is polluting the streams. Using the data of the animal DNA present in specific water samples along with the extracted data from our county's Nutrient Management Plans (NMPs), we can use ArcGIS to map and pinpoint potential sources of contamination.

This Project Focuses on Developing a Short Term Passive Sampling Method for Airborne Polychlorinated Biphenyl Congeners Using a Nanofiber Material.

Aoife Cleary

Polychlorinated biphenyls, also known as PCBs are a class of chemicals that were produced in the United states between 1930-1977, when their production was banned due to evidence that they cause health side effects to both humans and animals such as endocrine disruption, neuro developmental and reproductive issues, and cancer, in addition to evidence that they both persist and accumulate in the environment. This project focuses specifically on airborne polychlorinated biphenyl congeners, with our primary objective being to develop a novel passive sampling device. This project used a nanofiber mat of the polymer polyacrylonitrile (PAN), which was created by electrospinning a solution of PAN dissolved in N-Dimethylformamide. The majority of passive samplers currently available require weeks to months in order to be effective, and reflect only an average quantity over that time period. The preliminary results from our research indicate that there is potential to create a 12 hour sampling protocol with PAN, which could be effective in measuring the quantities of airborne PCBs in the environment. This allows for more specificity when analyzing PCB quantities in the air and makes identification of specific days or windows of time where PCB quantities are spiking possible.

Synthesizing Metal Thiolates As Soluble Precursors for Multinary Sulfides

Isabel Panicker

Multinary metal sulfides are an interesting class of materials many different properties. Of particular interest are the semiconducting metal sulfides which can be used for solar energy generation. While much research has been devoted to solution processing these materials when they contain late- and post-transition metals, but very little has been done for materials that contain alkaline earth metals. Solution-based approaches are desired because they can offer low energy, low cost, and scalable manufacturing processes. Therefore, this project focuses on creating a procedural basis for synthesizing multinary sulfides that contain alkaline earth metals (binary, tertiary and quaternary) with soluble metal thiolates precursors. The main procedure begins with a metal precursor that is reacted with a thiol in order to introduce a sulfur source. An appropriate solvent is selected, such as an amine, in order to form a precursor solution. Various metal precursors, thiols, and solvents are tested, ranging in polarity and size. If a metal thiolate solution is formed, then decomposition into the desired metal sulfide is performed by drop coating and heating the solution into films and then analyzing with XRD. Through this procedure, various multinary sulfides have been synthesized including Cu2BaSnS4 and the chalcogenide perovskite BaZrS3. This pathway opens up the possibility for widespread usage of multinary sulfides that contain alkaline earth metals in the fabrication of solar cells that can further sustainable energy goals.

A Review of Additive Usage in Polymer Manufacturing: Phenolic Antioxidants

Derek Orndoff, Jeffrey Seay, Mary Ternes

As the world comes together to address plastic waste and pollution, and the March 2, 2022, United Nations Resolution 5/14 to End Plastic Pollution with a Legally Binding Instrument by 2024, many strategies are being considered for how to best address this immense problem. One of the most significant arguments is the simplification of the additives and polymer universe. As time has gone on plastics have gotten much more variable so even within the main 7 types of plastic, there are thousands of different formulations for plastics. This variability can make post-consumer plastic relatively unusable. With the simplification of these additives and formulations, recovery methods could become much more effective with a much more valuable end product. As such, simplifying the plastic universe could make for an effective strategy for creating a more circular lifespan of these plastics. Based on these foundations, this presentation will be focusing on the additive class of phenolic antioxidants and the similarities identified within. It identifies different possible methods to take moving forward that take into consideration the regulatory aspect as well as the chemistry to limit variability and additives that are problematic (persistence, bioaccumulation, and toxicity) while still maximizing the function of the polymer.

Blending Silver Nanoparticles in a Membrane to Reduce Biofouling during Filtration

Amanda Bastian, David Lu, Isabel Escobar

Abstract

Membranes have been effectively used in water treatment for the separations of contaminants. During filtration, rejected materials are sent to the waste stream or accumulate on the surface of the membrane, with the latter leading to fouling. When microorganisms, such as bacteria, accumulate on the membrane, biofouling occurs. Silver nanoparticles have anti-microbial behaviors due to the reported release of reactive oxidative species. Reactive oxidative species are free radicals that are created after penetration to the cell, which cause toxicity in the microbial cells. By embedding silver nanoparticles in a membrane, biofouling build up on membranes can be minimized or even prevented. The membranes in this research were formed from a dope solution comprised of polysulfone dissolved in bioderived co-solvents Rhodiasolv® PolarClean and gamma-valerolactone (GVL).

Once the silver nanoparticles are added to the membrane, the silver can potentially leach into the water during filtration. Silver is not allowed in water, as the regulations require the silver concentration not to exceed 0.1 mg/L. Therefore, adding a polymer to coat, stabilize, and prevent the leaching of the silver nanoparticles was the goal of this study. The coating polymers studied were polyvinylpyrrolidone (PVP), polyethylene glycol (PEG), and polydopamine (PDA) due to their charge attraction. Characterization testing was then performed to ensure the polymers were coating the nanoparticles and attaching to the membrane polymer. After the coated nanoparticles were created, Fourier transform infrared (FTIR) was used to confirm the coating process was complete. There were no significant differences between the polymer and coated nanoparticles. Dynamic Light Scatter (DLS) was then used to measure the coated nanoparticle size and zeta potential or charge of the particle. The coated nanoparticles size increased and there was less of a charge comparably to the pristine nanoparticles. Then, UV-vis spectrophotometry was used to measure the particles' light absorption. The results showed the coated nanoparticles had absorbed less light than the pristine nanoparticles. After characterization testing was performed, x-ray photoelectron spectroscopy (XPS), Drop Shape Analyzer, and scanning electron microscope (SEM) were employed to measure the different characteristics of the coated nanoparticle membranes.

Arsenic Contamination in Groundwater of Nawalpur District

Nirajan Shrestha, Saragya Gyawali

Arsenic is a chemical element that is widely distributed throughout the environment in the air, water, and earth's crust. It is a confirmed carcinogen and highly toxic in its inorganic form. Groundwater is the primary source of water for drinking, irrigation, sanitation, feeding livestock, and industrial purposes in the Terai region of Nepal. Multiple studies have shown that the groundwater has been contaminated with arsenic (As) in districts such as Siraha, Dhanusha, Rautahat, Bara, Nawalpur, and Rupandehi. The effects of *As* contamination range from human health issues such as arsenic poisoning, skin diseases, and cancer to soil and crop contamination. Thus, groundwater contaminated with arsenic (*As*) has been a severe water issue for people residing in these districts. This study encompasses the determination of arsenic contamination in groundwater over different administrative wards of Gaindakot Municipality in the Nawalpur District. A limited number of 52 initial samples were pretested using the Hack As-Test kit on-site, which used mercuric bromide-coated film as a tester agent, while detailed analytical analysis was conducted using atomic absorption spectroscopy (AAS) and a UV-Vis spectrophotometer. Among the 52 samples tested, 7.69% had arsenic concentrations in groundwater above standard levels (National standards: 0.05 mg/L and WHO standards: 0.01 mg/L). Arsenic detected in the groundwater samples was of geogenic origin, as confirmed by various past research studies. The Arsenic Biosand Filter (ABF), made from locally available materials that proved to be up to 96% efficient in removing arsenic, was recommended to be installed in drinking water supplies at each administrative ward and in public places such as schools, parks, and hospitals around the municipality.

Modifying Pyrolysis Conditions to Increase CO_2 Adsorption of Biochar

Abigail Lin, Wesley Wong, Amanda Simson

Climate change is a pressing issue driven by excessive greenhouse emissions, particularly carbon dioxide. The atmospheric CO_2 level has significantly increased from pre-industrial levels of 280 ppm to a record high of 424 ppm in May 2023. While advancements in emissions reductions are the priority, carbon removal solutions such as direct air capture (DAC) should also be considered. Biochar, made from pyrolyzing biowaste, has been credited as a promising CO_2 adsorbent for DAC. The objective of this study is to examine if pyrolysis conditions impact biochar's CO_2 adsorption. Walnut, sawdust, coconut husk, and corncob feedstocks were pyrolyzed at various temperatures to see if the pyrolysis temperature affects adsorption. As expected, increased pyrolysis temperature led to increased adsorption capacity. However, the extent of the increase depended on each feedstock, suggesting that the influence of temperature is feedstock-dependent. Pre-drying the feedstock before pyrolysis was also studied to determine if it impacted biochar performance. Pre-drying walnut resulted in 40% less CO_2 adsorption, and similar results were observed for all but the corncob feedstock where pre-drying was not shown to influence the resultant biochar's capacity measurements. In addition to pre-drying, the effects of nitrogen flow rate during pyrolysis were tested to see if it influenced CO_2 adsorption capacity. Walnut had a threshold flow rate of 200mL/min, where, above this, there was no difference in capacity at higher flow rates but below it, a significant decrease of up to 24% was measured. Coconut followed a similar trend while there was no impact on sawdust or corncob. Not only does this show that higher nitrogen flow rates increase adsorption, but it also emphasizes the importance of mass transfer development during pyrolysis. These findings indicate that pyrolysis conditions, particularly pre-drying and nitrogen flow rate, significantly impact the adsorption of the subsequent biochar, and further, that these effects are feedstock dependent.

Evaluation of Pyrolysis Systems and Biomass Feedstock

Gloria Hernandez, Arianna Guillory, Catherine Brewer

Pyrolysis systems are technologies that convert biomass into chars, bio-oils, and combustible gases depending on the reaction conditions used and the desired primary products. The goal of this project is to identify and compare commercial pyrolysis systems for application at small-to-medium biomass processors. The candidate systems will be sorted based on technology complexity, scale, biomass suitability, and cost, and the needs of the community using the pyrolysis system. For the down-selected pyrolysis systems, vendor technical information and/or quotes will be used for techno-economic analysis and other decision metrics. Research into available pyrolysis systems will begin with information compiled by the U.S. Biochar Initiative (USBI) and the International Biochar Initiative (IBI). That system information will then be combined with biomass feedstock quantification and quality analysis, and then application testing for the produced biochars.

The focus of this research is on woody residues from forest burn scare reclamation and a local lumber mill. The purpose of burn scar remediation is to identify uses for materials that are too damaged to be used for lumber or fibers in order to provide revenue for site clean-up. Lumber mill wastes vary in particle sizes and include sawdust, bark, and defective wood. Among the different biochar applications to be considered during evaluation of pyrolysis systems are energy (heat), carbon credits through carbon sequestration, and adsorbents to remove arsenic (or other heavy metals) from water.

Investigating the Use of Waste Hemp Derived Biochar Adsorbents for Contaminants in Oil

Sable Phillips, Hema Ramsurn, Suriya Narayanan Ramasubramanian

Crystallization and precipitation of minerals (Ca^{2+}, Ba^{2+}, CO_3^{2-}, SO_4^{2-} etc.) present in water can cause organic scaling, which directly interferes with oil and gas production, impedes fluid flow, and adds costs due to treatment and removal processes. Scaling also obstructs the contact between metal and corrosion inhibitor, intensifying local corrosion of pipes. Therefore, it is important to reduce if not eliminate scale formation in pipelines. Our overall goal is to find a more cost effective and environmentally friendly way to prevent this issue by using different carbon-based adsorbents to remove scaling ions, in particular Ca^{2+}. The in-house prepared adsorbents tested were: Base Coke, Coke Hemp 1, Coke Hemp 2, Hemp 1, Activated Hemp 1, Cellulose 300, Activated Cellulose 300, Hemp 2, Activated Hemp 2, and Refluxed Hemp 2. As preliminary work, different Ca^{2+} concentration solutions ranging from 10-50 ppm were prepared and 24-hour adsorption tests were set for each of the adsorbents. A Flame Atomic Absorption Spectrometer (FAAS) was used to analyze and detect the Ca^{2+} ions in the solutions. After 24 hours, the adsorbents were separated from the solution which was then analyzed using FAAS to determine the amount adsorbed. While analyzing the Ca^{2+} results, it was found that activated cellulose and cellulose biochar gave the best results for Ca^{2+} adsorption. While hemp on its own (whether as biochar or activated) is not very successful in trapping the calcium, when mixed with coke, it successfully adsorbed Ca^{2+}. To gain an understanding of these results, it was decided, in the first instance, to look at the ash content of the bio-adsorbents. Coke and cellulose had negligible ash content (<1%) which aided in the adsorption process of calcium, but hemp biochars had a very high ash content (23-60%). It is suspected that the high ash content of hemp is due to the growth environment (soil, fertilizers added) and the ash may be blocking effective adsorption sites. Thus, an acid reflux was set up to see if the ash content of hemp biochar could be lowered to improve the adsorption of calcium.

Fluorinated Hydrogels for Remediation of Per- and Polyfluoroalkyl Substances from Water

Jacob U'Wren, Pranto Paul, J. Zach Hilt

Contaminated water is a growing issue as our world only industrializes more. Some common man-made contaminants are per- and polyfluoroalkyl substances (PFAS), otherwise known as "forever chemicals". These chemicals are commonly used as stain resistant or waterproof agents because of their hydrophobicity and extremely strong intramolecular forces due to their many carbon-fluorine bonds. The strength of the bonds within PFAS cause them to not break down naturally, leading to their accumulation in the environment (e.g., water sources) and human exposure and potential health consequences. One approach to this issue is to use hydrogels to absorb PFAS from contaminated water. Hydrogels are crosslinked, hydrophilic polymers that can be functionalized for targeted contaminant removal from water. We studied the use of fluorinated hydrogels to sorb PFAS from contaminated water. Fluorinated hydrogels will attract PFAS through fluorous interactions. Hydrogels were synthesized using free radical polymerization. Comonomers 2,2,3,3,4,4,5,5,6,6,7,7-dodecafluoroheptyl acrylate (DDFHA) and acrylamide (Aam), cross-linker N,N'-methylenebisacrylamide (NNMBA), catalyst tetraethylmethyl-ethyldiamine (TEMED), and initiator ammonium persulfate (APS) were used for the hydrogel synthesis. Swelling studies were conducted on hydrogels of varying DDFHA composition to confirm the DDFHA incorporation in the hydrogel. It was found that increasing DDFHA composition led to less swelling of water by the hydrogels. Binding studies of select PFAS compounds were conducted in water over 24 hours. A significant amount of PFAS was found bound to the hydrogels. These findings point toward a promising solution for remediation of PFAS-contaminated water.

Finding Trends in Biochar Types to Maximize Carbon Dioxide Adsorption

Harsh Reshamwala, Amanda Simson

Carbon dioxide (CO_2) emissions are a prominent factor in driving global climate change, intensifying concerns over global temperatures and socio-environmental instability. In addressing this worldwide challenge, biochar – a byproduct of biomass pyrolysis and a soil amendment widely used in facilitating agriculture and gardening – has gained favor for its ability to sequester carbon in the soil. In addition to its use as a soil amendment, it has been shown particularly with activation procedures to be a good adsorbent for various toxins and more recently for CO_2 from the air. Nonetheless, a significant knowledge gap remains regarding the identification of activators, both physical and chemical, that can be used to optimize the adsorption capacity. This study reviews 24 papers in the literature to uncover activators with high CO_2 adsorption capacities. Amongst the 11 activation processes evaluated in this review, specific activators, namely Potassium hydroxide (KOH) + CO_2, Aluminum (Al), Steam, and Iron (Fe), emerge as leaders for their potential for enhancing carbon capture when used to activate biochar. The study also investigates feedstock selection and the conditions under which biochar is produced. An analysis of feedstock was conducted to identify the most effective biochar types. From this subset of high-performing feedstocks, patterns in relation to 20 associated variables were analyzed. Among these, 2 specific parameters – Nitrogen content, and Carbon content – were recognized as potential predictors of feedstocks with a strong likelihood of yielding high-performing biochar after pyrolysis. This study forms the foundation for future investigations, including the development of a machine-learning model for identification of trends in CO_2 adsorption.

Desulfurization of Liquid Hydrocarbon Fuels Using Food Waste Activated Carbon

Caitlin Noonan, Henry Sokol, Julia A. Valla

Sulfur compounds present in crude oil contribute to air pollution, acid rain, and ozone depletion when sulfur-contaminated hydrocarbons are combusted in engines. Increasingly stringent environmental legislation limits sulfur concentrations in fuels; however, current methods used in refineries cannot meet these requirements without high energy and material costs. Simultaneously, food waste is a significant environmental problem that requires management solutions. The objective of this research was to convert food waste into activated carbon to investigate sulfur removal capabilities from fuels and to compare food waste activated carbons to more commonly studied uniform precursors to activated carbons such as walnut shells. Few previous studies examine food waste as a precursor to activated carbons for the removal of dibenzothiophenes (DBT) from fuels. Food waste and walnut shells were each pyrolyzed and physically activated with steam. Copper nitrate (5%, 1%, 0.25% weight (wt.)) was added to food waste prior to pyrolysis to study the effects of metal precursors on sulfur adsorption capacities for improved selectivity while limiting production steps. Pyrolysis (650°C for 2h) and activation (950°C for 4h) conditions were chosen to maximize the biochar yield and activated carbon microporosity. 0.25%wt. of copper nitrates resulted in the greatest amount of dispersion of copper on the biochar and was selected for activation. Computational studies showed that copper on carbon contributes to greater selectivity of DBT over naphthalene. Food waste activated carbons (3.0 mg S/g adsorbent) and 0.25%wt. food waste activated carbons (2.5 mg S/g adsorbent) were found to remove more sulfur than walnut shell activated carbons (2.0 mg S/g adsorbent) when all achieved similar yields after activation. This research demonstrates the promise of food waste activated carbons to offer a green solution for removing sulfur from fuels.

Biochar Produced from Corn Stover Affects Water Retention and Mechanical Properties of Soil

Aiden Brown, Bratoljub Milosavljevic, Tinglu Yang

Biochar can be mixed with soil to increase the water absorption capacity and decrease the evaporation rate, which could in turn increase crop yields particularly under prolonged drought conditions. In this project, the biochar was produced by pyrolyzing a sample of corn stover, a widely available resource in the U.S. The torrefaction procedure applied consisted of heating at 350 °C for 130 minutes in a nitrogen atmosphere. At the end of the pyrolysis, the biochar mass produced was 34% of the initial stover mass. The absorption and emission spectra corresponding to the corn stover pyrolysis product collected from the long quartz tubing by washing it using 95% ethanol indicates that a variety of (aromatic, potentially useful) compounds has been produced.

To assess the hygroscopicity, biochar sample was left to equilibrate in water for 2 weeks; the amounts of water absorbed was determined gravimetrically to be 72%. A TGA experiment was carried out at heating rates of 5, 10, 15, 20, and 25 °C/min on the sample hydrated for two weeks. The activation energy for the desorption of water was determined to be 54 kJ/mol using the Friedman method and 52 kJ/mol by using the Kissinger-Akahira-Sunose method. Another TGA experiment was performed to assess the effect of biochar on the dehydration rate at the temperatures relevant for agriculture. For example, at 35 °C, the presence of biochar results in more than tenfold decrease in the dehydration rate in comparison to that of free/bulk water. To further assess the potential use of biochar in agriculture, the dehydration rate of soil was compared to that of the soil and biochar mixture using a TGA technique. Due to a small sample size and variation in the soil particle size the data variance was very large. To overcome this problem, the hydrated 50 weighed percent biochar/soil mixture was placed in a cylindrical vessel with an open top and its mass is being measured as a function of time in order to obtain the dehydration rate. The results to be obtained will be compared to that corresponding to the hydrated soil. During the course of the experiment it was noted that the addition of biochar improves the mechanical properties of hydrated soil and the quantification of the effect observed is in progress.

Small-Scale Pyrolysis Reactor Modifications for Carbon Balance Measurements

Sivan Spiel

As of 2023, the US produces 120 billion pounds of foodwaste that ends up in landfills each year, releasing greenhouse gasses that contribute to the ongoing climate crisis. Currently the atmospheric carbon dioxide has been growing over 2 ppm each year, with a global average of 417.06 ppm recorded in 2022. One method of preventing CO_2 from being released into the atmosphere is using pyrolysis to convert biowaste into biochar. Pyrolysis is a method of sequestering carbon by heating biomass to high temperatures (~300°C-800°C) in the absence of oxygen. This process creates a solid carbon material called biochar, which farmers have been producing for millenniums to improve their soil. Although farmers use biochar in a purely agricultural sense, this process has the potential to be carbon negative. This study attempts to quantify the carbon sequestered and released during pyrolysis in a laboratory setup under a chemical hood. Thus far, two semi-batch reactors (with recirculating syngas) have been built that are partially able to pyrolyze biomass. These prototypes used propane as the energy source and wood shavings as the biomass. The first reactor, which contained an average of 171.25 g of biomass, had inefficient heat transfer, with a biochar yield ranging from 4-11% of the initial mass (w/w%), and a max temperature of ~314°C. The second prototype was much smaller, with an initial mass of 31 g. A flow sensor was also attached to measure syngas. This reactor had improved the heat transfer and data collection due to the change in size and shape, with a biochar yield of 65% of the initial mass (w/w%), and a max temperature of 330°C. It was found that the carbon emitted outweighed the carbon mitigated, with results showing 926.9 g of CO_2 was emitted and 73.3 g of CO_2 was sequestered with a reaction time of ~51 minutes and no measurable syngas flow. The reactor is still inefficient, the chemical hood worsens the heat transfer that reaches the biomass. To optimize the CO_2 emitted:stored as well as increase the max temperature, the next prototype will be powered electrically.

References:

1) "Food Waste in America in 2023: Statistics & Facts: RTS." *Recycle Track Systems*, www.rts.com/resources/guides/food-waste-america/. Accessed 28 Sept. 2023.

2) Lindsey, Rebecca. "Climate Change: Atmospheric Carbon Dioxide." *NOAA Climate.Gov*, www.climate.gov/news-features/understanding-climate/climate-change-atmospheric-carbon-dioxide#:~:text=In%20May%202023%2C%20carbon%20dioxide,people%20are%20burning%20for%20energy. Accessed 25 Sept. 2023.

3) "Biochar Then & Now." *Biochar Then & Now | US Biochar Initiative*, biochar-us.org/biochar-then-now#:~:text=Biochar%20has%20also%20been%20created,in%20some%20places%2C%20toxic%20soils. Accessed 30 Sept. 2023.

Mass Production of Hydrogel Beads for the Sustainable Biological Treatment of Emerging Contaminants in Soil and Groundwater

Katherine Bandettini, Conor Harris, Lewis Semprini, Skip Rochefort, Kaitlin Fogg

Chlorinated solvents are commonly found in drinking water supply aquifers and groundwater plumes throughout the United States[1]. The Centers for Disease Control and Prevention lists all chlorinated aliphatic hydrocarbons (CAHs) detected in groundwater samples as probable human carcinogens as well as having other hazardous health effects towards humans and the environment[2]. A permeable reactive barrier system has shown to be an effective *in situ* bioremediation technology for treating chlorinated compounds found in groundwater with microorganisms immobilized in hydrogel beads with orthosilicates as slow-release compounds (SRCs) to support aerobic cometabolism[3]. Rhodococcus rhodochrous ATC 21198 bacteria strain has successfully treated common chlorinated solvents and can be used as a passive bioremediation method for treating contaminated groundwater plumes. Thus, a lab-scale bead generator was built to generate mass quantities of hydrogel beads. Laminar jet flow breakup in addition to coaxial air flow and a vibrating nozzle allowed for bead parameters such as size and circularity to be controlled. Different concentrations of sodium alginate were flown through the bead generator at varying Reynolds numbers to characterize the optimal flow parameters of the polymer fluid. A correlation between the Reynolds number (inertial forces/viscous forces) versus Ohnesorge number (viscous forces/inertial forces and surface tension) was generated to find the flow regime for a laminar jet stream, and found that different concentrations could be binned into different ranges of Reynolds numbers. Images were used to capture and measure the bead perimeter and circularity. Data collected was used to optimize the independent variables (Reynolds number, vibrational frequency, and coaxial air flow), on the dependent variables (bead perimeter and circularity) for each with a design of experiments approach.

(1) Grindstaff, M. *Bioremediation of Chlorinated Solvent Contaminated Groundwater*; 1998. https://citeseerx.ist.psu.edu/document?repid=rep1&type=pdf&doi=caddf817ab5b9320dc7d1e91cbfc21b025ade36d (accessed 2023-08-10).

(2) *Organic Solvents | NIOSH | CDC*. www.cdc.gov. https://www.cdc.gov/niosh/topics/organsolv/default.html.

(3) Rasmussen, M. T.; Saito, A. M.; Hyman, M. R.; Semprini, L. Co-Encapsulation of Slow Release Compounds and Rhodococcus Rhodochrous ATCC 21198 in Gellan Gum Beads to Promote the Long-Term Aerobic Cometabolic Transformation of 1,1,1-Trichloroethane, *Cis*-1,2-Dichloroethene and 1,4-Dioxane. *Environmental Science: Processes & Impacts* **2020**, *22* (3). https://doi.org/10.1039/c9em00607a.

Streamlined LCA Tool for Sustainable Polymers

Ava Dahnke

The development of a streamlined Life Cycle Analysis tool for the purpose of guiding sustainable polymer design is presented. The NSF Center for Sustainable Polymers has an established Sustainable Polymer Framework which classifies sustainability considerations into four key stages: Feedstock, Process, Intended Use, and End-of-Life. The challenge lies in interlinking these framework components to evaluate potential trade-offs during the implementation or modification of polymer design and manufacturing processes. In response to this challenge, a tool is currently under development with the aim of facilitating informed decision-making within the polymer industry, while also being non-debilitating in the design phase. The tool will be open source, harnessing available data and models to create a collaborative community resource, and will be based in Microsoft Excel for ease-of-use. It is designed to address sustainability by estimating essential metrics such as energy intensity, water intensity, and greenhouse gas emissions at each stage of the Sustainable Polymer Framework.

Feedstock: The stage offers guidance on the selection of feedstock materials, which is often a challenge due to the multitude of available options. Leveraging data from the GREET (Greenhouse gases, Regulated Emissions, and Energy use in Technologies) model, outputs from this stage include a comparison of environmental impacts of producing a molecule of interest from different feedstocks when yield data is available.

Conversion: This stage addresses critical choices in polymer manufacturing processes, including the selection of solvents, reaction conditions, yield and selectivity, separation techniques, and the management of co-products or waste.

End-of-Life: In this phase, the tool assesses the environmental impacts associated with various recycling technologies such as mechanical, chemical, and biological methods based on literature-derived ranges. Data in this stage is standardized using the GREET model and allocation techniques are employed to quantify the impacts per kilogram of waste polymer.

The development of this streamlined LCA tool will enable scientists to quickly estimate environmental impacts when making decisions about sustainable changes in polymer design before they are even implemented. By providing an open-source platform that leverages existing data and models, this tool also seeks to foster collaboration and innovation, promoting a more sustainable future for the polymer industry.

Influence of Emulsion Properties on Poly(caprolactone) Microparticle Size

Ashbey Manning, Claire Rowlands, Hope Saindon, Brittany Givens Rassoolkhani

The role of polymer science in drug delivery is increasing in significance. Biodegradable, biocompatible polymers, polymeric micro- and nano- particles are promising advancements in targeted drug delivery. Poly(caprolactone) (PCL) has largely been successful in implantable devices and as tissue engineering scaffolds, due to its high biocompatibility and slow degradation time relative to other polymers. We hypothesize that these same properties make PCL a promising candidate for nano- and micro-particle drug delivery. Administration through the bloodstream, such as in tumoral drug delivery require nanoparticles below 300 nm whereas administration through the nose, such as lung disease treatment utilizes particles in the 1–5 um size range. In this work, we outline 16 unique preparation conditions for PCL particles to produce a range of particle sizes for each target application. An emulsion solvent evaporation method was used with varied solvent choice, polymer mass, polymer concentration, surfactant concentration and external water volume to synthesize particles between 300 nm and 1.7 micrometers. The particles were measured for size using scanning electron microscopy (SEM) and ImageJ. Particles were evaluated for cytotoxicity in HEK293, A549, and Ishikawa cell lines. We were able to determine that the particles met the size requirements for a range of healthcare applications and were not detrimental to cell viability at low concentrations.

Design and Assembly of VLP-Based Broadly Protective Flu Vaccines

Alexandra S. Seesee, Kathryn E. Loeffler, Daniel Kim, Ravi S. Kane

Influenza A is a genus of contagious RNA viruses whose potential to cause severe illness makes it a major public health concern; globally, it is responsible for hundreds of thousands of deaths every year. Influenza A is capable of infecting humans using the hemagglutinin (HA) glycoprotein spikes on its viral envelope, which have two domains: a head domain that helps the virus attach to the cell membrane and a highly conserved stalk domain that enables membrane fusion. Existing influenza vaccinations that aim to reduce its spread elicit an antibody response primarily focused on the head domain, but the frequent emergence of novel viral strains through mutations in the HA head domain demands the use of numerous influenza strains in vaccines, as well as annual vaccination. The goal of this project is to demonstrate the design and assembly of a broadly protective flu vaccine using virus-like particle (VLP) technology. The VLP platform studied consisted of an engineered capsid protein core (MS2) that displayed HA on its surface by biotin-streptavidin affinity interactions. The use of multiple HA proteins on each VLP also made them polyvalent, potentially amplifying immune response. In addition to regularly oriented HA VLPs, inverted HA VLPs were assembled in an effort to elicit a broad immune response to the highly conserved stalk domain. Steric hindrance by the immunodominant HA head domains in regularly oriented HA makes inversion—the attachment of HA to the MS2 core by the head domain, with the stalk pointing outwards—a promising alternative to improve the HA stalk's accessibility to immune cells. The VLPs were made by transforming E. Coli to express MS2, functionalizing it with biotin, producing HA in a mammalian expression system, and performing site-specific biotinylation of the AviTag at the C-terminus of the purified HA protein; in the case of VLPs displaying inverted HA, additional steps were necessary to express an HA-head-binding antibody fragment (Fab) to attach HA to the VLP in an inverted orientation. The VLP's components were characterized using size exclusion chromatography (SEC) and SDS-PAGE. After assembly, SEC and dynamic light scattering (DLS) supported the successful assembly of VLPs with both regular and inverted HA. These results demonstrate the successful design of VLP-based vaccines with the potential to provide broad protection against seasonal influenza A strains; however, more research still needs to be done regarding these vaccines' effectiveness *in vivo*.

The Inhibition of Topoisomerase IIα with (1*E*,4*E*)-1,7-Bis(4-hydroxyphenyl)Hepta-1,4-Dien-3-One)-Based Natural Products

Braden Scheidecker, Andreea Marinescu, Kiana Haynes, Owen Reid, Madison Taylor, Dr. William Carroll

DNA Topoisomerase enzymes are found in throughout all life on earth and are an essential part of DNA strand management of living beings. This unwinding is essential to ensuring continuous DNA replication of the strand, which cannot occur if one of the following is restricting said DNA replication. Seeing as these enzymes play a crucial role in a DNA's continuance, inhibiting cellular/DNA replication in cancer cells while limiting other cellular exposure has been a drive for finding chemicals that can properly dock and prevent the function of topoisomerase enzymes. This project served to use build from naturally occurring inhibitors of topoisomerase II α, optimize their physiochemical properties, evaluate their simulated chemical docking in topoisomerase II α enzymes, synthesis target the target molecule, and test their ability to inhibit topoisomerase II α. Current synthesis targets are based on a natural product isolated from mistletoe, (1*E*,4*E*)-1,7-bis(4-hydroxyphenyl)hepta-1,4-dien-3-one. This natural product was evaluated for its interactions with topoisomerase II α using the MOE-2020 computational suite. A series of candidates was evaluated and the best binding of them was undertaken as a synthetic target. Here we report on the design and synthesis of these candidates in advance of their testing.

Investigating Intrinsic Fluorescence Mechanisms in Amyloid Fibers

Aubrey Carey, Coral O'Brien, Leah Spangler

Fluorescent tags are minimally invasive fluorophore markers that can be used to identify proteins and monitor protein interactions for imaging and structural analysis. One advantage of fluorescent tagging is that they are largely non-toxic, which means they can be used in live cells. For example, the artificial fluorescent tag Thioflavin T (ThT) is used for identifying amyloid fibers, protein aggregates that are often associated with diseases such as Parkinson's, Alzheimer's, and Type II Diabetes. Recently we have found that amyloid fibers have an intrinsic fluorescence that can be used instead of an artificial dye like ThT. Here, we investigate this excitation-dependent fluorescence signal in fibers formed from Hen Egg White Lysozyme (HEWL). Batches of amyloid fibers were grown from HEWL, with each batch varying in external conditions. We find that these lysozyme amyloid fibers have a previously unidenfied absorbance peak at 350 nm separate from the UV signal given off by tryptophan at 280 nm. We also observe the excitation-dependent fluorescence from in the range of 420-500 nm with a Stokes shift of ~90 nm. By changing the growing conditions of the fibers, we have uncovered a potential correlation between fiber morphology and peak wavelength. We correlate these changes in intrinsic fluorescence to adjustments in amyloid structure caused by the change in external growing conditions. By better understanding the correlation between fiber structure and their fluorescence pattern, fluorescence can be used to identify proteins without the addition of dyes.

Use of a Subtilase By Commensal Microbiota to Suppress the flg22-Induced Immune Response in Arabidopsis thaliana

Kaeli Ficco, Samuel Eastman, Jonathan M. Conway

Plants have an innate immune system that is triggered in response to microbe-associated molecular patterns (MAMPs). Flg22 is a MAMP that varies by species and is found in the FliC subunit of bacterial flagella. An immune eliciting epitope of flg22 is present in the FliC protein of *Dyella japonica* MF079, an *Arabidopsis thaliana* Columbia-0 (Col-0) root commensal. Yet, the presence of this bacterium does not trigger the *A. thaliana* Col-0 immune system. Furthermore, both live bacteria and cell-free culture supernatant from *Dja* MF079 can suppress the plant immune system in response to free canonical flg22. *Dyella* Immunosuppressive Subtilase 1 (Dis1) was identified as an actor in this immune suppression using knockout genetics. Both bacterial and supernatant GUS assays have shown that Dis1 is important in plant immune suppression but not the only cause of immunosuppression by *Dja* MF079. Furthermore, by incubating flg22 in Dja MF079 wild type and ΔDis1 supernatant we show Dis1 alters flg22 such that it is not perceived by the plant. Additionally, homology was utilized to identify and knockout Dis1 homologs in other root commensal bacteria. These homologs, *Ralstonia* Immunosuppressive Subtilase 1/2 and *Luteibacter* Immunosuppressive Subtilase 1, were also found to contribute to immune suppression in strains from the respective genera. These data suggest Dis1 and homologous subtiliases in plant microbiota are an important mechanism by which commensals suppress activation of the flg22 immune response in plants.

Yeast-Based High-Throughput Screening to Isolate Inhibitor-Resistant Sars-CoV2 Protease Variants

Cassidy Simas, Samantha Martinusen, Sagar Khare, Carl Denard

RNA viruses have a rapid evolution rate (between anywhere from 10–4 to 10–6 substitutions per nucleotide site per cell infection), giving rise to drug-resistant mutants. This concerning evolutionary rate has been seen most recently in the years following the SARS-CoV-2 viral outbreak as researchers attempt to predict the next iteration of the virus. The key actor in the virus' ability to rapidly proliferate is cleavage of various sites in the viral polyprotein by the cystine protease 3Clpro, making the protease a focal point of many research studies. SARS-CoV-2 is often treated with drugs that covalently bind to 3Clpro and block its viral replication mechanism. Current in vitro studies seek out variants of 3Clpro that may be of concern due to this rapid proclivity to evolve, possibly towards resistance against such covalent inhibitors. However, present screens lack the ability to detect drug-resistant variants with low enzymatic activity (1-10% of the wild-type enzyme's activity). To address this issue, we built a high-throughput yeast-based surface display screen that can 1) identify low activity 3Clpro variants by tuning co-expression of 3Clpro and its biologically relevant substrate in the endoplasmic reticulum (ER) of the displaying yeast and 2) comprehensively profile resistance of these mutants to Nirmatrelvir, a known covalent inhibitor of 3Clpro. Adjusting our endoplasmic retention signal to retain one or both protease and substrate in the ER for differing periods of time gives each low resistant variant a dynamic potential to be detected in our screening system, with and without introduction of a drug inhibitor. In this way, our comprehensive screen provides a novel means to characterize previously undetected drug-resistant SARS-CoV-2 protease mutations.

Optogenetic Regulation of Transcription Factor Dynamics and Downstream Gene Expression in Single Cells

Amy Leister

Transcription factor p65, a subunit of the NFκB family, is involved in the regulation of cell immune response, inflammation, and carcinogenesis. The downstream gene expression from p65 activation is regulated by localization of the transcription factor into the nucleus of mammalian cells. Localization of p65 is triggered by external stimuli in the cell and is regulated by IκB proteins and the endogenous Nuclear Exporting Signal (NES) and Nuclear Localization Signal (NLS) within p65. For precise manipulation of the translocation of p65, a synthetic approach utilizing an optogenetic system was employed to deliver the stimuli. The endogenous NLS and NES were mutated, allowing for the localization of the transcription factor to be completely dependent on the synthetic, optogenetic signal, and to be unaffected by any naturally occurring signal in the cell. Variations of the mutant p65 plasmid were cloned to compare how deviations of the NLS and NES sequence affect the localization of p65 and the corresponding gene expression output. Using a fluorescent tag, quantification of the abundance of p65 within the nucleus can be achieved through microscopy. By implementing this synthetic approach, precise control over gene expression can be employed, and naturally induced and synthetically induced gene expression can be compared to better understand p65-implicated disease progression.

Use of Small Molecules in the Aid of Solubility of E. coli P3340 during Lysis Stage of Purification

Tausif Tamim, Radmila Janjusevic

Protein SFBNYU_003340 (P3340) is a surface protein of mice-specific segmented filamentous bacteria (SFB), gut bacteria distinguished for their role in development of Th17 and Th22 immune cells. These two subgroups of immune cells are known for their role in fighting pathogens. Protein P3340 homologous protein, P00774 from human-specific SFB, shares 53.6% similarities with P3340 on amino acid level. Protein P3340 also carries two epitopes that were immunodominant in the stimulation of antigen specific mucosal Th17 cells. Still, there is no known target for P3340 protein in intestinal epithelial cells (IEC) or mucus covering IEC. In order to study the function of P3340, we have expressed and purified this protein from E. coli in a N-terminal double His6-tag fusion and without a signal sequence (first 33 amino acids). Although we have produced the protein in our lab from synthetic DNA that was codon-optimized for E. coli and optimized expression and solubility in several buffers, yields were still very low. In this work, we wanted to test different chemicals generally known to increase the solubility of different proteins. One study has shown that adding of certain chemicals at the lysis stage of purification, when cell walls are broken down to release the soluble protein into the solution, significantly increased the yield of some of tested proteins. Before this publication, various molecules were typically added to already purified proteins to increase their solubility and stability prior to protein concentration and storage. In our study, different chemicals (trehalose, glycine betaine, mannitol, L-arginine, potassium citrate, copper (II)-chloride, proline, xylitol, CTAB, glycerol, and dipotassium phosphate) were separately added into the cell lysates of E. coli BL21(DE3) bacteria producing 2xHis6-P3340. Proteins from each lysate were then purified on Ni-NTA, concentration of protein 2xHis6-P3340 was measured in Bradford assay, and purity checked on SDS-PAGE gels.

Engineering Biomimetic Intestinal Models to Investigate Host-Microbiome Interactions

Aviral Misra

Biomimetic *in vitro* models of organ systems can help elucidate complex biomedical processes and structure-function relationships. For example, the human gut is home to a diverse range of cells which enable the gut to carry out its physiological functions. The gut's function is further enhanced by a diverse and largely uncharacterized microbiome. Currently, there are few systems which have been developed to study the interactions between human tissue and the microbiome. Further, these existing models like organ-on-chips and organoids are limited due to their lack of cell-type diversity and incorrect spatial configuration. In this project, a new biomimetic system which captures the colonic-crypt micro-architecture, the diverse epithelial cell types, and provides easily accessible luminal and basal compartments is used. With these features, the obtained system can be used to investigate colon development by applying precisely controlled gradients of chemical signals and morphogens that replicate *in vivo* development. In addition, this system provides an avenue for mimicking the behavior of the intestinal epithelia-microbiome interface by allowing anaerobic co-cultures of bacteria that inhabit the gut. This presentation will describe investigations with the bacteria *Candidatus Cibiobacter qucibialis* (the most common uncharacterized gut microbe) and discuss the optimization of co-culture conditions, including length of co-culture, bacterial growth behavior, and optimal seeding density. The effect of the bacteria on the gut epithelia is characterized using bulk RNA sequencing to determine differential gene expression between gut tissue with and without the bacterial co-culture.

Behavior of Myelin Lipid Compositions at the Air-Solution Interface

Mara Manolescu

The myelin sheath is a multilamellar lipid membrane wrapped around axons of neurons in the central nervous system (CNS) that facilitates rapid nerve impulse transmission, which is essential for the proper functioning of the CNS. Myelin destruction, called demyelination, causes Multiple Sclerosis (MS), a neurodegenerative disorder that results in the loss of sensory and motor functions due to nerve impulse leakage. In healthy subjects, the amphipathic lipids that compose myelin form a remarkably stable membrane. Changes in lipid composition have just recently been identified as a signature of demyelination; however, the details of these changes and the role they play in MS is not currently understood. We hypothesize that changes in lipid composition impact the packing of the film, which in turn impacts its stability. This work addresses the relationship between the structure and function of myelin to develop a biophysical model of the myelin sheath. Using a Langmuir-Pockels trough, we measured the interfacial properties (surface pressure (SP) and compressibility modulus) exhibited by monomolecular films of different lipid compositions when subjected to dilatational interfacial stresses. Major lipid components of myelin were used in this work: L-α-phosphatidylcholine (Brain PC), L-α-phosphatidylethanolamine (Brain PE), L-α-phosphatidylserine (Brain PS), and cholesterol. "Healthy" and "sick" brains were experimentally modeled by varying lipid ratios based on the lipid profile found in "healthy" and "sick" mice brain extracts. While little difference was observed between models in the absence of cholesterol, the addition of cholesterol promoted major differences in the mechanical properties of the interfacial film. Changes in SP with surface area and compressibility modulus suggested differences in packing due to differences in the amount of cholesterol in the model mixtures mimicking "healthy" brain compared to the "sick" brain.

Copper Oxide as an Anti-Neoplastic Agent for Endometrial Carcinoma

Jordan Berezowitz, Claire Rowlands, Brittany Givens Rassoolkhani

Carboplatin, a platinum-based drug, is commonly administered to treat endometrial cancer; however, efficacy is limited by chemotherapeutic resistance. Chemotherapeutic resistance is one of the greatest difficulties to overcome in cancer therapy, resulting in a vital demand for new anti-cancer drugs and drug-delivery devices. In addition to chemotherapeutic resistance, unequitable access to treatment options for endometrial cancer patients demonstrates the need for a non-invasive form of endometrial cancer therapy. A proposed solution to these challenges involves using copper oxide nanoparticles as an alternative metal to platinum-based drugs in a biodegradable polymeric nanoparticle to combat the issues of chemotherapeutic resistance and unequitable access to care.

Copper oxide nanoparticles pose a potential as an anti-cancerous compound due to Cu^{2+} ions demonstrating the ability to induce reactive oxygen species (ROS) leading to double stranded DNA breaks ultimately resulting in cell death. Polycaprolactone (PCL, Sigma) was used to encapsulate copper (II) oxide nanoparticles (CuO NPs, Sigma) in a polymeric nanoparticle using the double emulsion solvent evaporation technique with nanoprecipitation. Cell viability after 24 hours of treatment with 60-minute sonicated CuO NPs was assessed through MTT assays.

AN3CA, HEK293, A549, HEC-1A, KLE, and Ishikawa cell lines all showed sensitivity to CuO NPs; however, each cell line experienced a different response to CuO NP treatment. HEK293 cells showed a low sensitivity to CuO NPs with >50% viability at 100 µg/mL. KLE MTT assay results did not show a direct correlation between cell viability and CuO NP concentration. Ishikawa and A549 cells demonstrated a negative correlation between cell viability and concentration.

Compared to sonicating for 30 minutes and no sonication, CuO NPs sonicated for 60 minutes in a water bath resulted in effective dispersion and reduced hydrodynamic size. Sonication resulted in minor changes in zeta potential of CuO NPs in PBS. Zeta potential measurements of CuO-PEI nanoplexes were similar across three different wt% PEI used to create the complexes. Additionally, the zeta potential measurements of CuO NPs, PEI, and CuO-PEI nanoplexes were all different, indicating that CuO-PEI complexes were being formed. PCL shell CuO-PEI core in water nanoparticles were successfully synthesized using the proposed synthesis method. SEM imaging revealed that particle diameter was greater than 1 µm. MTT assay results confirmed that CuO NPs result in effective cell death in all human cell lines tested at or below 100 µg/mL. All cell lines showed differing trends and results for viability when treated with CuO NPs.

Dynamic Modulation of Matric Stiffness Regulates Apoptosis in Breast Cancer Cells

David Soares, Chinmay Sankhe, Esther W. Gomez

Breast cancer is the second leading cause of cancer death in women , with it accounting for 1/8th of cancer cases diagnosed worldwide. An increase in the stiffness of the tissue in the surrounding area of breast tumors is characteristic of breast cancer progression. Apoptosis or programmed cell death is known to be blocked in cancer cells leading to enhancement in tumor progression. Previous studies have shown that increased extracellular matrix stiffness reduces apoptosis in cancer cells and is potentially a mechanism by which cancer cells evade chemotherapeutic drugs. Most literature reports involving monitoring the impact of tumor mechanics on cancer cells have been performed on substrates with static mechanical properties, however, substrates with dynamic control of the mechanics provide an approach for monitoring how temporal changes in matrix mechanical properties impact apoptosis and tumor progression. Here, dynamic hyaluronic acid-based hydrogels that mimic the stiffness of normal and tumorigenic breast microenvironments were synthesized and characterized. MDA-MB-231 breast cancer cells were then cultured on the substrates and immunofluorescence staining was conducted to monitor the levels of cleaved caspase-3, an apoptosis marker. Quantification of cleaved caspase-3 staining revealed that when the modulus was replicative of a tumorigenic breast environment, the percentage cleaved caspase-3 positive cancer cells was reduced in comparison to when cells were cultured on hydrogels mimicking the mechanical properties of normal breast tissue. Furthermore, the percentage of cells staining positive for cleaved caspase-3 dynamically increased when the modulus of the hydrogels gradually softened over time. These findings indicate a role for matrix mechanics in regulating apoptosis in cancer cells. Overall, this hydrogel system provides a platform for examination of how dynamic modulation of matrix mechanics impacts cell phenotype.

Effects of Methyl Jasmonate and Jasmonic Acid Loaded Nanoparticles on Defensive Secondary Metabolites in Hairy Root Cells

McKenna Clinch, Poornima R. Sunder, Barbara Knutson, Stephen Rankin, Bert C. Lynn

Secondary metabolites in plants help with defense mechanisms against pathogen infection and abiotic stresses, including drought, extreme temperatures, etc. While an increase in these metabolites will benefit plants, they also provide potential benefits to humans as possible therapeutics. Methyl jasmonate (MeJ) and jasmonic acid (JA) are both known to elicit defensive secondary metabolites when delivered to plant cell cultures. Amine functionalized mesoporous silica nanoparticles (MSNPAs) loaded with these elicitors are hypothesized as carriers for delivery into the plant cell. This work focuses on the delivery of methyl jasmonate and jasmonic acid into *Lobelia cardinalis* hairy root cells using high surface area MSNPAs (with 2.5 nm diameter pores). Synthesis and amine functionalization of these nanoparticles are performed first, resulting in a porous particle with a positive surface charge, confirmed by zeta potential. Using evaporative loading techniques, MeJ/JA is loaded into the pores and quantified using thermogravimetric analysis (TGA). JA is expected to bind more strongly to the cationic MSNPA vector than MeJ. Isotherms will show the relative binding strength and capacity for both elicitors. Hairy root cell cultures are then treated with one of five different treatments of elicitor and particle combinations. The control group contains hairy root cells in plant media. Other treatments include root cells with elicitor in media, nanoparticles in media, elicitor loaded nanoparticles in media, and unloaded nanoparticles and elicitor suspended in the same media. Ultra high-performance liquid chromatography (UHPLC) coupled with electrospray mass spectrometry (ESI-MS) is used to analyze the metabolites synthesized with each treatment. Focusing on jasmonic acid as the elicitor, data manipulation is performed to determine which secondary metabolites are elicited in significant amounts. T-tests are then performed to determine which treatments are statistically significant from the control group. 1,808 metabolite features were found in common between all treatments. 133 features, or 7%, were found to be statistically significant with either a greater than 20% upregulation or downregulation within JA loaded MNSPAs compared to the control. Current work focuses on analysis of each metabolite and determining which treatment showed the greatest impact on area counts.

Investigating the Effect of Hydrogen Sulfide (H2S) on Hydrogen Peroxide Treated Hepatocytes

Kaleb Chan, Emma Henderson, Neeti Gandhi, Ishani Sarkar, Padmavathy Rajagopalan, John Matson

The liver is the largest visceral organ in the human body and plays a major role in multiple metabolic functions, including biotransformation, protein synthesis, and cholesterol metabolism. Hydrogen sulfide (H2S) may play a vital role in mitigating the effects of liver inflammation. Previous studies have shown that a deficiency in two H2S producing enzymes may lead to inflammation in the liver.

Amino acid-based C-substituted N-thiocarboxyanhydrides (NTAs) are H2S releasing-agents. Breakdown of the NTAs results in carbonyl sulfide (COS), which is converted to H2S by carbonic anhydrase, an enzyme that is naturally produced in eukaryotic cells. Different NTAs have various H2S releasing times and the remaining byproduct is the original amino acid from which it was derived.

The goal of this investigation is to determine the effects of H2S on mitigating liver inflammation after administration of hydrogen peroxide (H2O2), a known inflammatory agent. To investigate the effects, we assembled collagen Type I gels and cultured primary rat hepatocytes (primary liver cell type). To induce reactive oxygen species, we administered H2O2 to our cultures at varying concentrations. To enhance inflammation within our cultures we deprived the cells of media prior to H2O2 administration, based on previous reports. After H2O2 administration, 2 NTA solutions (Valine-NTA and Alanine-NTA) were administered to the cultures at two concentrations (100 μM and 250 μM) to investigate how H2S affected the cells. To determine the effects, we collected media every 24 hours for analysis and lysed cells for DNA quantification. Our future work includes investigating how the NTAs affect other hepatic functions after H2O2 administration.

Thermal Shock Eradication of Single-Species and Polymicrobial Biofilms of *Pseudomonas Aeruginosa* and *Staphylococcus Aureus*

Colin Houts, Josiah Power, Parham Parnian, Paraskevi Konstantina Zoga, Eric Nuxoll

Bacterial biofilm infections on medical implants cost the healthcare industry billions of dollars each year. Bacterial biofilms are highly resistant to antibiotic treatment, often requiring surgical removal of the medical implant. A new device must then be surgically implanted with twice the likelihood of infection. The surgery and recovery process are painful, time-intensive, and often ineffective. Recent research has highlighted the potential of controlled *in-vivo* thermal shocks as a biofilm eradication strategy. *In-vitro* thermal shocks have been shown to reduce or eliminate bacterial biofilms of *Staphylococcus aureus* and *Pseudomonas aeruginosa*, bacteria common in medical implant infections. Research thus far has focused on the thermal eradication of single-species biofilms; however, many implant infections are comprised of multiple species. More specifically, polymicrobial biofilms of *P. aeruginosa* and *S. aureus* have been linked to worsening patient outcomes in medical implant infections. The interactions of multiple species in a polymicrobial infection may impact the efficacy of a thermal shock. This study compares the thermal shock resistance of mono-cultured (single-species) and co-cultured (polymicrobial) biofilms of *P. aeruginosa* and *S. aureus* as a strategy for biofilm elimination. Results confirm thermal shocks at varying time and temperature combinations can eliminate bacterial biofilm infections of *P. aeruginosa* and *S. aureus*. Within mono-cultured biofilms *P. aeruginosa* was found to have significantly lower thermal resistance than *S. aureus*. When co-cultured the presence of *P. aeruginosa* in the polymicrobial biofilms significantly decreased the thermal resistance of *S. aureus*, while the thermal resistance of *P. aeruginosa* was not significantly impacted. Subsequent re-incubation of thermally shocked co-cultured and mono-cultured biofilms confirmed their elimination at clinically relevant time and temperature combinations. These results indicated no post-shock die-off of *P. aeruginosa* or *S. aureus* biofilms with the potential for regrowth even when bacteria population was dropped below the detection limit. Complete elimination of *P. aeruginosa* and or *S. aureus* was required to prevent biofilm regrowth. The results from this study can be utilized to develop more effective treatment for single-species and polymicrobial medical implant infections.

Extracellular Vesicle Production from hTERT-MSCs Embedded in Hydrogel

Gabriella Faircloth, Rachel Moen, Ethan Lippmann, Jamey Young

Extracellular vesicles (EVs) are lipid membrane-bound nanoparticles secreted by cells that contain nucleic acids, proteins, and metabolites derived from their cells of origin. While there is much to be learned about their structure and function, they are known to play key roles in intercellular communication and are involved in many physiological and pathological processes. EVs have wide therapeutic potential given their abilities to to cross membrane barriers and deliver cargo from their cells of origin to other locations throughout the body. Although they have much potential as therapeutics, current GMP that focus on the large-scale biomanufacturing of EVs is very limited. The use of hydrogel microspheres has been recognized as a potential way to scale up the culture of adherent cell lines, and thus the production of EVs, as they would provide a 3D model to culture cells at an increased seeding density and allow for extended culturing time. Though the pores in hydrogel should be large enough to allow for the secretion of EVs, in order to further confirm that hydrogel is a medium sufficient for cell growth and EV production, hTERT mesenchymal stem cells (MSCs), a cell type known to produce therapeutically relevant EVs, were cultured in 3D hydrogel layers so their EVs could be purified, counted, and compared to EV counts of parallel experiments done in traditional 2D cultures. Gelatin methacrylate (GelMA) and GelMA-Cad were the two types of hydrogel used for these experiments. From EV counts done on the Zetaview Nanoparticle Counter, the concentrations reflected an increase in EV secretion in hydrogel cultures while maintaining similar size distributions. EV protein characterization revealed that Flotillin-1, a common MSC EV marker, was not present in the EV counts for experiments done in hydrogel. Staining for other common EV surface markers should be done to better characterize the EVs and ensure their quality. Additional replicates of EV counts and EV protein characterization need to be done in hydrogel to further confirm the presence of EVs in hydrogel cultures. Ultimately, the goal of this project is to develop a scalable platform to produce EVs from tissue-specific cell types.

Toward the Synthesis of Potential Antiviral Carbocyclic Nucleosides

Heather Binion, Kyle Watson

Abstract: Human immunodeficiency virus, most commonly referred to as HIV, is a persistent problem in the United States. While therapies are becoming more widely available to prevent and treat symptoms of this virus, there is currently no cure for HIV.[1] One of the first steps taken to transform this otherwise fatal virus into a manageable ailment was the creation of nucleoside reverse transcriptase inhibitors. HAART, highly active antiretroviral therapy, utilizes nucleoside reverse transcriptase inhibitors as well as HIV-1 protease inhibitors in order to suppress the viral load of HIV-1.[2] The primary focus of this research is to synthesize a library of potential antiviral carbocyclic nucleosides that could further be tested for antiviral efficacy. Carbocyclic nucleosides are analogs of naturally occurring nucleosides where carbon replaces the endocyclic oxygen. Commonly referred to as carbanucleosides, these metabolically stable compounds could exhibit antibiotic as well as antiviral potency.[3]

References

1. Center for Disease Control and Prevention. https://www.cdc.gov/hiv/basics/statistics.html.
2. Gosh, A. K.; Osswald, H. L.; Prato, G., Recent Progress in the Development of HIV-1 Protease Inhibitors for the Treatment of HIV/AIDS. National Library of Medicine 2016, 59 (11), 5172-5208.
3. Li, F.; Brogan, B. J.; Gage, J. L.; Zhang, D.; Miller, M. J.; Chemoenzymatic Synthesis and Synthetic Application of Enantiopure Aminocyclopentenols: Total Synthesis of Carbocyclic (+)-Uracil Polyoxin C and Its α-Epimer. JOC 2004, 69 (4538-4540).

Biophysical Study of Novel Phlip-Stinga with Membrane

Dana Allababidi, Craig Klumpp, Hannah Visca, Anna Moshnikova, Oleg Andreev, Yana Reshetnyak

Immuno-oncology (IO) is a rapidly growing field that utilizes the body's own immune system to recognize and attack cancer cells. The tumor microenvironment (TME) plays a crucial role in the effectiveness of IO treatments and can be separated into two major categories: cold and hot tumors. Cold tumors contain M2 tumor-associated macrophages (M2-TAMs), cancer-associated fibroblasts (CAFs), myeloid derived suppressor cells (MDSC) and these tumors do not respond to IO treatment. On the other hand, hot tumors contain T-cells and M1 macrophages and typically respond well to IO treatments. STING agonists (STINGa) are a ligand for the STING protein, which activates the STING pathway, a fundamental pathway of innate immunity that is designed by nature to induce an inflammatory response, thus turning "cold" tumors into "hot" tumors. The non-targeted systemic use of STINGa is dangerous, since it can induce inflammation everywhere. To address these limitations, pH (low) insertion peptide (pHLIP) can be used to ensure the targeted delivery of STINGa. A biophysical study was done to assess the interaction of two novel pHLIP-STINGa constructs with membranes of liposomes, which were used to mimic the cell membrane. We monitored changes of tryptophan fluorescence and circular dichroism (CD) of pHLIP-STINGa agents in steady-state and kinetics modes to establish pK of insertion, and rates of membrane entry and exit. Obtained results indicate to a desirable pK of insertion and fast membrane insertion, which are important predictive parameters for in vivo performance of the agents.

Breakthrough Curves for Ettmp Purification with Alumina Oxide

Keira Korzeb, Brandon M. Vogel, Erin L. Jablonski

Biomaterials can be used in the advancement of targeted drug-delivery methods. Hydrogels are one of these useful drug delivery vehicles suitable for a range of drugs and therapeutic proteins. However, ultraviolet-cured and thermally-cured hydrogels suffer from drawbacks when prompted to control drug release in vivo because they need to be implanted. Degradable injectable hydrogels that cure in-situ are an emerging alternative that make clinical sense. Albeit there are some issues with these systems because there needs to be further research conducted to better understand them before broader adoption. In this poster, there's an exploration of the dependence of the curing kinetics with a range of factors for a thiol conjugate Michael Addition reaction used in the hydrogel system containing Poly(ethylene glycol) diacrylate (PEG-575-DA) and Ethoxylated trimethylolpropane tri-3-mercaptopropionate (ETTMP, Thiocure 1300).

The synthesis and curation of these gels are based on using a calculated volume of precursors resulting in an equimolar stoichiometric ratio of thiol and acrylate groups. Varying precursors were used to determine the most effective and consistent hydrogel curing. However, precursor impurities have a large effect on the cure time. This is relevant because the material, as shipped, degrades over time releasing mercaptopropionic acid (MPA). Subsequently, this impurity must be removed for any curing to take place at a pH of 7.4. Correlating MPA concentration to cure time, we can remove excess MPA effectively utilizing basic alumina. There is a correlation between the height of the column and ratio of alumina to ETTMP to remove MPA. When purifying in a 20 mL syringe, the height needs to be a minimum of 1.5 in., and a 2:1 ratio of alumina to ETTMP was found to remove MPA to trace levels. We attempted to determine breakthrough curves for MPA in an alumina column using titrations to gauge the pH of the purified ETTMP. Multiple runs of purification were tested with varying heights of alumina to determine breakthrough curves. It was found that the column slowly reacts with the ETTMP and eventually reduces the flow rate to zero before the MPA breaks through. Therefore, with low MPA concentrations, a hydrogel cures in about 1.5 min. Another number of tests were performed to analyze the mechanical properties of the hydrogel during its curation period. Dynamic rheology was used to determine the exact gelation time, which was confirmed by a frequency sweep to ensure the data was within the linear viscoelastic range.

Designing Peptide-Polymer Conjugates for Cell-Mediated Scaffold Degradation

Juliana Forero Arevalo, Natasha K. Hunt, Lesley Chow, E. Thomas Pashuck, Warren Grayson, Srujan Singh

Introduction: Biodegradable polycaprolactone (PCL)-based scaffolds for tissue engineering do not degrade at the same rate as new tissue formation. For example, bone regeneration typically occurs within 6 months, while PCL persists for over 1-3 years in vivo. This significant mismatch can lead to voids because tissue formation is completed before the remainder of the scaffold degrades. Ideally, scaffold degradation occurs concurrently with and in response to biological events that occur during new tissue formation, such as cells secreting proteases to remodel the extracellular matrix (ECM). This project focuses on developing solid peptide-PCL conjugate scaffolds with protease-sensitive peptides within the conjugate backbone to direct cell-mediated degradation.

Materials and Methods: Peptide synthesis and conjugation—A fast-degrading peptide, (CKRVKRRLLMETK(biotin)C) and non-degrading scrambled peptide with the same amino acid residues (CLTRLMEKRRKVK(biotin)C) were synthesized using Fmoc solid phase peptide synthesis. Fmoc-Lysine(Mtt)-OH and biotin were manually coupled to the resin before adding the remaining amino acids using a microwave-assisted automated peptide synthesizer. Another sequence, CGGGRGDSK(biotin), was synthesized using the same techniques. Peptides were cleaved from the resin and purified using high-performance liquid chromatography. Conjugate synthesis—PCL (25 kDa) was modified with amine-poly(ethylene glycol)-maleimide using a carbodiimide reaction to form PCL-mal. The fast-degrading and scrambled peptides were reacted with the PCL-mal via Michael addition to make PCL-peptide-PCL. PCL (14 kDa) was modified with p-Maleimidophenylisocyanate to form PCL-mal. The RGDS peptide was reacted with the PCL(14kDa)-mal to make a peptide-PCL-peptide RGDS-PCL conjugate. Nuclear magnetic spectroscopy was used to confirm each conjugation step. Scaffold fabrication—The PCL-peptide-PCL conjugate and the RGDS-PCL conjugate were co-dissolved in hexafluoroisopropanol at a total of 200 mg/mL with 5.13% by weight RGDS-PCL conjugate, and cast into flat disks onto 12 mm diameter glass coverslips. Degradation study—Scaffolds were seeded with fibroblasts and incubated for up to 22 days at 37°C and 5% CO2 . Media was collected and replaced with fresh media every two days. A colorimetric biotin assay was used to measure the amount of biotin released from the scaffolds normalized to the PCL-RGDS group to quantify scaffold degradation.

Results: NMR confirmed successful PCL-peptide-PCL and peptide-PCL-peptide conjugate synthesis. Over the 22-day period, small holes formed in the center of the fast-degrading scaffolds while the scrambled and PCL-RGDS scaffolds were unaffected cumulative biotin release from PCL-peptide-PCL scaffolds normalized to PCL-RGDS showed that the fast-degrading scaffolds released more biotin compared to the non-degrading scrambled scaffolds. Together, these results demonstrate that fibroblasts degrade the fast-degrading scaffolds at a higher rate compared to the scrambled and control scaffolds.

Conclusion: The scaffold with a fast-degrading peptide sequence degraded more quickly in response fibroblasts compared to the scaffold with a scrambled version of the same peptide. These results demonstrated our ability to tune degradation response. This platform introduces a versatile approach to tailor peptide sequences and conjugates for specific cells and cellular events.

Designing Synthetic Hydrogels to Use in Cell Therapy

Janine Ramtahal, Darren Miller, Israt Jahan Duti, Rachel Letteri

Hydrogels are polymeric networks characterized by their hydrophilic nature, which enables them to retain their structural integrity while exhibiting a high capacity for water absorption. They have many applications in biology and medicine, including medication delivery, tissue engineering, and wound healing. Peptides and polymers are regarded as very promising materials for hydrogel production due to their inherent advantages, such as biocompatibility, tunability, and reactivity to diverse stimuli.

Peptides, consisting of short sequences of amino acids, can undergo self-assembly and form diverse nanostructures, including hydrogels. Peptide hydrogels possess several notable advantages in comparison to conventional hydrogels. These advantages include their capacity to replicate the extracellular matrix and their potential for customization by the incorporation of specific amino acid sequences for cell adhesion or drug administration.

Polymers consist of elongated chains composed of recurring units, which possess the ability to undergo crosslinking processes, resulting in the formation of hydrogels. Polymer hydrogels frequently exhibit superior mechanical strength compared to peptide hydrogels; they may exhibit reduced biocompatibility. The integration of peptides and polymers enables the synthesis of hybrid hydrogels that exhibit a synergistic combination of the superior characteristics inherent to each constituent material.

The primary objective of this study is to investigate the advancement of peptide and polymer hydrogels in the context of biomedical applications. Our research focuses on the development of hydrogels for use in cell therapy. In this study, L and D peptides were synthesized and subsequently crosslinked to form stereocomplexed hydrogels. In the course of synthesizing the peptide LAKLA, we observed that the implementation of a double coupling strategy for lysine residues during the synthesis process resulted in a decrease in the occurrence of lysine deletions. In addition, the peptide was subjected to various analytical techniques for characterization, including matrix-assisted laser desorption/ionization (MALDI), circular dichroism (CD) spectroscopy, nuclear magnetic resonance (NMR) spectroscopy, infrared (IR) spectroscopy, and high-performance liquid chromatography (HPLC).

It is posited that peptide and polymer hydrogels can fundamentally transform the biomedical engineering domain. The synthesis of hydrogels with biocompatible characteristics can be achieved by the integration of the distinctive attributes of peptides and polymers. Hydrogels possess the potential to facilitate the advancement of novel therapeutic interventions targeting diverse diseases and medical conditions.

Proton Motive Force in *E. coli Persisters*

Isabella Schmitz, Annie Lee, Pushkar Lele

Persisters are bacterial cells that can survive lethal doses of antibiotics despite lacking genes responsible for antibiotic resistance. Persisters are assumed to be dormant with a low metabolic activity that protects against antibiotics. Once the antibiotic stress decreases, the persister cells resurrect and repopulate, contributing to chronic infections and antibiotic resistance. Understanding of persister physiology, though, is currently limited. We focused on the proton motive force (PMF) in 'induced' persisters as the PMF fuels several cellular processes, including the uptake of antibiotics. The previous notion in the field was that a low PMF promotes persistence by inducing dormancy. However, our recent findings suggest that the PMF varies considerably in populations of Escherichia coli persisters. Those conclusions were drawn from motility studies, which indicated that a part of the persister population swims robustly. As PMF powers motility in E. coli, this suggests the existence of two sub-populations within the persisters: persisters with low PMF and persisters with high PMF. We quantified the heterogeneity in PMF at a single-cell level using dyes that label cells as a function of their membrane potential, which is the major contributor to the PMF. Briefly, we treated cells with an antibiotic, rifampicin, which induces persistence by inhibiting transcription. We subsequently exposed the cells to lethal concentrations of ampicillin to concentrate persisters and eliminate the remaining population. We then labeled the cells with Thioflavin T (ThT) dye and used photon-sensitive cameras to record fluorescent signals from single cells. We observed a bimodal distribution in the intensity of fluorescence signals in persister populations, indicating that some persisters retained high energy levels, whereas the rest were metabolically dormant. Future experiments are anticipated to help determine if the so-called persisters employ PMF-dependent efflux pumps or decrease their membrane permeability to tolerate antibiotic stress. Insights from this research are likely to help develop therapeutics against persisters.

Biodegradable Magnetic Nanocomposite Hydrogels for Remote Controlled Drug Release

Kai Oddo, Pranto Paul, J. Zach Hilt, Anastasia Hauser

Some diseases such as diabetes require treatments with constant doses similar to biological processes in conjunction with on demand spike release of a drug. To address this challenge, a novel drug delivery system was investigated through the exposure of magnetic hydrogel nanocomposite to alternating magnetic field (AMF). With AMF exposure, MNP incorporated in the hydrogel could generate localized heating that could enhance the drug release. Hydrogel made of poly(beta-amino ester) (PβAE), loaded with magnetic nanoparticles (MNP) in proportions of 0, 1, 5, and 10 wt% were synthesized. Fluoresceine as a model drug was covalently bonded with the hydrogel. In vitro drug release in PBS was conducted with and without the exposure of AMF at amplitudes of 5 and 12mT. The proposed 5 wt% MNP incorporated hydrogel fully degraded in 135 minutes and 150 minutes in AMF at 5mT and 12mT amplitudes respectively compared to the 25°C water bath trials of the same gel degrading in 195 minutes. Current hydrogel nanocomposites degrade too quickly, as such future experiments will attempt to extend degradation time by altering molar ratio of PβAE reactants.

Programmable Dynamic Self-Assembled DNA Nanostructures

Yuhan Wu

In addition to storing genetic information, DNA serves as a distinct and predictable nanomaterial capable of self-assembling into precisely programmed structures. Among various DNA technologies, DNA nanostructures can be employed to create DNA nanorobots with the ability to perform various tasks, including molecular transport and release. Self-assembled DNA nanostructures emerge as excellent candidates to overcome the limitations of conventional drug delivery methods, thanks to their biocompatibility and programmability. Efforts have been focused on manipulating these structures to enhance their reconfigurability, leading to groundbreaking developments of structures that facilitate and cease nanotube growth.

In this project, we introduce a novel structure that combines both functions: nucleation and termination. By adjusting the structures dedicated to two individualized tasks, we achieve an integration of two functionalities into one structure. One end of the structure facilitates nanotube growth and the other end terminates growth. Consequently, segments of DNA nanotubes can connect, forming more sophisticated structures. Additionally, we investigated the effects of free monomer concentrations on nanotube growth. Concentrations play a pivotal role in nanotube growth dynamics, and an ideal growth curve encompasses successful nucleation of nanotubes followed by consistent, sustained growth. By observing nanotube growth under varying concentrations, we gain important insights that further optimize our nanotube system. This project significantly improves the programmability of our existing DNA nanotube systems and enhances our understanding of the optimal conditions for nanotube growth. The advancements are promising to contribute more to the DNA nanotechnology toolkit.

Designing Unique Mechanical Gradient Patterns in Alginate Hydrogels

Zoe Ostrowski, Juntao Zhang, Juan Ren, Ian Schneider

Directed cell migration is an important aspect in wound healing. Cells take in and process many signals including the mechanical signal of tissue stiffness. Durotaxis is the process in which cells migrate towards stiffer regions of the extracellular matrix. Alginate, which comes from brown algae, is commonly used in wound dressings, especially hydrogels, due to its biocompatibility and capability to retain large amounts of water. Alginate hydrogels can be crosslinked with calcium to create a stiffer hydrogel. While alginate gels have been used in wound dressings, sophisticated gels with stiffness gradients that could direct cells to particular targets within the wound bed at a faster rate have not been used. Using custom stencils, we can control where the calcium diffuses into the hydrogel, giving us an ability to design any unique stiffness gradient in the hydrogel. As the density of crosslinking increases, a change in opacity is observed in the hydrogel over time. The hydrogel swells in water and float when crosslinked, creating problems with characterizing and controlling stiffness. Swelling was minimized by tuning the NaCl concentration. To avoid floating hydrogels, glass coverslips and alginate were functionalized to adhere the hydrogel to the coverslip. We show an ability to generate gradients with different spatial features. Furthermore, the elastic modulus, a measure of the stiffness, of the alginate hydrogel was quantified using atomic force microscopy (AFM). Opacity changes correlated both spatially and temporally with crosslinking and elastic modulus in the alginate hydrogel, demonstrating that opacity can be used as a proxy for local mechanical properties. Future work is geared towards functionalizing these gradient gels with cell-adhesive molecules to examine durotaxis in 3D environments. Engineering stiffness gradients that facilitate cell migration has the ability to advance wound dressings by creating an environment that can direct cells to migrate into the wound bed faster.

The Impact of Fiber Inclusion on the Injectability of Shear-Thinning Hydrogels

Juhi Khandelwal, Grace Schwarz, Gregory Jensen, Julianne Holloway

Shear-thinning hydrogels change from a gel to a liquid state under high shear stress, such as the shear stresses that occur during injection [1]. Thus, these systems offer a non-invasive route for otherwise invasive treatments [1]. Shear-thinning hydrogels can be combined with other components to enable multi-functionality, including short fibers or particles [2]. For the inclusion of fibers, electrospinning has gained momentum as a simple technique to create fibrous structures that mimic the fibrous nature of the extracellular matrix [3]. Combining electrospun fibers within a shear-thinning hydrogel has the potential to provide cells with essential physical cues while retaining injectability [4]. For the inclusion of particles, hydrogel microparticles, also known as microgels, can be made from biocompatible, hydrophilic polymers [5]. Microgels are beneficial for inclusion due to their highly tunable size and structure, ability to encapsulate bioactive factors, and controllable composition and stiffness [5]. In the future, hydrogel composite systems including fibers and/or microgels can be manufactured using shear-thinning, guest-host hydrogels as an ideal multi-functional, injectable hydrogel for tissue repair applications. Studies manipulating the hydrogel, fiber, and/or microgel formulation, as well as the needle gauge are required to provide insight into the relationship between material properties, injectability, and biocompatibility, which is a prerequisite for clinical implementation. Thus arises the question: what is the effect of composite components on the shear stresses generated during injection of shear-thinning guest-host hydrogels?

JMPPro DOE was utilized to create a material-saving full-factorial design of experiments based on the independent variables (needle gauge, hydrogel weight percent, fiber length, fiber weight percent) and dependent variables (break force, glide force, and shear stress). For hydrogel synthesis, hyaluronic acid was modified with either cyclodextrin as the host functional group (Cd-HA), or adamantine as the guest functional group (Ad-HA) [6]. Regarding fiber synthesis, polycaprolactone was electrospun into an aligned fibrous mats and then cut into short fibers (20-60 um in length) using a cryotome [7]. To form hydrogel composites, fibers or were suspended in Cd-HA dissolved in PBS before being combined with Ad-HA within a syringe [10]. Then, a rheology strain sweep at 10 Hz from 0% to 500% strain was conducted to demonstrate the shear-thinning behavior of the guest-host hydrogel. Next, injection force testing was conducted to measure the break and glide forces for different hydrogel composite formulations and needle gauges. Using the iterative JMP Pro DOE software, significant independent variables were further analyzed and tested to determine trends.

Preliminary Study on Derivation of Sympathetic-like Tissue Using 3D Hydrogel Cell Culture

Lina Abu-Absi, Selina Banerjee, Guohao Dai, Abigail Koppes, Ryan Koppes

The peripheral nervous system, composed of the autonomic and somatic nervous subsystems, is responsible for relaying information between the brain and body. Despite efforts to optimize nerve regeneration, peripheral nerve injuries have poor recovery. Researchers rely on *in vitro* nerve models to better understand nerve injury, however these models often fail to simulate *in vivo* structure and function, as they incorporate animal cells and lack 3D culture environments. Human 3D *in vitro* models, such as organ-on-chip devices, can address these shortcomings, especially if they incorporate the use of human stem cell-derived neurons to better reflect human physiology. Previously, our lab has cultured neonatal rat autonomic neurons in 3D (Soucy et al., Advanced Biosystems, 2020); this preliminary study expands this work toward the development of a 3D human sympathetic-like tissue model. To our knowledge, this is the first attempt to differentiate human sympathetic neurons in 3D.

Induced pluripotent stem cells were differentiated to generate autonomic sympathetic-like neurons following a previously published method (Takayama et al.,Scientific Reports, 2020). In contrast to this method, we re-plated our cells at an earlier stage (Day 13 vs. Day 6, respectively) into a 3D gelatin methacryloyl (GelMA) hydrogel to allow for growth into the gel before development into post-mitotic mature neurons. Cells were re-plated on Matrigel-coated plates (2D) or encapsulated in GelMA hydrogels (3D), and neurite outgrowth was compared to a 2D group which was not replated. The number of neurites per overall cell count was evaluated in brightfield images taken on Days 7, 10, and 13. For the 2D cultures, there was an increase in neurites per total cell count, though most neurites for the re-plated group had developed prior to Day 10 whereas for the non-re-plated group more neurites grew post Day 10. Contrary to our hypothesis, the neurites per total cell count in 3D dropped from Day 10 to 13 and had less neurite growth than 2D cultures. These results indicate that the cells are maturing after re-plating, but are having difficulty growing into the GelMA matrix. Future experiments will incorporate the use of supportive glial cells to further encourage neurite formation in 3D, and morphology will be compared to rat sympathetic neurons in the same matrix to optimize growth conditions. Overall, this approach has shown promising results and can be further developed to become a robust 3D-cell-culture method for better understanding the relationship between autonomic nerves and their surrounding environments.

Structural Characterization of Scaffold-Free Tendons Constructs

Liza Rodriguez Gonzalez

Tendons are connective tissues that connect muscles to bones allowing for joint movement. Musculoskeletal injuries have become more prominent in the United States during the last years as they often occur in athletes and patients with moderate physical activity. Currently, tendon injuries are mostly treated by therapy or using another tendon source such as autologous or allogeneic grafts. The therapy treatment requires long periods of rehabilitation, and the success of the surgery is patient specific. Nevertheless, these replacements do not allow the tendons to regain its native properties. This concern generates a significant clinical interest in developing a tissue engineered tendon construct that can replicate the mechanical and biological structure of tendons. Previously developed engineered tissue replacements have proved to be unsuccessful because the traditional scaffold-based approach does not replicate the normal tendon development. In contrast, the scaffold-free approaches have succeeded to replicate early embryonic tendons. However, they fail to match the mechanical and structural properties of late embryonic tendons. A critical barrier stopping the development of tendon constructs is the lack of knowledge regarding the key mechanobiological mechanisms that influence the tendon maturation. This project aimed to establish a working lab protocol to build scaffold-free tendon constructs with the goal of characterizing the construct structural properties and compare them with a day fourteen chicken embryo tendon. For this, tendon constructs were formed using embryonic chicken tenocytes and relied on the cells' own self-assembly properties. Then, the constructs were subjected to analysis of structural properties through the use of confocal microscopy to visualize the cells' collagen deposition, alignment, and density. Approximately 72% of the tendon constructs exhibited high collagen alignment thus mimicking the 14-day chicken embryo showing a parallel organization along the direction of the long axis of construct. Consequently, scaffold-free tendons constructs were similar and comparable to the E-14 chicken embryo, suggesting an efficient lab protocol and practice. The outcomes from this study will be used to improve the in vitro maturation of tendon constructs through the activation of essential mechanotransduction signaling pathways, which play a crucial role in the development of embryonic tendons.

Controlling Peptide Binding Chemistry and Density within Gelatin Hydrogels

Samirah Salifu, Corinne Curry, Zander Schwartz, Ethan Lippmann

In an effort to test hydrogels without excessive animal testing, we are developing a microfluidic device to serve as a screening tool. These microfluidic devices will be used to grow vascular tissue in several hydrogel conditions. Each hydrogel has a gelatin base with different vascular promoting peptides chemically bound. My project has focused on synthesizing and quantifying these hydrogels across three different peptide factors: sequence, binding chemistry, and concentration. The three peptide activities being tested include QK (a VEGF mimetic peptide), N-cadherin (cell adhesion peptide), and HepPep (a heparin binding peptide). Two peptide-binding chemistries are being assessed in this experiment, click-chemistry via a maleimide-thiol reaction and EDC/NHS coupling. Finally, each peptide was bound at three concentrations, resulting in a low, medium, and high format of each. In total 18 distinct hydrogel formulations are being created for our screen. Additionally, to accurately quantify the amount of peptide in each dose of hydrogel, all variants will be duplicated with a fluorescently tagged peptide. Utilizing UV-Vis spectroscopy, the quantity of fluorescent peptide within each batch will be measured using Beer Lambert's law. By combining Nuclear Magnetic Resonance data with the fluorescent dataset, we can accurately quantify the peptide content within each hydrogel dose. Next steps include testing each hydrogel variation in the microfluidic devices to screen with vascular antibodies to determine the best hydrogel formulations. Subsequently, the best formats will progress to mouse fat pad injections for further evaluation and validation.

Chiral Modulation of Phenol Soluble Modulin Self-Assembly

Nicole Baalbaki, Mische Hubbard, Kody Whisnant, J. Scott VanEpps, Nicholas A. Kotov

Staphylococcus aureus biofilms have been implicated in numerous nosocomial and community associated infections worldwide. The virulence of S. aureus biofilms is dependent on the secretion of short, amphipathic peptides called Phenol-soluble modulins (PSM). PSMs provide a structural scaffold for S. aureus biofilms through fibrillation into amyloid-like fibers. Besides biofilm structuring, PSMs have other biological functions including triggering an inflammatory response, lysing mammalian cells, and show antibacterial activity towards other microbes. In this work, chiral carbon nanoparticles (CNPs) are explored for their enantioselective effect on the self-assembly process of PSM peptides. We assess the physicochemical properties of PSMa1-4 after co-incubation with chiral CNPs both as individual peptides and mixed together. With this, the fibrillation kinetics of nanoparticle-treated PSMs are assessed and the secondary structure is analyzed using circular dichroism. TEM images reveal asymmetrical hierarchical self- assembly of PSMs and chiral CNPs into supramolecular structures.

3D Modeling of *Plodia Interpunctella* silk Gland: Unraveling Silk Gland Dynamics for Advanced Manufacturing

Attisun Blackwell, Lauren Eccles, Whitney Stoppel

Silk fibers are produced by silkworms, such as the domesticated silkworm (*Bombyx mori*) and the pantry moth (*Plodia interpunctella*). Silk fibers are formed in an insect-specific organ, called the silk gland. Silk glands are complex structures, where the proteins that make up the silk fiber are produced. These proteins self-assemble within the gland, which utilizes pH shifts, enzymes, and shifts in dimensions to aid in silk spinning.[1-4] Aside from studies on the domesticated silkworm (*Bombyx mori*), very little is known about the structure and function of silk glands in other insects. Our laboratory utilizes silk fibers and purified silk fibroin proteins from *Plodia interpunctella* to form materials for applications in advanced manufacturing and healthcare. To investigate the way *Plodia interpunctella* silk fibers form, we utilize both experimental and computational tools to develop 3D models of the *Plodia interpunctella* silk gland.

First, we determined the size, structure, and architecture of the silk gland in a 5th instar *Plodia interpunctella* larva using nano-computed tomography (Nano-CT). Then, we used image segmentation and reconstruction to model the 3D architecture. Next, we performed post-processing of the surface meshes to ensure a solid object. Once we had this 3D model, we used it in two applications: 1) COMSOL™ modeling and 2) 3D printing.

We are using COMSOL™ modeling to gain insight into changes in fluid shear stress as a function of gland architecture (gland diameter changes) to understand the impact of shear forces on beta sheet formation, a necessary secondary structure to form insoluble silk fibers from liquid silk solutions. First, the 3D model was moved to AutoDesk Inventor to convert the surface mesh into a solid object. Then, we explored shear stresses at the walls based on viscosity and flowrate of the solution flowing through the gland. Results provide wall shear stress and pressure at predetermined intervals. To further visualize the silk gland and build a 3D microfluidic model, we printed the gland at 5X the original size in polydimethylsiloxane (PDMS). Preliminary results evaluate the impact of silk fibroin solution viscosity on flow through the gland.

Overall, this work represents the first steps in understanding the fundamental fluid dynamics of *Plodia interpunctella* silk fiber self-assembly. Moreover, this work sets the stage for designing artificial silk glands or silk spinning devices, which will aid future manufacturing of silk fibers for a wide range of applications.

<u>References</u>

1. Terry AE, Knight DP, Porter D, Vollrath F. pH Induced Changes in the Rheology of Silk Fibroin Soluon from the Middle Division of Bombyx mori Silkworm. Biomacromolecules. 2004;5(3):768-72. doi: 10.1021/bm034381v.
2. Dicko C, Vollrath F, Kenney JM. Spider silk protein refolding is controlled by changing pH. Biomacromolecules. 2004;5(3):704-10. doi: 10.1021/bm034307c. PubMed PMID: 15132650.
3. Domigan LJ, Andersson M, Alber KA, Chesler M, Xu Q, Johansson J, Rising A, Kaplan DL. Carbonic anhydrase generates a pH gradient in Bombyx mori silk glands. Insect Biochemistry and Molecular Biology. 2015;65:100-6. Epub 2015/09/15. doi: 10.1016/j.ibmb.2015.09.001. PubMed PMID: WOS:000364618800012; PMCID: PMC4628561.
4. Wu X, Wu X, Shao M, Yang B. Structural changes of Bombyx mori fibroin from silk gland to fiber as evidenced by Terahertz spectroscopy and other methods. Internaonal Journal of Biological Macromolecules. 2017;102:1202-10.

How Many Passes Does It Take? an Investigation to Determine the Optimal Number of Liposome Extrusion Cycles

Kasey Piper, Kenneth Mineart

Liposomes are lipid-based nanoparticles commonly known for their drug carrying capacity and value in membrane and nanoreactor research. One commonly used method of fabricating liposomes of particular size is membrane extrusion in which liposomes are pushed through a porous membrane and break apart to form smaller liposomes. However, the details pertaining to liposome extrusion, such as the number of passes, membrane pore sizes, and pre-extrusion sample preparation, vary between different labs. The specific goal of this project is to establish a relationship between the number of passes through a track-etched, polycarbonate membrane, and the average diameter and diameter distribution of extruded liposomes. To investigate, we extrude samples to several different amounts varying from 1 to 1,000 passes and measure the average hydrodynamic diameter and diameter distribution of each using dynamic light scattering (DLS). In addition to cross-experiment comparison, we compare liposome size data to the size distribution of membrane pores, which is acquired using scanning electron microscopy (SEM) and subsequent image processing. Cases in which liposome and pore size distributions are similar suggest that extrusion has been completed, and further extruding would not change the liposome size. These results provide relevant researchers with the optimal number of passes for liposome extrusion that will allow them to form ideal distributions in the shortest amount of time.

Engineering a Probiotic Bacterium for Utilization of Pyrolytic Sugars

Hailey Bates, Sanjeeva Kumar Murali, Thomas J. Mansell

Fermentable sugar obtained from lignocellulosic material exhibits substantial potential as a renewable feedstock. Upon pyrolysis, this biomass produces a major product called levoglucosan (LG), which is not fermented by typical microbes such as *Escherichia coli*. Previous reports suggest that engineering microbes with the levoglucosan kinase (Lgk) enzyme can enable the use of this sugar, directing carbon flux toward growth and energy metabolism. In this study, we have engineered the probiotic strain *E.coli* Nissle 1917 (EcN) for LG utilization with recombinant expression of Lgk enzyme from *Lipomyces starkeyi*. We demonstrated inducible and constitutive expression of Lgk in EcN, and increasing protein expression enabled use of LG up to 10% (w/v) as the sole carbon source in minimal media. This engineered strain also showed a competitive growth advantage when co-cultured with organisms lacking Lgk. Our current investigation involves constructing a metabolic pathway for the production of anti-inflammatory molecules like butyrate, secreting anti-bacterial peptides, and identifying transporters that will enable more efficient catabolism of pyrolytic sugars.

Effect of Egg White Addition on Setback Viscosity of Mayonnaise

Julianna Lopez, Tao Wu, Min Li

Food emulsions are inherently unstable, and over time will separate into the respective oil and aqueous phases. Consequently, it is important to identify conditions that prevent the breakdown of emulsions, like Mayonnaise, that can lead to an increase of food waste from its unappealing appearance. This study investigates the influence of egg yolk and egg white proportions on the stability and rheological properties of mayonnaise. Egg yolk acts as the emulsifier and plays a critical role in mayonnaise stability due to its ability to interact with oil and water phases. In addition, it's usually the most used emulsifier in the mayonnaise industry due to its reliable stability. However, egg whites are abundant in protein count and have been proven to be an economical option when it comes to processing purposes. Therefore, the impact of egg white addition on mayonnaise viscosity and emulsion stability was explored by testing different ratios of liquid egg yolk combined with liquid whole egg. Rheological properties, including flow behavior, were evaluated to understand the interaction between ingredients and its effect on mayonnaise stability. The findings confirm that higher proportions of egg yolk contribute to increased viscosity in mayonnaise, and that the regulation of oil and egg yolk temperature leads to more consistent results. More surprisingly, it has been noted that higher proportions of egg white contribute to better mayonnaise stability over time. This research provides valuable insights into optimizing ingredient ratios and processing conditions for enhancing the quality and stability of mayonnaise formulations.

Unmasking Adulteration of Commercial Avocado Oils: A Case Study Using Parameters from Novel Regulatory Standards

Daniela Hernández González, Héctor Plascencia Villanueva, Veronica Patino-Gonzalez, Diana Jessica Obispo Fortunato, Luis Martín Marín-Obispo, Carmen Hernández-Brenes

Daniela Hernández González, Héctor Plascencia Villanueva

Verónica Patiño González, Diana Obispo Fortunato, Luis Martin Marín-Obispo, Carmen Hernández-Brenes

Unmasking Adulteration of Commercial Avocado Oils: A Case Study Using Parameters from Novel Regulatory Standards

In a landscape marked by an expansion of options of Mexican avocado oil in the market, consumers must ensure that the product they acquire is of good quality and meets regulations. Due to the recent improvements in Mexican avocado oil regulations, this represents a significant change for consumers. Oils being 100% avocado not only benefits the consumer, but also reinforces a brand's dedication to transparency and integrity. The goal of this research was to verify if some avocado oil brands comply with the recently approved NMX-F-811-SCFI-2021 standard (NMX) and CXS 210-1999 Standard (CODEX) through the fatty acid profile. Avocado oils from 12 commercial brands from different batches were used for each evaluated brand in the random sampling (n=3). The determination of fatty acids was carried out according to the AOCS Method Ce2-66 in a gas chromatography coupled with a flame ionization detector. To be approved, a brand had to meet allowed fatty acids values in both CODEX and NMX standards; failure in NMX, even if it was within CODEX limits meant automatic disapproval. Out of the 12 brands, 58% failed to meet the stearic acid values as well as 42% that missed the palmitoleic acid values specified in both standards. When it comes to arachidic acid, 33% did not meet the values specified in both standards, and 25% didn't meet the NMX specifications. Similarly, for linoleic acid, 25% exceeded the values specified in both standards, while 16.6% didn't meet the NMX specifications. This research exposes the avocado oil brands according to NMX and CODEX through the fatty acid profile, and how the brands that don't meet the mandatory requirements indicate implications for the industry and consumers. This study contributes to ensuring that the food industry remains competitive and has rigorous standards that hopefully lead to a higher ratio of products that meet the required values of components present in Mexican avocado oil.

Comparing in Vitro Degradation Evaluation Methods for Lyophilized Silk Sponges: Discrete Vs. Continuous Sampling Approaches

Axianax Merone, An Nguyen, Jonathan Valentin, Nisha Kotta, Henry Lutz, Julie F. Jameson, Whitney Stoppel

Silk fibroin proteins purified from *Bombyx mori* silkworm cocoons are used to create biopolymer solutions, forming versatile biomaterials.[1] These proteins offer tunable mechanical properties, limited inflammatory response activation, non-thrombogenicity, and controlled biodegradability.[2] Silk scaffolds are useful in tissue engineering and wound healing due to features like enzymatic degradation and the potential for cellular integration. Customizable silk fibroin sponges demonstrate adjustable characteristics such as pore size, elastic modulus, and crystalline content, showing promise as a scaffold to facilitate tissue regeneration. Adjusting fabrication parameters such as freezing temperature, water annealing time, and degumming time enables fine-tuning of enzymatic degradation rates of silk scaffolds.[3] This research quantifies the relationship between parameter alterations and enzyme degradation rates via a kinetic model. Previous work shows faster degradation rates with lower silk fibroin molecular weight and lower polymer concentrations. Experimental mass loss data over time supports this hypothesis. However, predicting *in vivo* scaffold degradation remains challenging. This study compares discrete and continuous sampling techniques for monitoring the degradation of sponges over time, varying formulation parameters like extraction time and silk concentration. While we have explored methodologies before, our results show that use of the oven to dry the samples quickly altered the structural integrity and crystalline composition of the scaffolds. We showed this by evaluating scanning electron microscopy images over time during the degradation process when using either discrete or continuous sampling methods, as shown in Figure 1. While the continuous method has the advantage of following the same sample over the course of the experiment, our data fit a modified first order reaction rather than the anticipated Michaelis-Menten kinetic model, leading us to further explore the role of structural changes on the resulting data. In this study, we hypothesized that utilizing the discrete method would eliminate changes in sponge structure due to the elimination of the repeated use of oven dried samples. We then confirmed that the data fit a Michaelis-Menten kinetic model. This approach improves the accuracy of degradation prediction and provides insight into the influence of low heat (40°C) on protein secondary structure in porous silk sponges.

Inventing a No-Knead, No-Stretch Pizza Crust

Anh Le, Margot Vigeant

Although pizza making has been mostly mechanically operated, kneading has been a challenging and labor-intensive step for chefs and machines. This research project aimed to revolutionize pizza crust preparation by providing a simple alternative method to eliminate the stretching step, taking advantage of the no-knead bread concept. In a moisturized environment, proteins can move around freely and make direct contact with each other to form gluten. However, without the help of kinetic movements, this process would require a longer waiting time, a higher percentage of water, and the dough being in the right shape from the beginning.

After conducting a series of experiments, the project determined the most efficient method could be achieved by increasing the water amount by 75% and maximizing the waiting time of the dough to be 72 hours in the refrigerator plus 3 hours at room temperature both before and after the 72-hour session. While further research is warranted, this research provides an alternative recipe that generates products attaining approximately 70% of the traditional method's result by volume. Consequently, further development might lead to an application to an industrial scale to decrease the cost of labor and the need for additional equipment.

A Kinetic Model for Single-Yeast Sour Beer Fermentation

Sofi Jeffrey, James E. Maneval

During the fermentation process in which glucose and or other sugars are metabolized by yeast to form ethanol, certain strains of yeast, called 'sour yeasts', will first metabolize sugar to produce lactic acid, after which, the mechanism switches to produce only ethanol. We sought to adapt an existing mathematical model of a fermentation to describe this lactic acid fermentation.

After understanding and implementing several papers, we decided that the math model contained in [1], was ideal to adapt to describe a lactic acid fermentation. The key aspect of the model is a switch function, which we replicated to demonstrate the change in metabolism when a sour yeast ceases to produce lactic acid during a fermentation. This model is also limited to glucose as the sugar source. Utilizing the existing rate equations and stoichiometric coefficients within the model, we were able to construct a rate expression for lactic acid and integrate it into the model proposed [1].

These rate equations should model a lactic acid production coinciding with cell mass growth causing an early pH decrease to about 3.2. We accomplished this through the switch function model approach [1] controlling lactic acid production with a single parameter which we established by comparing our model to data [1]. Our final pH was achieved at around 5 hours.

We hope that our model could be used in the industrial production of sour alcohol. Improvements include describing more realistic worts, such as maltose and maltotriose, as well as modeling a greater range of beverage production settings, as described in [2]. Fermentation modeling can be implemented to improve the ease of production and decrease the waste involved in experimentation.

[1] *González-Hernández, Y.; Michiels, E.; Perré, P. A Comprehensive Mechanistic Yeast Model Able to Switch Metabolism According to Growth Conditions. Fermentation 2022, 8, 710. https://doi.org/10.3390/fermentation8120710*

[2] *J. Gee DA, Ramirez WF. Optimal temperature control for batch beer fermentation. Biotechnol Bioeng. 1988 Feb 20;31(3):224-34. doi: 10.1002/bit.260310308. PMID: 18584597.*

Evaluation of Biosynthetic Approaches Toward the Production of Non-Natural Psilocybin Derivatives

Jessica Flower, William Gibbons, J. Andrew Jones

Psilocybin is the psychoactive prodrug of psilocin that has been shown to lessen the negative effects of PTSD, depression, and addiction. While there are multiple established process for the synthesis of psilocybin including extraction methods and traditional chemical synthesis, they have proven to be tedious, time-consuming, and expensive. However, my lab has championed the production of psilocybin with a fermentation-based approached using genetically engineered E. coli. Psilocybin treatment is, however, not perfect. It has been shown to be safe and nonaddictive but there is one significant side effect that cannot be ignored...a roughly 6-hour hallucinogenic experience immediately after consuming the drug. Neuroscientists are beginning to unravel the mechanism of mental health treatment by psychedelics and have suggested that the presence of a hallucinogenic experience is not necessarily connected to the positive mental health impacts. Here, we evaluate four different approaches towards the production of non-natural psilocybin derivatives to create a larger drug candidate pool with diverse efficacies and psychedelic properties. The first approach expands upon our previously reported psilocybin production pathway to develop a route for the *de novo* co-culture biosynthesis of this compound. Incorporating a P450 monooxygenase, PsiH, to catalyze the hydroxylation of tryptamine intermediate compounds enables both a route to *de novo* psilocybin biosynthesis as well as a study of the substrate promiscuity for the full psilocybin biosynthesis pathway through supplementation of a library of single substituted indoles. A second approach will explore the production of psilocybin derivatives through supplementation of disubstituted indoles, bypassing the low-yield PsiH catalyzed step. These starting substrates will be supplied to a previously optimized psilocybin production strain in monoculture. A third approach will expand the current psilocybin biosynthesis pathway through use of a new type class of enzymes to the Jones Lab, halogenases. Specifically, we will focus on site-specific chlorination enzymes which have previously been shown to have activity towards tryptophan substrates in prokaryotic hosts. Finally, the last approach will combine the first and third approaches into a single study, enabling *de novo* psilocybin derivative production by expansion of the psilocybin biosynthesis pathway to include both P450 monooxygenase and halogenase modules.

Development of a Rechargeable, Antimicrobial, Reactive Textile Dye

Nabeel Hadad

The process of imparting antimicrobial properties into fabrics is a topic that is extremely beneficial in the medical field, military applications, and commercial uses due to the proximity of these fabrics to the human body. We focus on using N-halamines, which are recognized for their rechargeable and powerful microbial contact-killing properties. Inspired by reactive dye chemistry, a compound was designed using cyanuric chloride as a binding agent and cysteine as the linker to the intended N-halamine built off 5,5-dimethylhydantoin. This compound is comparable to a dichlorotriazine (DCT) dye, possessing water solubility and reactivity suitable for the industrial textile industry. The synthesized compound can be applied to fabric in a single reaction, avoiding complex processes and making it industrially viable.

We confirmed the synthesis of dye using NMR, FT-IR confirmed the attachment of dye to the fabric, and an iodometric titration was used to measure the uptake of chlorine. These validations ensure the effective chlorine loading, the capability of rechargeability, and overall storage stability of developed fabrics. A specialized antibacterial assay is utilized to affirm the contact-killing efficacy of the compounds, neither the hydantoin nor chlorination alone imparts any antibacterial activity to the fabric. This project is meant to dye fabrics to ensure bacteria growth is inhibited and can be re-energized multiple times, ensuring it's practical, durable, and effective in various real-world applications such as medical linen and athletic apparel.

Development of a Method to Determine Regional Bacterial Eradication of Dry Powder Aerosols.

Olivia Clark, Christopher Vidmar, Jennifer Fiegel

Lung infections caused by the bacterium *Pseudomonas aeruginosa* are challenging to eradicate due to their formation of biofilms. These bacterial colonies excrete a polysaccharide matrix that provides protection to the bacteria from the immune system and antibiotics. To be effective, antibiotics must be delivered directly to the site of infection to provide the high drug concentrations required for eradication. Localized delivery can also minimize side effects experienced throughout the body with other routes of delivery. Dry powder aerosols containing antibiotics can be inhaled at high doses, thus providing a method to kill lung infections. The size and deposition of such aerosol particles can be characterized by impaction methods, such as with the Next Generation Impactor (NGI). This impactor, however, cannot support bacteria viability studies, meaning it cannot determine how well a dry powder aerosol can eradicate bacteria at various locations in the lungs. The 6-stage Viable Andersen Cascade Impactor (vACI) is commonly used in environmental studies to collect microbes, including bacteria, via impaction onto petri dishes. In the vACI, aerosolized material will deposit on the six stages based on aerodynamic size. The purpose of this research is to adapt the vACI to deposit antibiotic-containing dry powder aerosols to bacteria-laden petri dishes and quantify *P. aeruginosa* eradication as a function of aerodynamic size. This study aimed to determine the correlation between the deposition of dry powder aerosols in the NGI and vACI. This research will expand with the formulation of dry powder aerosols at various concentrations of ciprofloxacin to characterize how concentration of ciprofloxacin affects particle size, deposition, and lethality.

Nanopore Sequencing in the Regenerating Axolotl Limb

Connor Powell

In the landscape of modern genomics, RNA sequencing (RNA-seq) has become a cornerstone for deciphering intricate molecular compositions. However, the field's progress has been challenged by the costs and limitations inherent in traditional short-read sequencing methods. Oxford Nanopore Technologies (ONT) has rapidly emerged as a transformative force in the field of RNA-seq by utilizing nanopores - nanometer-scale channel proteins - as artificial transport mechanisms. This innovative approach presents an exciting long-read sequencing alternative that counters the limitations of conventional short-read sequencing methodologies, such as the occurrence of multi-mapping reads and difficulty of generating de novo assemblies. In addition, ONT provides a series of distinct library preparation alternatives allowing for additional epigenetic and transcriptomic analyses.

Here we present an optimized, cost-effective, and streamlined protocol for ONT RNA-seq that maximizes the potential of this technology in diverse research settings. We use the axolotl salamander (*Ambystoma mexicanum)* to test ONT technology in a regenerative biology setting, developing a pipeline which can be applied to tissues to characterize differential gene expression in varied contexts. To validate our characterization pipeline, we compared our axolotl intact limb and blastema expression data to established short-read RNA-seq datasets and observed consistent differential expression of well-established markers. Additionally, we characterized a dataset derived from mutant axolotls to identify misregulated genes by immunohistochemistry in tissue samples.

Our data indicate that ONT technology represents a convenient and low-cost approach which can generate in-house agnostic differential gene expression data for various tissue sets. We hope to further broaden our understanding of long-read sequencing by including a wider range of epigenetic and transcriptomic analyses that are currently unavailable in traditional short-read RNA-seq alternatives.

Platinum Nanoparticles' Mechanism of Action in Triple Negative Breast Cancer

Ashish Kokkula, Aida Lopez-Ruiz, Kathleen McEnnis

Nanotechnology has emerged as a potential therapeutic for cancer treatment. Metallic nanoparticles, such as gold and silver, have gained significant importance in nanomedicine. More importantly, platinum nanoparticles (PtNPs) have shown promising anticancer activity while maintaining low toxicity towards healthy cells. Triple negative breast cancer (TNBC) is an aggressive form of breast cancer that metastasizes rapidly and is currently only treated with toxic chemotherapeutics. Cisplatin, in particular, is a platinum-based therapy that works by entering the cell, binding to the DNA, and breaking the DNA, effectively disabling the cell's ability to proliferate. Cell viability tests using PtNPs treatments for TNBC, double-positive breast cancer (DPBC), cisplatin-resistant uterine sarcoma, and cardiac fibroblasts indicate that PtNPs works similarly to cisplatin, but only kills TNBC, leaving healthy cells unaffected. Additionally, the acidic environment of cancer cells can lead to more platinum ions, increasing PtNPs efficacy at killing TNBC. The mechanism of action of PtNPs needs to be more fully understood and to do so, ion release studies and cellular uptake studies were conducted. Acidic environments promote more ionization and the data shows that PtNPs in MES, a buffer at pH 5, has variably equal and higher release of platinum ions. Uptake of PtNPs by cancer cell lines and healthy fibroblasts was studied. Cancer cells are prone to devour more nutrients and resources than healthy cells, attributing to their increased proliferation rates. As such, it was expected that the cancer cell lines would have higher PtNPs uptake, especially TNBC. Surprisingly, the healthy fibroblasts were observed to have higher uptake levels than the cancer cell lines, with TNBC having the lowest uptake of PtNPs, implying that TNBC uptake PtNps through a different mechanism. Ongoing mice studies are being conducted as well to determine the efficacy of PtNPs *in vivo*. Preliminary data does show non-toxicity to the mice, further supporting PtNPs low toxicity towards healthy cells. **References:** Shim et. al 2017; Lopez et. al 2022. **Acknowledgements:** This research was funded by the New Jersey Health Foundation and NJIT's Undergraduate Research Innovation Grant. This research was also advised by Dr. Aida Lopez-Ruiz and Dr. Kathleen McEnnis. PtNPs formation was conducted by Prova Joy Foundatoin, Darshan Danak, and Pranav Prabhu.

Generating Multilayer Polyelectrolyte Coating Libraries on Protein Nanoparticles for Mucosal Delivery

Sarah Jenison, Thomas Pho, Julie Champion

The efficacy of nanoparticle therapeutics is greatly limited by biological barriers that hinder access to intended sites. One biological barrier that has yet to be overcome is the mucus layer, which covers epithelial surfaces throughout the body. Mucus is a viscoelastic gel that restricts movement of foreign matter in the body, including beneficial material such as nanoparticles. Conventional therapeutic nanoparticles have adhesive interactions within the mucus layer that hinder mobility. To enable mucus-penetration, polyethylene glycol (PEG) is commonly used to achieve a hydrophilic and neutral surface coating for improved diffusion across the mucus barrier. However, we have demonstrated that PEG coatings on protein nanoparticles elucidate off-target immune responses through the production of anti-PEG antibodies in a murine model upon intranasal administration. In this study, we developed a library of protein nanoparticles with different multilayer-polyelectrolyte coatings that modulate diffusion in mucus as an alternative to PEG coatings. We formulated over 30 unique nanoparticles and screened them for enhanced diffusion coefficient using Nanoparticle Tracking Analysis (NTA) in mucus. NTA revealed that certain formulations had higher diffusion coefficients. In addition, we screened four different cell lines derived from colorectal, lung, cervical epithelial tissue, and dendritic cells using flow cytometry to assess their uptake profile with a mucosal barrier. We identified formulations with improved interaction with different cell lines, linking unique polyelectrolyte coatings with specific cell types. These experiments indicate that certain multilayer polyelectrolytes formulations have the potential for improved diffusion and increase cellular uptake delivery in mucus enriched cells.

Immunomodulatory Nanoparticles for Inflammatory Arthritis

Ryan Jones

DMARDs, or Disease-Modifying Antirheumatic Drugs, are frontline therapeutic drugs used in treatment of autoimmune diseases such as autoimmune arthropides.[1] These drugs operate by suppressing the hyperactive immune response responsible for inflammation and joint damage in these diseases.[1] Aside from their beneficial effects, there are noteworthy challenges associated with DMARDs, such as the potential for adverse side effects ranging from gastrointestinal problems to increased susceptibility to infections.[1] Additionally, not all patients respond equally well to DMARDs, making it essential to explore alternative therapies or combinations of medications.[1,2] In the research presented here, I explore the anti-inflammatory and immune modulating effects of a poly lactic(co-glycolic) acid (PLGA) nanoparticle (NP) encapsulating the immune modulating compound all-trans retinoic acid (ATRA) on T-cells and dendritic cells (DCs) cultured *in vitro.* At a glance, it was observed that the ATRA loaded NPs significantly decreased the amount of proinflammatory cytokine IL-17 and increased the amount of anti-inflammatory FoxP3 expressed from T-cells in conditions set to polarize Naïve T-cells into Th17s. Additionally, the nanoparticles significantly decreased the expression of key markers CD80, CD86, and MHC II in dendritic cells; indictive of ATRAs suppressive properties on DC maturation.

Use of Gold Nanoparticles for Electrochemical Detection of Dopamine

Seejal Padhi, Omowumni A. Sadik

Dopamine is both a neurotransmitter and a hormone. It has been used to analyze cognitive function. Due to its significance in neurodegenerative conditions, a comprehensive understanding of how this neurotransmitter affects the brain is needed. Previous research has demonstrated success in novel therapeutics to treat neurodegenerative diseases via the use of dopamine, including using L-DOPA and carbi-DOPA to manage Parkinson's disease (Iversen et al 2007). Electrochemical techniques, specifically cyclic voltammetry, have shown great promise in understanding the reactivity of its molecular complexes and detection in neural tissue. Gold nanoparticles have been used as markers in the ultrasensitive detection of Dopamine in tissue samples. In this study, a nanostructured electrochemical biosensor is being developed to detect Dopamine in neural tissue sample.

Dopamine solutions were prepared using dilution methods at varying concentrations (0 μM, 20 uM, 40 uM, 60 uM, 80 uM, and 100 uM) with phosphate buffer solution. Cyclic voltammetry was conducted using silver/silver chloride (Ag/AgCl) as the reference electrode. Platinum electrode was employed as the counter electrode. Glassy carbon electrode was used as the working electrode. An initial set of experiments were run using the bare glassy carbon electrode. Gold nanoparticles were adsorbed onto the glassy carbon electrode overnight for the second set of experiments

A series of cyclic voltammetry experiments were carried out at scan rates of 50 V/s, 100 V/s, 150 V/s, 200 V/s, and 250 V/s for varying molar concentrations of Dopamine in phosphate buffer pH 7.2. Figure 1 below demonstrates the reaction of Dopamine using a bare glassy carbon electrode. A second set of experiments was performed using a glassy carbon electrode with gold nanoparticles. There were little to no differences in the peaks for oxidation and reduction, indicating that the dopamine present is not detected or not actively participating in the electrochemical reaction. Another set of experiments was performed using a glassy carbon electrode with gold nanoparticles.

Previous studies show great promise in using Dopamine to better understand the neurochemical changes that occur during the onset of neurodegenerative diseases. However, further research is needed to devise a method to more accurately detect Dopamine. Future experiments will be conducted using gold nanoparticles to determine the effect of gold on detecting various dopamine levels in biological fluids such as sweat and blood using biosensor devices. These experiments will be conducted via cyclic voltammetry at various concentrations to see what can most effectively detect dopamine.

Electrochemical Strategy for Eradicating Multi-Drug Resistant *Candida Auris* Using Low Level Direct Current and Echinocandins

Colin Landis, Camila Cue, Maria Geraghty, Sricharani Balmuri, Tagbo Niepa, PhD

The growing threat of antimicrobial organisms compromises even the most impervious healthcare systems. *Candida auris*, an emerging pathogen as recognized by the CDC, has proven to be not only highly lethal, but also extremely resistant to drug treatment. Various treatment methods have been proposed in addition to antifungals from essential oils to nanoparticles. One treatment form already proven effective at eradicating *Candida spp.* is electrochemical therapy (ECT) used in conjunction with drug treatment. At the highest current density tested (70μA/cm^2), CFU studies reveal almost complete log killing of *C. auris* cells. Echinocandins are a novel drug class that due to their targeting of the (1,3)-β-D-glucan synthase have made resistance harder to develop. Therefore, their increased susceptibility makes them a useful tool for investigating antifungal diffusion properties and their potential for inhibiting cell growth. In this study, ECT is applied to *C. auris* cells and the effect of that treatment as well as the synergy between ECT and antifungal treatment on cell viability, function, and morphology are studied. Caspofungin, a commonly prescribed echinocandin for candidiasis, when applied to *C. auris* cells demonstrates rapid cell envelope adhesion, localizing to the cell wall within fifteen minutes. When caspofungin is incubated with the yeast cells for just as long, it enters the cells via endocytosis in vesicles. Upon further incubation, these vesicles rupture releasing the drug into the cell. Applying ECT lowers the energy barrier for endocytosis to occur, allowing for greater drug uptake. Visualizing fluorescently tagged caspofungin using confocal microscopy, these diffusion properties are analyzed. SEM images of the ECT-treated cells offer a look at how the electrical current affects cell morphology. Overall, this study demonstrates that ECT is a promising alternative strategy for effectively killing persistent yeast infections.

Biomimetic Nanoparticles for the Homotypic Targeting of Triple-Negative Breast Cancer

Grace Azevedo, Sara Aboeleneen, Emily Day

Triple negative breast cancer (TNBC) accounts for 15-20% of all breast cancer diagnoses, and is characterized by its lack of expression of estrogen (ER), progesterone (PR), and human epidermal growth factor receptor 2 (HER2) receptors, the three most common cell membrane receptors present on other subtypes of breast cancer. This lack of receptor expression hinders the development of effective targeted treatments for TNBC, resulting in higher recurrence and mortality rates among TNBC patients. Currently, the only available treatment options for TNBC patients are surgery, radiotherapy, chemotherapy, or a combination of two or more of these treatments, all of which have adverse side effects and lack the ability to specifically target the tumor without damaging healthy tissues. Therefore, there is an urgent clinical need to develop effective treatments that can target the tumor and improve patient outcomes. Recent studies suggest that coating nanoparticles (NPs) with cancer cell-derived membranes can improve their tumor delivery *via* homotypic targeting. Homotypic targeting is thought to be achieved by the the presence of special cell membrane proteins and adhesion molecules that include "markers of self" proteins that hide the NPs from the immune system, prolonging their circulation time in the blood and thus increasing their chances of accumulating at the tumor, as well as "self-recognition" proteins that enable cancer cell-specific binding of the NPs. Thus, we wrapped light-sensitive gold nanoshells (NS) with 4T1 TNBC murine cell membranes, forming 4T1-NS to enable the NS to target the tumor *via* homotypic targeting. Here, we present results of *in vitro* studies confirming the successful synthesis and characterization of cancer cell membrane-wrapped gold nanoshells that demonstrate greater targeting and uptake in TNBC cells over non-cancerous breast epithelial cells. We also investigate the expression of TNBC cell membrane adhesion molecules *via* immunocytochemistry to better understand homotypic targeting. This work provides a promising biomimetic platform to improve the targeting and delivery of nanoparticles to TNBC for improved treatment outcomes.

Mesoporous Nanoparticles As Carriers for Increased Effectiveness and Welfare for Ovarian Cancer

Caleb Dang, Marian Olewine, Paulina Naydenkov, Angelea Maestas-Olguin, Rita Serda, Achraf Noureddine, C. Jeffrey Brinker

Ovarian cancer is the deadliest amongst cancers of female reproductive system. It is estimated that in the United States for 2023, over 13,000 women will die from ovarian cancer. In advanced ovarian cancer, common treatments involve intraperitoneal injections of high doses of chemotherapeutics which cause serious side effects such as neurotoxicity. Many chemotherapeutics (such as paclitaxel and oxaliplatin) and targeted inhibitors (such as afatinib) used in the treatment of cancers have low bioavailability due to their hydrophobicity and quick removal from blood circulation. This can lead to multiple treatment doses in order to affect the cancer. My work focuses on the implementation of a carrier that increases the bioavailability of such therapeutics in the blood stream. A carrier that protects the therapeutics and increases bioavailability can reduce the frequency and intensity of treatments while providing comparable or better outcomes. My work aims at implementing and optimizing nanoparticle features to administer a combination therapy (targeted inhibitors supplemented with chemotherapy) that can, in the long term, reduce side effects by lowering chemo doses. In this project, I am evaluating both lipid-based and silica based nanoparticles and using the state of the art characterization techniques to assess drug loading and nanoparticles quality. My presentation will include in silico, in vitro, and in vivo evaluation of the efficacy of drug-loaded nanoparticles against BR5 ovarian cancer cells. In vitro cytotoxicity assay showed that an afatinib/oxaliplatin combination killed up to 95.5% of the cells and an afatinib/paclitaxel killed up to 93.8%. A preliminary in vivo study indicated that a combination of afatinib and oxaliplatin is comparable to pure oxaliplatin and paclitaxel.

Engineering Nanobody-Drug Conjugates for Logic-Gated Immunotherapy

Jonah Finkelstein, Blaise Kimmel, Neil Chada, John T. Wilson

The stimulator of interferon genes (STING) pathway is an emerging and clinically relevant target for cancer immunotherapy. The innate pathway functions to detect cytosolic DNA through the activation of cyclic GMP-AMP Synthase (cGAS). Activated cGAS generates cGAMP, which then activates ER-resident protein STING, leading to the downstream production of immunostimulatory molecules. Natural cyclic dinucleotide agonists suffer from rapid clearance and poor stability, whereas powerful synthetic diamidobenzimidizole (diABZI) agonists can cause immune-related adverse effects when administered systemically. To combat these delivery problems, we developed STING-activating nanobody conjugates that target specific cell populations. We have previously shown that macrophages constitute at least 20% of tumor volume in a murine breast cancer model. Macrophages within the tumor microenvironment, also called tumor-associated macrophages (TAMs), are divided into two classes: M1 and M2. M1 macrophages are proinflammatory, drive increased immune infiltration in solid tumors, and have high CD80 expression. M2 macrophages are anti-inflammatory, promote tumor growth, and have high CD206 expression. CD206, also known as the macrophage mannose receptor (MMR), is involved in pathogen recognition and resolving inflammation. Studies have shown that STING activation can reprogram M2 TAMs to M1 TAMs. By targeting M2 TAMs with a diABZI STING agonist conjugated to an anti-MMR nanobody, we propose that we can elicit the reprogramming of M2 TAMs while avoiding the systemic inflammatory effects of free STING agonists.

To test our hypothesis, we generated an anti-MMR nanobody by transforming a plasmid vector containing the nanobody DNA sequence, an enzymatic conjugation tag, and a 6xHistidine affinity tag into *E. coli*. We grew and induced large bacterial cultures, then lysed the cells by sonication. The lysate was centrifuged, then purified by nickel affinity chromatography and an endotoxin-removal column. Endotoxin-free protein was then purified using size exclusion chromatography. An azide-containing linker was conjugated to the nanobody using an engineered sortase enzyme. A DBCO-functionalized diABZI STING agonist was then covalently conjugated to the nanobody through click chemistry. Through ELISA, we found micromolar binding affinity for murine MMR, which is unusually poor for nanobodies. However, we showed increased STING activation *in vitro* in M2 polarized versus M1 polarized RAW264-Dual murine macrophages. Future work involves testing the ability of anti-MMR-diABZI conjugates to reprogram polarized bone marrow-derived macrophages *in vitro* and assessing the biodistribution *in vivo* in tumor-bearing mice. We aim to assess the therapeutic potential of reprogramming TAMs with STING agonists and display the modular ability of nanobodies to target immune cell populations.

Protein Corona Impacts Surface Properties of Layer-By-Layer Nanoparticles

Alfonso Restrepo, Bhuvna Murthy, Tamara Dacoba Gomez, Simone Douglas-Green, Paula T. Hammond

When introduced into biological mediums, a layer of proteins known as the protein corona (PC) adsorbs to the surface of nanosystems. In recent years, the effects of PC in the therapeutic outcome of drugs and drug carriers such as nanoparticles (NPs) has become more evident, impacting their stability, cellular uptake, circulation time, and immunotoxicity. NPs constructed via the layer-by-layer (LbL) method show promise as a modular theragnostic platform that uses a variety of charged polymers, giving rise to tunable surface properties that can influence targeting capacity and biodistribution. We hypothesized that PC is likely affected by this variable surface chemistry and may contribute to the performance of LbL NP *in vivo*. However, PC formation on LbL NPs remains to be characterized.

Here, we report a reproducible method to isolate and characterize the PC on LbL NPs, then an evaluation on how the presence of PC impacts *in vitro* cellular uptake. First, we prepared a library of LbL NPs with different outer polymers (polypeptides, polysaccharides, and synthetic hydrocarbon backbones) based on their therapeutic potential in different disease models, together with PEGylated liposomes as the gold standard. These NPs were incubated in a biologically relevant medium, then subjected to an optimized ultrafiltration method to remove unbound protein for characterization of the isolated PC. Particle size and surface charge before and after PC isolation, quantified by dynamic light scattering, showed moderate size increases, while also maintaining surface charges. Differences in the amounts of proteins associated with LbL NP were observed, with polypeptides showing lower protein concentration than PEGylated liposomes. Further studies revealed differences in cellular uptake in both an ovarian cancer cell line and in macrophage. In particular, for the best performing layer, a pre-formed PC increased the uptake by the cancer cell line, while decreased the uptake by macrophages. These findings indicate that further characterization can and should be carried out for the PC on LbL NPs to determine how protein corona influences the fate of these NPs, providing a key step toward accelerating their clinical translation.

Towards Greener Production of Complex Molecules: Enhancing Triterpenoid Production in Yeast Via PAH1 Gene Mutation

Nicole Babineau, Samuel Scott, Lucas Busta

Triterpenoids, one of the largest classes of phytochemicals, are valuable polycyclic molecules that can be applied as pesticides, pharmaceuticals, and more. Typically, the main production mode for triterpenoids is direct extraction from plants, however yeast cell factories may prove to be a more cost-effective alternative. The objective of this project is to optimize a strain of *S. cerevisiae* for triterpenoid production through alteration of the PAH1 gene. PAH1 codes for the enzyme phosphatidic acid phosphatase, which converts phosphatidic acid to diacylglycerides within a cell membrane. Disrupting synthesis of the PAH1 enzyme has been shown to alter the phospholipid composition of the cell membrane, increasing its permeability and facilitating the export of triterpenoids into growth media to prevent the toxic buildup of triterpenoids inside the cell. The CRISPR-Cas9 gene editing system was used to target and mutate the PAH1 gene, first by using a knock-out method (non-homologous end-joining). Resulting mutant strains of *S. cerevisiae* were isolated and sequenced, which revealed varying frameshift mutations. The location of base pair deletions was consistent across strains, though the number of missing base pairs differed. The chemical profile of both mutant and wild type strains was analyzed via GC-MS to determine whether the mutant strains produce higher levels of triterpenoids than the wild type. To create a genetic disruption more likely to cause a phenotype and to optimize the yeast engineering process, the same CRISPR-Cas9 system is being used to perform a gene knock-in (homology-directed repair) with donor DNA containing numerous stop codons. Overall, we demonstrate that gene editing with CRISPR-Cas9 is effective in *S. cerevisiae*, paving the way for further genetic modification and the continued optimization of triterpenoid production with yeast cell factories.

Scalable Expression of LbCas12a Proteins in *E. coli*

Jenna Kolbe, Zach Hetzler, Qingshan Wei

CRISPR Cas proteins such as Cas12 and Cas9 proteins offer great opportunities for a wide range of sensing applications but at a high cost. When obtaining proteins for research commercially, it can cost the lab up to $1,000 for only 0.5 mg of LbCas12a. The purpose of my project over the summer was to determine if we could produce needed proteins via a low-cost bacterial expression method utilizing the equipment and materials available in our lab. If successful, it would also allow for the purchase of plasmids that would produce other Cas proteins that are not commercially available, giving our research lab a wider range of CRISPR protein options. The overall process involved 1) plating the plasmid containing *LbCas12a* from the agar stab, 2) purifying it via Qiagen Miniprep, 3) transforming the plasmid into *E. Coli*, and 4) cell growth and protein expression via fermentation. After fermentation, the cells were pelleted and lysed. Half of the cells were lysed via freeze/thaw and sonication methods. The other half of the cells were lysed by chemical lysis to compare the efficiency of each method. LbCas12a was purified and multiple BCA Assays were completed to determine yield. We had roughly 2 mg of protein after purification from a one-liter fermentation. Next steps include characterizing the efficiency of the protein through CRISPR diagnostic assays as well as determining protein purity with SDS-PAGE. An additional fermentation will also be conducted to enlarge the amount of culture and protein we can use.

An Integrated Approach for Exploration of Plant Rhizomes' Specialized Metabolism and Low-Cost Production of Related Valuable Compounds

Catherine Xu, Justin Seyedmoomenkashi, Codruta Ignea

Plants can synthesize an incredible myriad of specialized metabolites, providing one of the richest sources of bioactive molecules with applications in pharmaceuticals, agriculture, cosmetics, and even biofuels. Unfortunately, only a small percentage of these chemicals have been harnessed because extraction from natural sources and chemical synthesis of these metabolites is frequently challenging, low-yielding, and economically unviable. Engineering metabolic pathways into heterologous hosts, such as yeast and bacteria, has received increased attention as a cost-effective and environmentally sustainable alternative for generating large amounts of valuable compounds. However, heterologous production requires complete knowledge of a molecule's biosynthetic pathway, which is missing for most plant specialized metabolites. Additionally, the RNA-seq and transcriptomic data needed to elucidate these pathways is sparse. Plant rhizomes are of particular interest because they are a specialized stem tissue that often contain unique metabolites with medicinal properties. Classic examples include ginseng (ginsenosides) and turmeric (curcumin), which have been heavily studied. Accordingly, extensive transcriptomic profiles of these species are available, and the corresponding biosynthetic pathways are broadly known. However, rhizomes are one of the most difficult tissues for nucleic acid extraction due to their high polyphenol and polysaccharide content, limiting RNA sequencing and thus transcriptome analysis. To overcome this challenge, additional extraction steps and optimization are required to obtain high-quality RNA. The rhizomes of *Tacca sp.* are particularly valuable but lack transcriptomic data. These modified stem tissues produce a novel class of anti-cancer compounds called taccalonolides that can circumvent resistance to current chemotherapeutic drugs, such as taxol. With an interest to engineer low-cost and sustainable production of taccalonolides in yeast, we set out to elucidate the biosynthetic pathway of these pentacyclic steroid-type compounds and reconstruct it in host cells. Considering the lack of transcriptomic data, we designed an integrated bioinformatic, metabolic and analytical workflow targeting rhizome specialized metabolism. This encompasses RNA extraction and sequencing, transcriptomic analysis combined with rhizome metabolomics, candidate library screening, and heterologous expression of biosynthetic enzymes in a microbial host. Our ongoing approach will facilitate exploration of the untapped diversity of rhizome-produced metabolites and can be readily expanded to other plant tissues for the discovery of novel bioactive compounds. From a chemical perspective, it will guide elucidation of unknown pathways for biosynthesis of other valued terpenoids, such as withanolides or pentacyclic triterpene saponins. Further engineering of dedicated microbial production platforms will enable scaled up precision fermentation for sustainable and low-cost biomanufacturing of important molecules.

Development of a Cell-Free System for Producing Vaccines Against *Streptoccocus pneumoniae*

Regina Fernandez, Derek Wong, Katherine Warfel, Peivand Sadat Mousavi, Michael C. Jewett

Streptoccocus pneumoniae is a leading cause of pneumonia, sepsis and meningitis worldwide and kills over 300,000 children under 5 globally per year [1]. Glycoconjugate vaccines, such as Prevnar 13 and Prevnar 20, have proven safe and effective at preventing infection from the most prevalent serotypes of *S. pneumoniae*. Composed of a pathogen-specific glycan antigen attached to an immunogenic carrier protein, conjugate vaccines train the immune system to recognize the glycans that are natively displayed on the outside of bacteria. Current methods, however, for producing conjugate vaccines require expensive equipment, growing the pathogenic bacteria of interest, and cold-chain distribution. Cell-free systems circumvent many challenges posed by centralized manufacturing by utilizing crude extracts derived from bacteria to produce any protein of interest via DNA template. Recent work has developed cell-free glycoprotein synthesis (CFGpS) which modifies CFPS for producing glycoconjugate vaccines. In CFGpS, a lysate enriched with both the glycan of interest and the oligosaccharyltransferase (OST), which enzymatically transfers glycans onto proteins, is used to produce the carrier protein of interest [2]. In this work, we modify the CFGpS platform to produce a candidate conjugate vaccine against *S. pneumoniae* serotype 4. We began by improving glycosylation efficiency by using a mutated PglB (OST) to transfer CPS 4. We then further improve glycosylation efficiency by using extract derived from a glycoengineered strain of *E. coli* for increased production of the CPS 4 glycan [5]. Lastly, we demonstrate that by incorporating these two changes, we can efficiently transfer the *S. pneumoniae* CPS 4 glycan onto an FDA approved carrier protein, Protein D. Future work will focus on expanding the CFGpS platform to other serotypes of *S. pneumoniae* to enable production of a protective, multivalent pneumococcal vaccine. Overall, these promising results open the door to increased access to vaccines against *S. pneumoniae* in low resource settings.

Continuous Manufacturing of Nanomedicines for the Treatment of Breast Cancer Brain Metastases

James Kurdi, Federica Carnamucio, Victoria Garcia, David Boyd, Emily Zboril, Aaron Valentine, Chuck Harrell, Sandro da Rocha

250,000 new cases of breast cancer are diagnosed each year in the United States, 15-20% of which are the triple-negative (TNBC) subtype. TNBC is highly aggressive, with patients demonstrating higher incidence of recurrence and metastasis. Selinexor, an FDA-approved multiple myeloma drug, targets nuclear export protein XPO1, which is commonly overexpressed in TNBC compared to normal tissue. Interestingly, Selinexor is also capable of crossing the blood-brain barrier, which may enable it to treat the brain metastases that approximately one-third of TNBC patients experience. So, Selinexor presents a promising anticancer treatment for TNBC. Selinexor has demonstrated toxicity at normal to high doses in clinical trials. Selinexor's toxicity could be addressed by engineering a delivery system to improve its pharmacokinetic profile. Liposomal nanoparticles have shown success in slowing the release of other chemotherapies, thus increasing their tolerability. If the plasma concentration of Selinexor can be more tightly controlled, it may be administered at effective doses without exceeding a patient's tolerated limit. To mitigate Selinexor's toxicity while eyeing clinical translation, we first needed to produce and characterize liposomes with an appropriate quality target product profile and assess their efficacy in vitro. We began by creating and characterizing a liposomal formulation and testing its kill ability in relevant tumor models. We hypothesized that liposomal Selinexor would potently kill cells in a genetically relevant TNBC model. We created multiple batches of liposomal Selinexor via thin film hydration to ensure compatibility of the structural lipids with each other and with the drug. Then, we moved on to a continuous manufacturing approach that utilized a microfluidic system, the NanoAssemblr Ignite, to produce liposomes at a milliliter scale in a fraction of the time. We characterized key physical parameters such as zeta potential, hydrodynamic diameter, and encapsulation efficiency using dynamic light scattering and HPLC-UV. The thin film batches require further optimization to increase encapsulation efficiency, but the continuously manufactured sample demonstrated comparable physical attributes to the quality target product profiles indicated in literature. Next, we assessed the efficacy of free Selinexor in killing MDA-MB-468 cells, an immortalized TNBC cell line. We administered our liposomes at the same concentrations to compare efficacy. The liposomal formulations showed high biological activity, efficiently killing the TNBC cells. Moving forward, we seek to repeat experiments for replicability, control for potential cytotoxic effects of the liposome nanocarriers, and test our formulations in vivo.

Characterization of Phzmpp and Directed Evolution for Novel Phenazine Production

Rebecca Garcia

Phenazines are planar, nitrogenous heterocyclic compounds that are natively produced by Pseudomonads. These chemicals are redox-active pigmented compounds identified for use in antibiotics, dyes, and electrochromic redox mediation. Two enzymes (PhzM and PhzS) are known to decarboxylate phenazine-1-carboxylic acid (PCA) in tandem to produce pyocyanin, a blue-colored phenazine. Recently, a strain of pseudomonas was reported to methylate PCA to 5-methyl-phenazine-1-carboxylic-acid (5MPCA) with a previously undiscovered methyltransferase. We identified and recombinantly expressed the methyltransferase (PhzMPP) in *E. coli* and characterized its native activity. We performed directed evolution experiments PhzMPP, screening for mutants with the ability to convert phenazine to 5-methyl-phenazine, a non-native metabolite. Kinetic rate constants for both the wild type and the best mutant PhzMPP were determined. New mutant enzymes that can readily produce non-native phenazines will be useful for a number of applications including emerging energy technologies where organic redox mediators can be used for large scale energy storage.

Increased Recombinant Protein Expression of Petase Enzyme in *E. coli*

Christopher MacFarlane, Luke Carter, Jason Berberich, Jason Boock

In 2015, more than 380 million metric tons of plastic were produced, with a large quantity being polyethylene terephthalate, or PET, that is commonly used for disposable water bottles. PET is engineered for durability, leading to its slow natural breakdown in the environment. A recently discovered enzyme, PETase, speeds up depolymerization of PET; however, it's turnover rate is still too slow for practical recycling application. Many groups have sought to engineer PETase for faster catalysis and elevated stability. Considerably less effort has been focused toward expressing large quantities of the protein, which is necessary for large-scale application. Our previous work proved that making PETase in a disulfide bond promoting strain of *E. coli,* SHuffle T7 Express, results in higher active expression of the enzyme compared to standard *E. coli* production strains, as PETase is known to require disulfide bonds to be functional. The goal of this work is to further enhance and scale-up production of PETase by studying different medias, carbon sources, induction times, and bioreactor feeding strategies. PETase activity was measured in clarified lysates by quantifying catalysis of p-Nitrophenyl acetate. Using the common laboratory media, 2xLB, PETase was produced in a shake flask with an activity of 10 units/L (unit = umol pNitrophenol produced/s). In a controlled bioreactor, using the same media, expression was increase to 15U/L. Switching to a complex media, with glycerol as its main carbon source, and early induction time, we observed a further 3-fold increase in expression, up to 45U/L, for a batch time of 48 hours. We hope these advances in PETase production will improve the economics to apply this industrially relevant enzyme.

Effects of Altering Purification Steps on the Enzymatic Activity of Glycerol Phosphate Dehydrogenase When Incorporated in a Recombinant Protein Vesicle

Joseph Mancini Jr., Jooyong Shin, Yeongseon Jang

Synthetic vesicles can be used for a variety of applications, ranging from serving as drug delivery vehicles to biosensing capability to studying nanoscale phenomena. In recent years, protein vesicles have showed advantageous qualities, namely increasing biocompatibility and specific bioactivities designed by protein components. In this work, we designed a protein vesicle through self-assembly by combining a recombinant fusion protein, glycerol phosphate dehydrogenase fused with a glutamic acid-rich leucine zipper (GPDH-Z_E), with a complimentary arginine-rich leucine zipper fused to an elastin-like polypeptide (Z_R-ELP). We hypothesized that GPDH-Z_E on the protein vesicle membranes will facilitate an enzyme cascade reaction of glycerol to dihydroxyacetone phosphate (DHAP), which is instrumental in lipid metabolism in the body. The GPDH enzyme provides an active site for glycerol-3-phosphate (G-3-P) to convert to DHAP, which is one of the reactions involved in the enzyme cascade. The GPDH in its commercially used state has successfully converted G-3-P to DHAP, however the enzymatic activity of this enzyme on the protein vesicle membrane of GPDH-ZE has not been sufficient to carry out the reaction in previous trials. My research aims to identify how altering purification steps in the expression of the GPDH fusion protein may affect the enzymatic activity of the GPDH-ZE vesicles. The research will involve incorporating different buffer solutions, different concentrations of those buffer solutions, and the subsequent measure of the GPDH vesicle's enzymatic activity after these changes have been applied.

Psilocybin Pathway Optimization Via Gene Species Variation Toward Improved Strain Production

Madeleine Keller, Madeline McKinney, Andrew Jones

Psilocybin is the hallucinogenic compound found in the recreational drug known as magic mushrooms and is currently being studied as a treatment for PTSD, anxiety, and depression. The psilocybin biosynthesis pathway as reconstituted in *Escherichia coli* contains three exogenous genes: PsiD, a decarboxylase, PsiK, a kinase, and PsiM, a methyltransferase. Previous efforts in the Jones Lab have shown high-titer psilocybin production, up to 1.16 g/L, in *E. coli* using gene sequences originating from the most common recreationally used psilocybin mushroom, *Psilocybe cubensis*. Here we attempt to expand our biosynthetic toolbox by testing PsiD, PsiK, and PsiM variants from 3 additional mushroom species: *Gymnopilus dilepis, Panaeolus cyanescens*, and *Psilocybe cyanescens*. These variants were identified by sequence homology to the original *P. cubensis* variant. The gene variants were cloned randomly into a biosynthetic library consisting of 64 (4^3) pathway combinations, four total species variants of three genes. Additionally, a smaller library of only 2 genes, PsiD and PsiK, was assembled and screened for norbaeocystin production. Both libraries yielded top producers that outperformed the production metrics of the original *P. cubensis* control in small scale studies. The top gene combinations were subjected to promoter optimization to further improve biocatalytic performance. Building upon this work, scale up shake flask studies were performed to fully evaluate the potential of the genetically superior strains developed in this work.

Development of a Robust Pipeline for Synthesizing Various Components of CRISPR/Cas

Sara Olsen, Lilia Yang, Piyush Jain

CRISPR/Cas systems have emerged as powerful gene editing tools. With CRISPR/Cas we have the power to correct genetic mutations and thereby cure many as of yet incurable diseases. However, it has been difficult to find a reliable and efficient way to deliver CRISPR/Cas complexes in mammalian cells, as the large ribonucleoprotein complexes make the delivery challenging. The current best method is to deliver Cas proteins, is by transfecting cells with a plasmid, however they are very large with uptake issues. However, by delivering an mRNA coding for CRISPR/Cas proteins some of the current problems can be alleviated such as the size issue, as mRNAs are much smaller than plasmids (1). Furthermore, the mRNA delivery approach with lipid nanoparticles has been shown to be safe and effective with vaccines during the COVID-19 pandemic (2). To investigate the efficacy of this method of expressing Cas proteins, a DNA fragment that encodes for enhanced green fluorescent protein (EGFP) was transcribed into an mRNA, then transfected into HEK 293T cells, and monitored for protein expression using fluorescence microscopy. We tested three different mRNA constructs: 1) created straight from the purchased DNA fragment, 2) synthesize from that purchased fragment that underwent polymerase chain reaction (PCR), and 3) stabilized with pseudo-uridine made straight from the purchased gene fragment. It was found that the EGFP mRNA construct that was stabilized with the pseudo-uridine produced the brightest fluorescence after just 17 hours with ~90% of the cells glowing green. We also found that by day 8 almost all the green fluorescence disappeared, and by day 11 it had completely vanished, indicating that the EGFP protein had fully degraded. In conclusion, we have demonstrated potential efficacy of transiently delivering a Cas protein as an mRNA in mammalian cells by showing success in expression and degradation of the EGFP transfect. We are currently applying this approach to various CRISPR/Cas proteins.

References:

(1) Stadelmann C. Molecular Therapy – Nucleic Acids. 2022; 28; 47-57

(2) Chaudhary N. Nat Rev Drug Discov. 2021; 20(11): 817-838

Efficient Skin Delivery of Long-Acting Contraceptive from a Microneedle Patch

Nicolas Kelhofer, Gülçin Arslan Azizoğlu, Mark R. Prausnitz, Steven P. Schwendeman, Avantika Dalvi

Introduction: Currently available contraceptive methods are not able to meet the needs of all users, leading to high rates of unintended pregnancy each year. Our team has been developing a novel contraceptive technology that utilizes microneedle (MN) patches[1] aimed at providing effective, long-term contraception, enhancing user accessibility and adherence through self-administration, and ensuring cost-effectiveness for global usage. An important feature of MN patches with biodegradable MNs is the ability of the MNs to rapidly detach from the patch backing upon insertion. In this work, we evaluate the effect of MN height on the detachability of MNs and demonstrate sustained contraceptive hormone delivery from MNs optimized for efficient needle detachment.

Methods: MNs were fabricated by casting a solution of poly (lactic-co-glycolic acid) (PLGA) and a contraceptive hormone, levonorgestrel (LNG), onto polydimethylsiloxane molds to form drug-loaded MNs, the mold allowing for a maximum MN height of 600 µm. A second casting solution of polyvinyl alcohol (PVA) and sucrose was used to form the rest of the MN patch. MN patches were imaged, and the height of drug-loaded MNs was measured. The ability of MN tips to detach upon insertion was evaluated by *ex vivo* pig skin insertion. An *in vitro* release study was conducted to determine the rate of LNG released from the MNs. We then evaluated MN detachment and drug release profile *in vivo* using a rat animal model. Liquid chromatography–mass spectrometry (LC-MS) was used to monitor the plasma concentration of LNG.

Results: More than 90% MN detachment was achieved in pig skin insertion using one cast only either with 8.5 or 10 µL casting amount indicating that the MN height should not exceed more the 2/3 of the total MN shaft height. A total of 6 MN patches were applied to each rat and an average of 90% MN detachment was achieved *in vivo*. MN patches showed release for 21 days *in vitro* and more than 17 days *in vivo*.

Conclusion: MN height is an important factor in the efficient detachment of biodegradable polymer MNs. By determining the optimal MN height, we developed MN patches with a high MN detachment rate which achieved long-term drug release both *in vitro* and *in vivo*.

Acknowledgments: This work was prepared under sub agreements funded by Family Health International (FHI 360) under Cooperative Agreements No. AID-OAA-15-00045 and 7200AA20CA0016, funded by the U.S. Agency for International Development (USAID). The content of this presentation does not necessarily reflect the views, analysis, or policies of FHI 360, or USAID, nor does the mention of trade names, commercial products, or organizations imply endorsement by FHI 360, or USAID. Mark Prausnitz is a co-founder and has a significant financial interest in Micron Biomedical. This conflict of interest is managed by the Georgia Institute of Technology.

References:

(1) Li W et al. J Control Release 2022, 347:489-499.

Development of β-Lactamase Responsive Microneedles for Treatment of Diabetic Foot Ulcer Bacterial Biofilm Infections

Anna Li, Alec Mccall, Nina Hernandez, Anita Shukla

Diabetic foot ulcers (DFUs) often lead to infections, with 60% of those infections associated with biofilm formation. Biofilms represent three-dimensional microbial communities embedded within a matrix of extracellular polymeric substances (EPS). Bacterial biofilms exhibit various antibiotic-resistant mechanisms, including the production of β-lactamases (βLs), which inactivate many common β-lactam antibiotics via hydrolysis. The EPS barrier limits the effectiveness of current topical and systemic treatments, diminishing the efficacy of traditional topical and systemic treatments, thus elevating the risk of antibiotic resistance. A novel antibiotic-loaded microneedle (MN) patch offers an alternative to biofilm infection treatments by using miniaturized needles (<1 mm in height) to puncture the EPS, releasing an antibacterial cargo into the biofilm. To overcome antibiotic resistance, we developed a βL-responsive MN patch containing a βL-responsive crosslinker that can undergo free radical photopolymerization and degrade in the presence of βLs. The MNs will be compared to a non-responsive formulation using poly(ethylene glycol) diacrylate as the crosslinker. To assess the mechanical stability, compression testing and *in vitro* puncture studies were conducted with non-responsive MNs formulated with and without fluorescent nanoparticles as a drug cargo model. At a pre-load of 0.1 N, the compression testing results indicated an exponential increase in compression force at a displacement of ~0.3 mm of the non-responsive MNs. Puncture studies indicated that non-responsive MNs punctured through 2 parafilm layers, with a 47% decrease in holes created between the first and second layers. Visual confirmation of MN formation and its deformation under mechanical stress was provided by scanning electron microscopy and confocal microscopy. Current work is using similar approaches to characterize the βL-responsive MNs.

Non-Invasive Delivery of V_2O_5 Nanowires for the on-Demand Synthesis of Antimicrobials

Shuxian Wen, Sophie Liu, Rong Yang

Otitis media (OM), a prevalent middle ear infection, has been traditionally treated with antibiotics, raising concerns about the rise of antibiotic resistance in children after repeated treatments. The goal is to develop a safe and effective alternative for OM treatment. In previous works, we demonstrated the potent antibacterial properties of vanadium pentoxide nanowires against OM pathogens in a chinchilla animal model, specifically the bacteria *Streptococcus pneumoniae*. However, the effectiveness of these nanowires is hindered by their inability to penetrate the middle ear non-invasively. Their average size over 400 nm poses a challenge for crossing the eardrum — a barrier that is structurally similar to the skin and impermeable to most molecules. This project aims to confer tissue-penetrating capability to the antibacterial vanadium pentoxide nanowires through the conjugation of select tissue-penetrating peptides. By attaching these peptides to the nanowires, we anticipate the enabling of minimally invasive delivery of the nanowires across the eardrum. 1-Ethyl-3-(3-dimethylaminopropyl)carbodiimide / N-hydroxysuccinimide (EDC/NHS) crosslinker chemistry was used to conjugate the vanadium nanowire and the membrane-penetrating peptide. The successful crosslinking was analyzed and validated using X-Ray powder diffraction crystallography (XRPD) and Fourier-transform infrared spectroscopy (FTIR). *Ex vivo* study was performed to examine the membrane-penetrating activity of the peptide-nanowire construct across the eardrum, by infecting the chinchillas (animal model) with OM pathogen and retrieving the bulla bone which contains the ear, to perform time-dependent penetration study of the peptide-nanowire construct. Samples were analyzed using inductively coupled plasma mass spectrometry (ICP-MS) and other analytical methods. The conjugation of peptides to vanadium pentoxide nanowires represents an innovative approach to address this issue by designing a single-dose, antibiotic-free nanomedicine.

Formulation and Characterization of Sting Agonist Implants to Overcome Barriers to Delivery

Julia Treese, Shicheng Yang, Jillian Jurczyszak, Srinivas Sridhar, Needa Brown

Targeting of the stimulator of interferon genes (STING) pathway has emerged as a valuable combinatorial approach for patients with aggressive BRCA-mutated tumors such as ovarian and breast cancers. Combination of STING agonists with conventional clinical therapies such as poly-ADP-ribose polymerase (PARP) inhibitors have shown promise in pre-clinical settings, however translation to the clinic has been limited due to STING agonist drug delivery barriers. Clinical trials of STING agonists require multiple intratumoral injections to deliver a high dose while reducing associated systemic inflammatory responses; however, this delivery method limits treatment options and increases patient burden. We propose an alternative approach using a biodegradable poly (lactic-co-glycolic) acid (PLGA) implant loaded with ADU-S100, a STING agonist, to sustainably release the drug over 28 days and maintain continuous tumor dose.

We have developed a solvent-based evaporation method to fabricate PLGA ADU-S100 implants. Characterization of drug loading and release was evaluated using high performance liquid chromatography (HPLC). To predict how the implants will degrade and release drug over time, a study was performed in phosphate-buffered saline (PBS) at 37°C and pH 7.4 to mimic *in vivo* conditions. Every seven days, buffer samples were collected, and the implants were transferred to fresh PBS. Initial HPLC analysis resulted in overlapping peaks due to similar retention times of free ADU-S100 and PLGA hydrolyzed products, necessitating HPLC optimization. Given that ADU-S100 has a greater affinity for the nonpolar stationary phase than the hydrolyzed products, the ratio of aqueous solvent in the mobile phase was modulated between 50% and 75%. Drug loaded implants were compared against negative control blank implants to validate separation. Increasing the aqueous solvent percentage increased the retention time of the ADU-S100 and resulted in improved separation from the PLGA hydrolyzed products.

Formulation of STING agonist expands the clinical benefit of a cumbersome treatment to previously unattainable patient populations. Further characterization of the implant formulation will be conducted using the optimized HPLC method to assess drug release over time in *in vitro* and *in vivo* models of BRCA-deficient metastatic ovarian cancer. Given that STING agonist is also known to enhance combinatorial drug efficacy, studies evaluating ADU-S100 implants alongside PARP inhibitor implants will also be performed in patient-mimicking metastatic ovarian cancer to study the impact of sustained drug delivery on tumor progression.

The Processing Space of the Spray-dried Mannitol-Leucine System for Pulmonary Drug Delivery

Riley T. Schweizer [1,2], Mani Ordoubadi [3], Cody A. Prather [1], Reinhard Vehring [3], Kimberly B. Shepard [1]

[1] Research & Development, Lonza Group AG, Bend OR 97703, USA;
[2] Department of Chemical, Biological, and Environmental Engineering, Oregon State University, Corvallis, 97331, USA;
[3] Department of Mechanical Engineering, University of Alberta, Edmonton, AB T6G 2E1, Canada;
* Correspondence: kimberly.shepard@lonza.com

Abstract: Designing spray-dried particles for inhalation aims at specific physicochemical properties including a respirable aerodynamic diameter and adequate powder dispersibility. Leucine, an amphiphilic amino acid, has been shown to aid in optimizing bulk powder properties. Mannitol, a model crystalline active and common bulking agent, was co-sprayed with leucine at several excipient ratios, ethanol/water ratios, and spray dryer outlet temperatures in order to experimentally probe the underlying particle formation mechanisms in this binary crystalline system. During the droplet drying of two crystallizing components, the material which nucleates first will preferentially enrich the surface. It is desired to have a completely crystalline leucine shell to improve powder properties, however, mannitol will compete with leucine for the surface depending on excipient concentration and manufacturing parameters. The resulting particles were studied initially and at a two-month timepoint via solid state characterization, visual analysis, and particle size analysis in order to detect changes in bulk powder properties. It was determined that, similar to systems where only leucine can crystallize, leucine saturation in the initial formulation dictates powder characteristics, regardless of ethanol concentration and outlet temperature.

Keywords: spray-drying; leucine; mannitol; polymorphism; powder stability; nucleation

Development and Characterization of Spray-Dried Quercetin-Loaded Nanocomposite Microparticles (Q-nCmP) for the Mitigation of Pulmonary Inflammation

Camila Cersosimo, Amanda Pepler, Daniel Perez Torres, Samantha Meenach

Pulmonary inflammation stimulates the production of reactive oxidative species (ROS) that often initiate chronic pulmonary disorders. Flavonoids are a class of compounds with highly effective ROS-reducing capabilities, and quercetin is a type of flavonoid that has demonstrated a superb ability to mitigate pulmonary inflammation. Unfortunately, quercetin is poorly water-soluble and exhibits low bioavailability, posing challenges in its administration. Therefore, there is a dire need for means to administer quercetin directly to the lungs. Dry powder nanocomposite microparticle (nCmP) formulations show great promise due to their ability to provide targeted delivery of poorly water-soluble drugs to the site of disease. Quercetin-loaded nCmP (Q-nCmP) involve quercetin-loaded nanoparticles (Q-NP) that are encapsulated into microparticles via spray drying. The parent Q-NP dissociate from the Q-nCmP upon deposition in the lungs, delivering the quercetin in a localized and targeted manner. This work aims to characterize Q-nCmP dry powder formulations for their size, homogeneity, antioxidant activity, and aerosol dispersion properties to investigate their potential as a novel pulmonary drug delivery platform. Preliminary results show that Q-NP have significant antioxidant properties comparable to that of unencapsulated quercetin. Future work will determine the aerodynamic properties and drug release profiles of these promising formulations.

Reactive Oxidative Species in Cancer Following Platinum Nanoparticle Treatment

Darshan Danak, Ashish Kokkula, Aida López Ruiz, Kathleen McEnnis

There has been increasing interest in using metallic nanoparticles, such as gold and silver nanoparticles, as therapeutics for cancer treatments. Specifically, platinum nanoparticles (Pt NPs) have become potential cancer therapeutics due to their significant anticancer effects as well as low toxicity to healthy cells (compared to other types of nanoparticles such as Cisplatin). Previous data has shown that Pt NPs are effective against triple negative breast cancer (TNBC) cells. TNBC is a very aggressive and metastatic form of breast cancer. Current literature claims that Pt NPs induce cell death by increasing concentrations of reactive oxidative species (ROS) in extracellular cancer environments. We proposed that, due to platinum's antioxidative properties, the addition of Pt NPs would decrease ROS levels. In this study, we measured the effect of several concentrations of Pt NPs on ROS concentrations in various cell lines. We conducted ROS assays on three separate cancer cell linnes, double positive breast cancer (DPBC) cells, triple negative breast cancer (TNBC) cells, and cisplatin-resistant uterine sarcoma cells, all with differing responses to Pt NPs.

In order to determine the effect of Pt NPs on reactive oxidative species (ROS) concentrations, we used several cell lines including TNBCs, DPBC, and cisplatin-resistant uterine sarcoma. We measured the ROS concentrations after either 1, 3, or 5 days for each cell line after treatment with Pt NPs. There was a significant increase in ROS after day 1 across all cell lines. Across all cell lines, ROS levels decreased after 3 to 5 days even at high concentrations.

The results of this study reflect our hypothesis that Pt NPs decrease ROS levels due to platinum's antioxidative properties. The decrease in ROS levels are indicative of platinum's antioxidant effects, which contradicts current literature. The results and data from this study can aid in the understanding of the mechanism of Pt NPs on ROS levels in TNBC, DPBC, and the cisplatin-resistant uterine sarcoma cells.

Future studies will continue to examine the mechanism of action for Pt NPs in TNBC cells as well as determine their efficacy in mice models.

References: Lopez, A. & McEnnis, K. (2022). Poly(lactic-co-glycolic acid) encapsulated platinum nanoparticles for cancer treatment. *Royal Society of Chemistry*

Acknowledgements: This research was supported, mentored, and overseen at the New Jersey Institute of Technology by Ashish Kokkula and Dr. Kathleen McEnnis.

Fabricating a Flexible Thermoplastic 3D Printed Wrist Support Device for Injury Immobilization and Healing

Sabrina Zhou, Kayla Lee, Amanda Blanca, Mili Shah, Jennifer Weiser

Wrist sprains, fractures, and contusions are common, with an estimated incidence of 896 hand and wrist injuries per 100,000 people yearly[1]. During recovery, most of these injuries require immobilizing the wrist for protecting the injury and promoting faster healing. Today, wrist braces are made using materials that are either too bulky or lack the rigidity needed to immobilize the injured area. Thus, there is a need to design and use a material for a lightweight, rigid brace that still gives the wearer the ability to do everyday tasks.

Inspired by ActivArmor™ , a custom 3D printed support device, the goal of this research is to fabricate a wrist brace using thermoplastic polyurethane (TPU), a 3D-printable polymer. The material shows promise in that it has a smooth finish, abrasion-resistance and is simultaneously rigid and elastic. When 3D-printed thinly, the material can bend and flex, but when printed above a certain thickness, the material becomes stiff and rigid. The TPU cast is created by scanning the wearer's forearm and hand, which is processed and refined in Autodesk Meshmixer to model and generate a cast with a vertical clamshell opening. This is then printed at a filament temperature of 225°C and 20% infill. The cast features a Voronoi pattern, a computer-generated tessellation pattern that removes the maximum amount of material from the cast body possible while still maintaining stability, for a more breathable and sustainable design. Cylindrical notches are present on both sides of the clamshell opening and connected across by O-rings to create a closing mechanism.

The lightweight, Voronoi-patterned TPU cast is flexible enough to allow the cast to be opened and put on while simultaneously being rigid enough to hold the cast's shape and the wearer's wrist in place once secured. Additionally, the cast's design allows for a one-handed open/close mechanism. For this specific cast, the use of a 3D scan allows for a custom fit around the wearer's forearm and wrist. Using a 3D print compatible material allows for cost-effective and environmentally friendly manufacturing, as the brace is removable, washable, and has the potential for long-term use.

[1] Gordon, A. M.; Malik, A. T.; Goyal, K. S. Trends of Hand Injuries Presenting to US Emergency Departments: A 10-Year National Analysis. The American Journal of Emergency Medicine 2021, 50, 466–471. https://doi.org/10.1016/j.ajem.2021.08.059.

Optimizing Natural Killer (NK) Cell Labeling Using Magnetic Particle Imaging (MPI) Tailored Tracers

Nahis Cruz Delgado, Bo Yu, Suzanne Lightsey, Blanka Sharma, Carlos Rinaldi-Ramos

Immunotherapy is one of the most promising cancer treatments available. Despite their incredible promise, clinically approved T-cell immunotherapies present considerable challenges such as accessibility limitations due to their autologous nature, as well as the potential treatment-related toxicities. As such, natural killer (NK) cells are being studied as an alternative with unique advantages. Being the major cytotoxic effectors in the innate immune system, NK cells provide immune surveillance and can induce an immune response without the need for priming. Additionally, NK cells do not carry a risk of graft-versus-host disease, making them attractive as an allogenic source and a more feasible alternative for clinical settings.

To better understand the mechanisms of NK cell therapies, there is a need to track the localization and persistence of these cells after their delivery. Magnetic particle imaging (MPI) is a technology that allows non-invasive visualization of superparamagnetic iron oxide nanoparticles (SPIONs), allowing tracking of SPION-labeled cells. This imaging modality does not use ionizing radiation and has proven to have signal intensity proportional to SPION's mass. The objective of this research was to investigate the feasibility of labeling NK cells using SPIONs tailored for MPI.

NK-92 cells, were labeled using RL-1 PMAO-FL. This RL-1 tracer is engineered for MPI sensitivity, coated with a charged polymer (PMAO) that enhances cell uptake and tagged with fluorseceinamine (FL) to allow their microscopy visualization. NK-92 cells were incubated with tracers at different incubation times, concentrations, and media conditions. Cell viability, as well as cell uptake were compared between conditions. Fluorescence imaging was also performed, to confirm uptake and determine intracellular localization of the tracers.

It was observed that the NK-92 cells incubated with RL-1 PMAO-FL for 2 hours at a concentration of 25 μgFe/mL in low-serum media had better uptake of the tracers. Cell viability was not significantly affected by the different tracer concentrations and incubation times studied. In contrast, observations showed that media conditions seemed to pose an inversely proportional relationship between cellular uptake and cell viability, an interaction that will have to be further studied. Microscopy images suggested tracer internalization in the cells. Microscopy images suggested tracer internalization in the cells. Together this study indicates that NK-92 cells can be labeled using RL-1 PMAO FL particles and traced using both confocal imaging and MPI. Future efforts will include measuring NK cell functionality after labeling, as well as experiments validating MPI tracking of NK cells *in vivo*.

Designing Nanoparticle-Loaded Aerosols for Controlled Interactions with Pulmonary Immune Cells for Treatment of Respiratory Diseases

Nicole Gill, Michael Trautmann-Rodriguez, Catherine Fromen

Approximately 544.9 million people globally are diagnosed with chronic respiratory diseases, which has increased 39.8% since 1990 (Lancet Respiratory Medicine, 2020). Aerosolized drugs that can yield controlled effects on the pulmonary immune system may afford new opportunities to treat such diseases. In this research, we investigate the role of nanoparticle (NP) formulations as respirable drug delivery carriers. We first investigate how the phagocytosis of NPs affects the intracellular signaling of pulmonary immune cells. Cell survival studies using bone marrow-derived macrophages (BMMs) dosed with inert polyethylene glycol (PEG) NPs were conducted to derive a relationship between NP uptake, cell secretion, and cell survival. We demonstrated that NP-dosed BMMs drive secretion of pro-survival signals that increases the lifespan of untreated BMMs. To deliver these PEG NPs to the lung, we next explored the physical effects of NP loading into nebulized aerosols to create respirable aerosols with tunable sizes. Formulations of different concentrations of NPs were aerosolized with an Aeroneb vibrating mesh nebulizer and characterized by laser diffraction. We find that at a critical NP concentration, the aerosol volumetric median diameter increased upwards of ~150%. On-going work includes combining these effects to study bio-active NPs for immune and gene delivery applications.

An Improved Strain for Aromatic Aldehyde Stability and Incorporation: Recoded RARE

Miyu Mudalamane, Neil Butler, Aditya Kunjapur

Aromatic aldehydes find widespread use as end products in fragrances and flavorings as well as intermediates in the synthesis of other valuable compounds. When incorporated into proteins via non-standard amino acids (nsAAs), aromatic aldehydes can also act as clickable handles for conjugation to hydrazine or hydrazide molecules. However, chemical synthesis of these compounds is complex and utilizes nonrenewable petroleum feedstocks. Although biosynthesis of aromatic aldehydes could help reduce reliance on petroleum, these pathways are limited by the instability of aromatic aldehydes within microbes due to endogenous oxidoreductase enzymes. In addition, nsAA incorporation in wild-type *E. coli* is hindered by competition with native translation termination systems. The reduced aromatic aldehyde reduction (RARE) strain ameliorates stability issues through translational knockouts of selected aldo-keto reductases and alcohol dehydrogenases in *E. coli* K-12 MG1655. Another engineered chassis, the C321 strain, has all amber UAG stop codons replaced with ochre stop codons UAA. This leaves the UAG codon free for translation by orthogonal aminoacyl-tRNA synthetases (AARSs)/tRNAs. The recoded RARE strain (RR3) combines the features of the RARE and C321 strains for improved incorporation of non-standard amino acids with aldehyde functionalities, along with targeted mutations to improve growth compared to C321. In this work, we sought to evaluate the recoded RARE strain's stabilization of aromatic aldehydes and incorporation of selected nsAAs. RR3 and C321 cultures were supplemented with 1 mM *para*- and *meta*-formylphenylalanine (formylPhe) and monitored over 24 hours to compare the stability of nsAAs in each strain. Subsequent analysis via high performance liquid chromatography revealed up to 5-fold improvements in the retention of formylPhe after 24 hours in RR3 compared to C321. Incorporation of a well-studied nsAA, pAzF, into a fluorescent reporter protein, GFP, was monitored via fluorescence to compare the incorporation efficiency of three strains: RR3, C321, and BL21, a widely used microbial host for protein overexpression. When compared to BL21, the RR3 strain showed an up to 20-fold increase in fluorescence and up to 8-fold increase compared to C321, indicating an improved incorporation efficiency. We further tested a library of AARSs within RR3 and C321 for activity on *para*- and *meta*-formylPhe. After identifying a variant with activity on *para*-formylPhe, we were able to demonstrate improved incorporation of the nsAA into recoded GFP within the RR3 chassis compared to C321. The recoded-RARE strain presents an improved bacterial strain for incorporation of aromatic aldehyde non-standard amino acids and derivatives, ideal for genetic code expansion applications.

Discovering Darpin Inhibitors Against Disease-Relevant Targets for Potential Therapeutic Development.

Ethan Slaton, Samantha Martinusen, Marian Pulgar, Carl Denard

Proteases are involved in several biological processes and, upon dysregulation, often contribute to disease onset and progression, making them critical therapeutic targets. Strategies for targeting protease therapeutics using small molecule inhibitors often rely on active site binding. Due to active sites being conserved across protease families, this strategy can lead to off-target toxicity, resulting in unintended consequences. Protein binders have been found to modulate proteases. Thanks to the large surface area of interaction, protein-based binders can modulate proteases in ways small molecules cannot. These include highly specific and selective inhibitors, activators, enhancers, and substrate-selective reprogrammers. However, isolating such modalities remains challenging because they are often discovered from high-throughput binding assays. To address this, our lab has developed a High-throughput Activity screen for the functional Reprogramming of Proteases (HARP). HARP is a yeast surface display system that allows us to measure the activity of a protease. HARP provides us with the ability to engineer and discover protein binders that can reprogram the activity of proteases. These discoveries will enable us to understand protease mechanisms, establish their roles in disease, and help develop new novel targeted therapeutics.

We have applied our approach to discover tight binding proteases inhibitory Designed Ankyrin Repeat Proteins (DARPins) from randomly generated libraries. DARPins are genetically engineered protein binders with exceptionally high binding affinity and specificity. DARPins against a protein target are typically isolated using high-throughput screens such as ribosome display, which could potentially lead to discarding strong protease inhibitors. We show that we can isolate TEV inhibitory DARPins from randomized libraries and perform affinity maturation of weak inhibitors via error-prone PCR followed by fluorescence-activated cell sorting (FACS). Using this approach, we will be able to identify DARPins that are substrate-selective against disease-relevant proteases to potentially create new therapeutics. Our approach can fully enlist the potential of DARPins as protease modulators and provide the necessary foundation for modulator design.

Dihydromyricetin (DHM) Increases Lipid Clearance and Mitochondrial Function in Mice Exposed to Long-Term Ethanol Consumption

Eva Jones, Jinah Kim, Karina Hiranandani, Samantha Skinner, Isis Janilkarn-Urena, Daryl Davies

Alcohol-associated liver disease (ALD) is one of the most prevalent liver diseases in the United States and is caused by long-term, heavy alcohol consumption. ALD is characterized by three major stages, progressing from steatosis to steatohepatitis/fibrosis, to finally cirrhosis, an end-stage liver disease. Unfortunately, there are no effective treatments for ALD besides abstinence, or in extreme cases, a liver transplant. Corticosteroids can be used to treat hepatic inflammation associated with steatohepatitis, but long-term use is associated with adverse side effects and risk of infection. The lack of effective ALD pharmacotherapies necessitates the need for novel therapeutics. Dihydromyricetin (DHM) is a bioactive flavonoid derived from a variety of plants, particularly *Holvenia dulcis*, that has demonstrated anti-alcohol effects. Chronic alcohol consumption has been shown to induce dyslipidemia and impair mitochondrial health. Promising evidence has shown that DHM produces hepatoprotective effects. In this study, we demonstrate the effect of DHM on lipid droplet accumulation and mitochondrial function. Mitochondrial proliferation and increased activity of critical enzymes involved in mitophagy can regulate healthy mitochondria function. Using an established ALD mouse model, the Lieber DeCarli diet, we measured the effect of DHM supplementation on mitochondrial health and lipid clearance. We observed and quantified the changes in lipid size and concentration and found that DHM supplementation led to an increase in lipid droplet clearance and heterogeneity. In addition, we looked at mitochondrial health and proteins associated with mitophagy. With this evidence, we determined that DHM supplementation improved mitochondrial metabolic function, with data suggesting an increase in mitophagy. Taken together, this data demonstrates that DHM can increase lipid clearance and mitochondrial efficiency as demonstrated by a decrease in lipid count and improved metabolic function and health. This evidence will support future research investigating DHM as a promising therapeutic to treat ALD by providing insight into DHM's role in mitigating mitochondrial damage and hepatic lipid accumulation.

High-Throughput Single-Cell Sequencing of *B. Fragilis* Populations in the Mouse Gut

Christine Stark, Ophelia Venturelli, Freeman Lan

High-throughput single-cell sequencing of *B. fragilis* populations in the mouse gut

[1] Christine Stark, Freeman Lan [2], Ophelia Venturelli [2]
1. North Carolina State University, 2. University of Wisconsin - Madison

Background:
B. fragilis is a gram-negative gut commensal with eight capsular polysaccharides which could be expressed depending upon promoter expression. Capsular polysaccharides are essential as they contribute to the harmfulness of many gut pathogens as these capsular polysaccharides are able to hide cell-surface components which could trigger an immune response from the host [1]. This prompts the question regarding which capsular polysaccharides are selected for when the host undergoes various stressors. Through the use of high-throughput single-cell sequencing, promoter inversion and show was determined to be diverse in *B. fragilis* populations due to many capsular polysaccharide promoter states [2].

Results:
On day 29 the mice were inoculated with communities, thus increasing competition and selection within the mouse gut; consequently, variation decreased. By day 33 compositions with initially high capsular polysaccharide A proportions begin to experience less selection. This indicates a chance that capsular polysaccharide A does not play a significant role in the mouse gut. Furthermore, by day 33 those with low capsular polysaccharide H proportions begin to experience increased selection; thus, indicating the plausibility of capsular polysaccharide H playing a prevalent role in the mouse gut. Furthermore, capsular polysaccharides B and E seem to have a relationship as both proportions individually are low in the mouse gut as time progresses; however, when capsular polysaccharides B and E promoters are turned on in the same composition the proportion is exceptionally high. These results were confirmed when sequencing of the biological replicate occurred. On day 43, ATC is introduced into the mice's water supply. Furthermore, the introduction of ATC induced promoter scrambling resulting in increased variation.

Conclusions:
When looking at the data as a whole there is a chance that skewing occurred in the analysis process as those compositions containing ambiguities either in the form of "X" or "U" were removed. This could greatly affect the proportions and could account for some of the unexpected results. The next steps for this project include running the DoTA-seq workflow on the other experimental groups. Once compositions are obtained for all the experimental groups big picture conclusions can be made in regard to how the state/health of the mouse gut impacts capsular polysaccharide composition proportions.

Reparative Mechanisms of STEM Cell-Based Therapy Using Extracellular Vesicles Derived from Human Placental STEM Cells for Necrotizing Enterocolitis

Michelle Ebu, Maryssa Ellison, Arion Dey, Victoria Weis

Necrotizing Enterocolitis (NEC) is a life-threatening gastrointestinal disease that primarily affects premature infants born at 34 weeks or less. It is characterized by the infection, inflammation, and death of intestinal tissues in the ileum of the small intestine. Though the etiology and pathogenesis of NEC remain unknown, several factors such as ingestion of formula instead of breast milk, hypoxic injury, and bacterial invasion, have been shown to contribute to the development of this disease. To date, therapy options have been limited to surgical and medical management, often leading to subsequent issues such as short bowel syndrome in the infant's future.

Recent cell therapy studies show that using the paracrine signaling mechanism, human placental derived stem cells (hPSC) therapy promotes epithelial healing of NEC intestinal damage. Although this has proven to be an effective therapeutic, the use of Extracellular Vesicles (EVs) produced from hPSCs provides a novel cell-free therapeutic approach for optimized clinical translation. In this study, we tested the efficacy of hPSC-EVs in ameliorating the NEC disease damage to proliferation, progenitor stem cells, LGR5+ active intestinal stem cells, and differentiated epithelial cells.

We used breastfed (BF), diseased (NEC), and treated (NEC-EV) ileal tissues from newborn Sprague-Dawley rat pups for our investigation. We utilized cell-specific markers through immunofluorescence staining to quantify changes in cell number using the GNU Image Manipulation Program (GIMP) and a newly developed ImageJ plug-in. We hypothesized that EV therapy will promote the inherent mechanism of the small intestine to repair epithelial damage from NEC disease, specifically within particular cell types. We quantified the following markers: Ki67 (proliferation), SOX9 (intestinal progenitor/stem cells), OLFM4 (LGR5+ intestinal stem cells), and Ezrin (enterocytes). BF, NEC, and NEC-EV ileums were immunostained with specific antibodies for each marker. The number and location of positively stained cells were quantified. From this analysis, we were able to identify changes within proliferation, progenitor stem cells, LGR5+ active intestinal stem cells, and differentiated cell types in the NEC and NEC-EV tissues. Additionally, we were able to quantify the distribution of each marker along the crypt-villus axis.

Based on our findings and past studies, we observed that EVs derived from human placental stem cells are an effective therapeutic for Necrotizing Enterocolitis. Future studies will focus on the additional immune modulation benefits of this EV therapy, providing a better understanding of the mechanisms of action for this therapeutic approach and guiding future translation in the NICU.

Controlling Disulfide Formation and Dimerization to Constrain Alternate Peptide Conformations

Clara Victorio, Nicholas Sawyer

Disulfide bonds form covalent bonds between distal regions of peptides and proteins to dramatically impact their folding, stability, and oligomerization. Given the prevalence of disulfide bonds in many natural products, considerable effort has been invested in site-selective disulfide bond formation approaches to control folding of chemically synthesized peptides and proteins. Here we show that careful choice of thiol oxidation conditions can lead to monomeric or dimeric species from fully deprotected linear bisthiol peptides. Examination of reaction conditions indicates that the use of disulfiram (DSF) under denaturing conditions leads to the formation of intramolecular disulfide bonds while reactions with dimethyl sulfoxide (DMSO) under aqueous conditions yields disulfide-linked, antiparallel dimeric species. Intramolecular disulfide formation using DSF and denaturing conditions proved to be sequence-independent. On the other hand, the formation of antiparallel dimers is sequence-dependent and requires partial α-helical folding of the linear peptide and the presence of a hydrophobic triad to encourage peptide association. Treatment with Proteinase K indicates that both disulfide species demonstrate enhanced protease stability compared to the linear precursor. The half-life increases by a factor of 100 for the dimer and even greater for the intramolecular disulfide. Additionally, both disulfide species are fully reduced by tris(2-carboxyethyl)phosphine (TCEP) within 15 minutes to regain the original α-helical structure for targeting proteins. These results provide an approach for using disulfide bonds to control peptide folding and oligomerization to better understand how these properties influence interactions with protein and membrane targets.[1]

1) Victorio, C.G. and Sawyer, N. (2023) Folding-Assisted Peptide Disulfide Formation and Dimerization. ACS Chem. Biol. 18, 1480. DOI: 10.1021/acschembio.3c00268

Silk Sponges As *in Vitro* Skeletal Muscle Tissue Model

Kayleigh Trumbull, Elizabeth Aikman, Asha Rao, Emily Fussell, Whitney Stoppel

Introduction: Limb Girdle Muscular Dystrophies (LGMD) encompass over 30 genetic mutations leading to degenerative muscle conditions. Limited patient data hampers LGMD study and treatment. Silk sponges are a biomaterial that can stimulate muscle cell growth *in vitro.*[1,2] We hypothesize introducing mechanical stimulation to myoblasts during growth will aid functional muscle tissue development. This approach holds potential for enhanced fundamental understanding and treatment strategies for diseases like LGMD.

Materials & Methods: Aqueous silk fibroin from *Bombyx mori* cocoons was processed using existing protocols for boiling and solubilization.[3] The silk solution, diluted to 3% and mixed with porcine muscle decellularized extracellular matrix (dECM), was frozen in an anisotropic manner using a dry ice and ethanol slurry.[4] After lyophilization (Labconco) and water annealing, the sponges were imaged via SEM (ThermoFischer Scientific) and mechanically evaluated via a tensile test using a rheometer (Anton Parr) with custom immersion cup to maintain hydration of samples.[5] Sterile sponges were seeded with healthy human skeletal muscle myoblasts (HSMMs, Lonza®), incubated in growth media (GM) at 37°C for 8 days and subsequently fixed for analysis. The sponges were wax embedded and sectioned prior to H&E staining and imaging (Keyence) to assess cell integration.

Results & Discussion: SEM images show the sponges exhibit uniform and aligned pores that resemble the structure of mature muscle tissue formed by natural mechanical forces. Tensile tests simulate these forces and strain rates, producing comparable results for sponges composed of only silk and made of silk and dECM in various media conditions. The Young's modulus of each sample type showed no significant difference at low strains, indicating that the sponges' viscoelastic properties for applications in skeletal muscle tissue engineering are unaffected by ingredients in GM. However, at high strain rates, the presence of salts within the media did impact elastic properties. Staining HSMM-seeded dECM sponges showed widespread distribution of cells within the sponges, indicating cell migration from the surface. Future work includes putting HSMM-seeded sponges on a bioreactor to mechanically stimulate the myoblasts and encourage their maturation. These results show promising progress toward generating a human primary cell-derived *in vitro* muscle tissue model that can be customized to accommodate diseased cells, thereby advancing knowledge and therapeutic strategies for rare muscle diseases such as LGMD.

Modeling the Impact of Pediatric Tonsil Size on Aerosol Deposition and Drug Delivery

Saurav Padhye, Catherine Fromen

Current pediatric applications of aerosolized medication consist of off-label prescription of adult medications to scale down the dosage for children; however, studies have shown that there are significant anatomical differences between the airways of pediatric and adult subjects, which could potentially lead to off-target aerosol deposition and decreased drug delivery. *In vitro* and *in silico* tools provide innovative ways to model aerosol behavior without the ethical concerns surrounding *in vivo* testing with pediatric patients. Recent papers from Kolewe et al. study the impacts of this changing anatomy using computational fluid-particle dynamics (CFPD); however, these studies focus on the geometric changes of the airway during development. Other key organs throughout the airway may also have significant impacts on flow development and aerosol transport. The tonsils are one such organ. Tonsil hypertrophy is a common occurrence during early development, whether it is related to tonsillar infections or tissue development. Understanding both flow and transport of aerosols at various levels of tonsil size would allow medications to more effectively target the tonsils or lower lobes of the lung depending on need, leading to maximal bioavailability at the intended tissue. This work aims to extend previous work from the Fromen lab in modeling pediatric aerosol behavior by adding variable tonsil sizes. This consists of 1) creating an anatomically-accurate in vitro model of the pediatric upper airway for each of the four tonsil grades in the Brodsky scale, 2) characterizing nebulized aerosol deposition patterns in the models, and 3) modeling the same aerosol deposition computationally to cross-examine apparent trends. Currently, a 6-year-old airway has been 3D printed using a Carbon 3D digital light synthesis printer, with a focus on modular design for printing and analysis. Pilot studies are also being conducted on using an Aeroneb nebulizer and a next-generation impactor to deliver aerosolized rhodamine b to the model and size the effluent aerosol particles. Thus far, resultant aerosol sizing aligns with the manufacturer-specified particle size, although this procedure is still being refined. Computational data has been recorded for this geometry, which will be used to benchmark the *in vitro* findings. Future work will include modeling different levels of tonsillar swelling and measuring the impact on deposition, as well as potentially expanding this study to a wider range of patients.

Determining Regions in USP15 That Regulate Its Stability and Turnover in Ovarian Cancer Cells

Pavankumar Umashankar, Noel Amadu, Achuth Padmanabhan

Due to the lack of early diagnostic markers and effective therapeutics, human ovarian cancer remains the most lethal gynecologic malignancy. This unfortunate clinical reality highlights the urgent need to identify novel therapeutic targets and more effective treatment strategies. Recent studies show that the deubiquitinase USP15 plays an important role in ovarian cancer progression and therapeutic response. Despite this, mechanisms that govern USP15 stability and turnover remain unknown. Our project aims to address this knowledge-gap by identifying regions in USP15 that regulate its stability in ovarian cancer cells. To this aim, we are cloning full-length USP15 and truncation mutants with either N- or C-terminal epitope tags into the mammalian expression vector, pcDNA3.1. Truncation mutants are generated by sequential deletion of USP15 domains, which will allow us to pin-point the regions in USP15 that are most crucial to regulating USP15 stability. Subsequently, cloned plasmids are transfected into human ovarian cancer cell lines and the half-life of the full length and truncation mutants is measured using the cycloheximide chase assay. Furthermore, a small-molecule drug called MCB-613 was previously identified to impact USP15 stability. We will also determine how MCB-613 impacts the stability of the full-length USP15 and the truncation mutants using western blot and mass spectrometry analyses. Currently, we have successfully cloned several of the USP15 mammalian expression constructs and are optimizing their transfection into ovarian cancer cell lines. The above structure-function studies will determine mechanisms and regions in USP15 that regulate USP15 stability, ultimately enabling the development of new therapeutic strategies that target USP15 in ovarian cancer.

Investigating the Impact of Titanium Dioxide Nanoparticles on Wound Healing

Briman Yang

Approaches in the study of wound healing attempts to promote and accelerate either of the four stages in wound healing: homeostasis, inflammation, proliferation and remodeling. An approach in developing wound dressing is titanium dioxide (TiO2) nanotechnology-based interventions, where its antimicrobial properties are utilized in creating nanocomposite wound dressings. While there has been many studies in accelerating the inflammation phase, little has been researched on promoting the proliferation and remodeling phase of wound healing. Thus, this study focuses on the TiO2 nanoparticles on human umbilical vein endothelial cells (HUVEC) angiogenesis and impact of exposed rutile TiO2 nanoparticles on human skin organotypic was studied on a cellular and tissue level.

To test for TiO2 cytotoxicity, cells were treated at a concentration of 0.4 mg/mL rutile TiO2 particles for 24 hours prior to the tests. Fibroblast and keratinocyte cell proliferation assays were then evaluated for 5 days after treatment. Data shows that the presence of TiO2 nanoparticles did not significantly impact the proliferate rate of either keratinocyte or fibroblast.

For HUVEC angiogenesis assay, after the TiO2 treatments, HUVECS were sorted through fluorescence activated cell sorting. Cells sorted with a high scattering angle window were assumed to have an increase in TiO2 uptake. HUVECs with and without TiO2 underwent 18 hours of vascular network formation on Matrigel. Quantification of network formation of HUVECs cells using EVOS microscope shows greater vessel network density in the treated sample, suggesting promotion of angiogenesis.

Lastly, wound healing in treatments of TiO2 nude mice skin grafts were studied. 8 millimeter diameter puncture wounds was inflicted posteriorly on both mice A) with transplanted vascularized autografts as control and mice B) 20 microliters of 0.4 mg/mL TiO2 placed onto the wound followed by the transplantation of vascularized autograft. Through post-autograft histology of both mice, TiO2 did not affect the fibroblast's ability to support the epidermis and there were nanoparticles aggregation near the stratum corneum for TiO2 treated skin equivalents.

In conclusion, the data demonstrated that TiO2 nanoparticles do not interfere with the skin wound healing process, rather promote angiogenesis. TiO2 not only did not have a negative impact on the functionality of the epidermis, but also showed evidence that TiO2 can be removed from tissue through epidermis differentiation.

Application of Mhetase-Displaying Whole-Cell Biocatalysts for Conversion of PET-Degradation Intermediates

Brian Ortiz, Siddhant Gulati, Eric Imhoff, Qing Sun

Enzymatic depolymerization is a promising solution to manage polyethylene terephthalate (PET) waste.[1] Complete PET enzymatic degradation to terephthalic acid (TPA) and ethylene glycol (EG) needs the synergistic action of two enzymes, PETase and MHETase.[2] MHETase facilitates conversion of MHET, an intermediate in the PETase-based enzymatic degradation process, to TPA, and prevents inhibition of PETase activity. Surface display of PET-degrading enzymes is an attractive approach to achieve improved activity and eliminate the need of protein purification, but current work has focused on the surface display of only one PET-degrading enzyme.[3] Thus, complete PET depolymerization has not been achieved. Our hypothesis is that an engineered bacterial community can be developed for complete PET depolymerization, with one part of the community displaying MHETase.

In this work, whole cell biocatalysts displaying PET-degrading enzyme MHETase on the surface of *Escherichia coli (E. coli)* was constructed. MHETase was surface displayed on *E. coli* using transporter proteins YeeJ[3] and AIDA-I[4] because of their demonstrated ability to efficiently surface display heterologous passenger proteins. After expression, activity of surface displayed MHETase was determined by incubating cells with dissolved MHET to confirm the activity of both AIDA-I-MHETase and YeeJ-MHETase via HPLC analysis. This work is an essential step in achieving complete PET depolymerization to its constituent monomers which can be upcycled, potentially by other members of the bacterial community or through other methods.

Editing a Zinc Finger Inducible Promoter Via Recombinases to Change Expression

Yunbeen Bae, Sneha Kabaria, Kate Galloway

Engineering mammalian cells provides key insight into translatable research, but it requires precise control of gene expression to recapitulate and manipulate biology. Furthermore, natural systems often exhibit expression changes over time which can be difficult to precisely control with current synthetic biology tools.

Toolkits like Composable Mammalian Elements of Transcription (COMET) provide design rules to vary the strength of a zinc finger (ZF) dependent promoter, but their features are often statically encoded. Furthermore, current methods of gene expression control are binary or transient, limiting the extent to which we can engineer and recast biological systems.

Therefore, we explore a new method that can introduce a stable change in gene expression between two non-zero levels upon a stimulus. To this end, we apply the COMET toolkit and the Cre-Lox recombinase system to edit the DNA length between the last ZF binding site and the TATA box to increase gene expression.

In transient transfection, excising the spacer causes various fold increases in the downstream reporter gene expression, which corresponds with the previously established COMET rules. We showed that longer spacer lengths start at lower levels of gene expression, but all reporters converge at the same expression level after excision.

Due to toxicity and challenges associated with delivering multiple plasmids, we sought to consolidate the ZF expression and the reporter gene into one cassette. Despite maintaining the same promoter architecture, we found that differences in DNA syntax affect expression profiles of both ZF and reporter gene. Tandem expression yielded the best expression pattern, with overall highest expression levels and no bimodal distribution seen with other orientations.

In addition to varying the spacer length, ZF expression can modify the output gene expression level. This modularity of our design allows multiple ways to alter the reporter gene expression, which can enable multiplexing a variety of inputs to confer downstream changes. Using multiple parameters can be useful in constructing gene circuits with editable setpoint levels. To demonstrate the utility of both ZF-expression mediated expression with editing spacer length, we hope to use these promoters to change the setpoint of a previously published antisense-based integral controller.

This work using recombinases to edit the DNA composition of a ZF-dependent promoter allows heritable changes to tune gene expression. Altogether, these designs can add to the synthetic biology toolkit by facilitating complex biological behaviors in mammalian cell engineering.

The Regulation of Histone H2A.X and DNA Damage Repair By Environmental PFAS Exposure and the Activation of ER Stress Sensor IRE1α.

Elaina Gouin, Kevin Chen, Sean Foster, Caleb Sandum, Sardar Murtaza, S. Patrick Walton, Christina Chan

Per- and polyfluoroalkyl substances (PFAS) are man-made chemicals found in household cleaning products and packaging. They enter the environment via waste from manufacturing facilities or metabolism in organisms. Thus, they have been accumulating in drinking water, food, and human bodies.

High concentrations of two PFAS chemicals, perfluorooctanoic acid (PFOA) and perfluorooctane sulfonate (PFOS), may be linked to cancer by inducing endoplasmic reticulum (ER) stress in cancer cells. In response to this stress, cancer cells utilize cellular survival mechanisms including the unfolded protein response (UPR) and DNA damage repair (DDR) to survive, metastasize, and resist treatment.

One signal protein of the UPR, inositol-requiring enzyme 1 alpha (IRE1α), mediates ER stress via X-box binding protein-1 (XBP1) mRNA splicing and regulated IRE1-dependent decay (RIDD). The DDR mediates genotoxic stress by repairing DNA damage like double stranded breaks (DSBs). An early marker of damage is the phosphorylation of histone H2A.x to γH2A.x. We hypothesized that the activation of the IRE1 pathway by PFOA and PFOS affects the DDR and cellular repair mechanisms.

The results have shown an increase in DNA damage repair in response to PFOA and PFOS induced DNA damage. This research is instrumental to understanding health risks caused by PFAS chemicals from the environment.

Development of Enhanced Transformation and Promoters to Engineer an Environmental-Isolate of *Priestia Megaterium*.

Paul Carter III, Yoseb Song, Elaine Reece, Kristala Prather, Jason Boock

Engineering microbes for sustainable bioproduction is a promising avenue to tackle problems in pollution, medicine, and energy. Traditional model organisms have received much attention due to their easily manipulated genetics and well characterized growth. However, there are a wealth of non-model microbes with attributes that are beneficial for biotechnological applications, but genetic tools must first be developed to engineer them by incorporating foreign DNA and characterizing promoters for controlled expression. In this work, we are looking to expand the genetic toolbox for an environmental-isolate of *Priestia megaterium*, called SR7, that can survive under high pressures of CO_2, simultaneously creating a sterile culture environment as well as extracting bioproducts as they are formed. For these reasons, SR7 was previously engineered to produce isobutanol, a second-generation biofuel, under supercritical CO_2.

Genetic manipulation of SR7 was previously possible using a protoplast-osmotic shock method, but was highly inefficient, necessitating the development of an enhanced transformation method. Plasmid methylation was controlled using a strain of *E. coli* which lacks native methylation (*dam, dcm mcrA, mcrC*) genes and contains a methyltransferase gene from SR7. Using the methylation strain as well as osmotic balancing agents, transformation efficiencies have been enhanced, permitting bioparts validation. We next sought to apply constitutive promoters to our bioproduction pathway by mining a transcriptome of SR7 grown in different environments. Several promoters were identified and characterized, finding particularly high expression from the native citrate synthase promoter. This has allowed us to begin experimentally testing gene expression levels which will lead to an optimized set of bioparts to produce isobutanol.

Building Phase Diagrams for Coacervate Systems with Low Volume Samples

Sydney Volheim, Whitney Blocher McTigue

Complex coacervation, a liquid-liquid phase separation phenomenon driven by electrostatics of two oppositely charge polyelectrolytes, is a topic of growing interest in the biomedicine and biological communities. A large part of this investigation focuses on where this complexation occurs, commonly illustrated through binodal curves outlining the two-phase region. With industrial polyelectrolytes, large sample volumes (mililiters) of both the dense liquid phase, the coacervate, and the surrounding dilute phase, the supernatant, can be generated and analyzed. When using biologically relevant polyelectrolytes, such as proteins, enzymes, peptides, etc., small sample volumes (microliters) are obtained. Single large volume samples may be made to generate a handful of points to create a binodal curve, though without the same degree of detail as binodal curves utilizing industrial polymers. Here, we are developing a method to utilize small sample volumes focusing on the supernatant phase with absorbance and conductivity readings to create a more detailed map of the binodal curve. We initially focused on using poly(styrene sulfonate sodium salt) (PSS) and poly(diallyldimethylammonium chloride) (PDADMAC) to form complex coacervates as a point of comparison between what has been done already in the literature and our method. This process will then be applied to more biologically relevant systems where large sample volumes are difficult to make, specifically poly(L-lysine) and poly(D,L-glutamate). By building these binodal curves, we will have a more accurate representation of where phase separation occurs and can more readily compare how different solution conditions (i.e., salt, pH) change the phase envelope.

CRISPR-Cipher: DNA-Based Data Storage and Hierarchical Encryption Using CRISPR-Cas9

Zoe Fang, Santosh R. Rananaware, Carlos Orosco, Vedant Karalkar, August Bodin, Piyush Jain

We are living in the information age, a time during which massive amounts of digital data are generated every second. Never before has it been more important to save, encrypt, and transmit these data in a way that is secure and efficient. Biological systems, particularly DNA, have very large information densities, presenting unprecedented potential for data storage. Current storage methods that use computers inside of data centers consume enormous amounts of energy, so storing digital information as DNA offers a more sustainable alternative. However, given that DNA could potentially be used to store highly sensitive data, such as medical records or proprietary information, robust encryption methods must be put in place to protect against potential security breaches and ensure data privacy. This is where the CRISPR-Cas9 (**C**lustered **R**egularly **I**nterspaced **S**hort **P**alindromic **R**epeats - CRISPR associated protein 9) system, renowned for its genetic engineering abilities, could be of use in an entirely new field. Using CRISPR-Cas9 to integrate DNA encryption directly into the storage medium represents a pivotal advancement in DNA-based data storage, guaranteeing the preservation of data integrity and confidentiality at a molecular level. Here, we introduce CRISPR-CIPHER (**C**RISPR **I**nduced **P**artitioning for **H**ierarchical **E**ncryption and **R**etrieval), a novel approach to encrypt DNA-based information. Our methodology leverages the precision and programmability inherent in the CRISPR-Cas9 system to partition DNA-encoded information into fragments. The resulting jumbled fragments effectively serve as encrypted messages that can only be correctly deciphered with the corresponding key composed of guide RNA sequences known only to the user. To achieve this, we coded an algorithm that takes alphanumeric data as input and transforms it into DNA format while generating a unique key. The encrypted message can be synthesized and partitioned using the Cas9 enzyme complexed with the RNA sequences outlined in the key. To decode the message, the fragments can be sequenced and decrypted by the same algorithm, when provided the key and sequence information. Our innovative approach merges the fields of synthetic biology and cryptography, forging a novel path towards secure, high-density data storage. Future applications of this work encompass a wide spectrum, ranging from secure transmission of sensitive information to the long-term archival of substantial data in a safe and confidential manner. CRISPR-CIPHER not only introduces a unique application of CRISPR-Cas systems but also pioneers a new genre of biological cryptography.

Reversible Intracellular Gelation Reduces MCF10A Spheroid Growth

Delaney McNally, Laura Macdougall, Bruce Kirkpatrick, Chima Maduka, Benjamin D. Fairbanks, Christopher Bowman, Kristi Anseth

In nature, some organisms can survive in extreme environments by inducing a biostatic state within the organism's molecular structure. Synthetic biostatic states in adherent mammalian cells have been previously achieved via intracellular network formation. This is attained by using bio-orthogonal strain-promoted azide-alkyne cycloaddition (SPAAC) reactions between functionalized poly(ethylene glycol) (PEG) macromers introduced to the cell by a lipofectamine aided transfection. These macromers are able to spontaneously crosslink within the cytosol. In this work, the effects of intracellular network formation in a complex 3D epithelial MCF10A spheroid model are explored. Transfected cells are encapsulated in a proteinous 3D matrix, Matrigel, and overall spheroid area is reduced by ~50% compared to controls. Interestingly, intracellular network formation also induced a change in cell cycle state. Network formation results in a higher quiescent cell population indicated by the loss of phospho-Rb and a gain in p21 expression within single cell analysis of spheroids. After lipofectamine-aided transfection with SPAAC macromers, the formed network reduces overall bioenergetic (ATP/ADP) levels and functional metabolic rates, while also inducing quiescence -like effect in the mitochondrial electron transport chain. These effects are confirmed by Fluorescence Lifetime Imaging Microscopy (FLIM) and Seahorse Cell Metabolic Analysis. To enable reversibility of the observed biostasis effect within the model, a photodegradable nitrobenzyl moiety is incorporated into an azide containing macromer. This allows the PEG network to experience photoinduced degradation. The degraded network allows for continued proliferation and a return to normal spheroid growth. Following light exposure at day 12, growth and metabolic rates return to control levels, while SPAAC treated spheroids that did not receive light exposure (i.e., spheroids containing intact intracellular networks) remain smaller and less metabolically active through this same period. These results demonstrate that photodegradable intracellular polymer networks in 3D spheroid culture is a novel method that can control metabolic states and induce a reversible quiescent state.

Spatial Mapping of the Embryonic Lung and Salivary Gland Development

Kylie Cyhn, Celeste Nelson, Michelle Chan, Pengfei Zhang

Branching morphogenesis is a key process during embryonic development that shapes epithelial tubes into the tree-like architectures observed in mature organs, including the airways of the lung and ducts of the salivary gland. Although both organs contain trees of epithelial tubes that form via a similar branching process, it remains unclear whether their development is regulated by common or divergent mechanisms. Exploring the similarities and differences may uncover information that could be used to improve overall respiratory and oral health (i.e., treatments for congenital lung diseases and salivary gland disorders). Furthermore, a better understanding of branching morphogenesis could also suggest innovative tissue engineering approaches to regenerate or replace lung or salivary tissue.

To this end, it is important to understand the spatial and temporal patterns of gene expression that take place within the developing lung and salivary gland, which can provide a clue to the underlying signaling pathways and patterns of differentiation in each organ. Thus, my project is focused on high-resolution spatial transcriptomic mapping of the murine embryonic lung and salivary glands, specifically focused on exploring patterns of gene expression at branch points during morphogenesis. I combine spatial transcriptomics with deterministic barcoding and computational analysis to explore the potential correlation between spatial patterns of gene expression and patterns of differentiation at branch points. My project will uncover the extent to which branching is regulated by a conserved signaling pathway in the embryonic lung and salivary gland.

Engineered Artificial Human Neutrophils Exhibit Mature Functional Performance

Sydney Hummel, Yun Chang, Xiaoping Bao

Neutrophils are a key innate immune component in the body, serving as powerful effector leukocytes by mediating opposing effects on tumor progression and ameliorating pathogen infections. They are the most prevalent leukocyte in the body but have a limited clinical application due to their short lifespan, resistance to genetic modification, and complex purification process. The ability to mimic neutrophil function in the tumor microenvironment (TME) holds promise for cancer therapeutics and presents an important challenge. An approach to optimizing the potential of neutrophils in the TME is the design of artificial neutrophils.

Artificial cells are built to mimic the function, shape, morphology, or behavior of naturally occurring cells and provide the ability to have greater control over cell behavior. Here, we combined genetic engineering technology with a nanodrug system and constructed artificial neutrophils that display similar functions as native neutrophils. K562 and HL60 human leukemia cells were used as the platform for construction of artificial cells because of their longer lifespans and ease of genetic editing.

Our artificial neutrophils strove to emulate neutrophil chemotaxis and bacterial phagocytosis. G-protein coupled receptors (GPCRs) play an essential role in neutrophil chemotaxis and serve as a chemoattractant receptor. Our study aims to demonstrate success of mimicking neutrophil chemotaxis in the artificial cells through targeting modified GPCR receptors that regulate the Gi signaling pathway, specifically the hM4Di receptor. GPCRs can also be engineered to be activated solely by a synthetic ligand, such as small molecule clozapine-N-oxide (CNO). These engineered GPCRs hold great potential to redirect the migration of modified cells under a controllable manner with CNO treatment.

As compared to the parental cells, engineered hM4Di-K562 and hM4Di-HL60 cells exhibited excellent chemotaxis ability towards clozapine-N-oxide (CNO) and superior bacteria phagocytic behavior. The antibacterial ability of the hM4Di-K562 cells was further enhanced by loading them with the glycopeptide vancomycin via mesoporous silica nanoparticles (Nano@Van), resembling the granule proteins of native neutrophils. Engineered hM4Di-K562 cells were able to effectively phagocytize the vancomycin-SiO2 nanoparticle complex and migrate to target bacteria in a CNO-responsive manner, reminiscent of many aspects of native neutrophils.

Our proposed artificial cell engineering platform provides a new avenue to investigate the physiological properties of neutrophils. By engineering these cells to recognize and respond to specific antigens, pathogens, or cancer cells, they can effectively simulate the immune system's response. Enhancing the immunomodulatory capabilities of artificial immune cells can lead to breakthroughs in disease treatment and prevention.

The Effects of Chronological Age on Cell Motility upon 2D Microenvironments

Anshika Agrawal, Mariah Snelson, Ladaisha Thompson, Pratik Kamat, Jude Phillip

Background. Aging is a physiological process characterized by an accumulation of damage and dysfunctions that increases the risk of disease[1,2]. Census data indicates that the US population 65 years or older is to increase from ~16% to ~20% by 2050[3]. There is a need to develop robust aging biomarkers, which are encoded within the propensity of cells to move in 3D substrates, that can be used to map baseline aging trajectories and improve precision medicine[4,5,6]. *We hypothesize that the observed decrease in cell motility in aged cell samples is not due to a population-wide decrease in cell motility but rather from a differential redistribution of cells among distinct cell motility patterns upon 2D microenvironments.*

Methods. To elucidate the notions of redistribution in cell motility states in cells during aging, we will create a high-throughput motility reference that connects distinct single-cell aging trajectories and motility characteristics. A panel of healthy human dermal fibroblasts were seeded upon Collagen-I coated plates and imaged using live-cell confocal microscopy for 8 hours. To evaluate the impact of stressors on the cell motility, senescence was induced on biological replicates of the cells and were also imaged. The movies are analyzed with commercial tracking software Metamorph and CaMI[7]. CaMI is a computational pipeline that utilizes single-cell motility data to identify and classify spatio-temporal behaviors of single cells.

Conclusion. Experiments conducted on 2D collagen surfaces have indicated the effect of aging on motility trends. Current data shows that there is a decrease in displacement of cells prior to their persistence with aging and occurs in a differential shift. We expect to gain a single-cell level analysis of the age-associated patterns of cell motility improving understanding of cellular determinants of aging and robust information regarding individual aging trajectories.

[1] Phillip, et al., *Annual Review of Biomedical Engineering* (2015) [2] Vigetti, et al., *Journal of Biological Chemistry* (2008) [3] U.S. Department of Health and Human Services (2020) [4] López-Otín, et al., *Cell* (2013) [5] Phillip, et al., *Nature Biomedical Engineering* (2017) [6] Phillip, et al., *Communications Biology* (2021) [7] Maity, et al., *bioRxiv* (2022)

Chemoenzymatic Functionalization of Quinolines with Carbene Transferases

Luis Burgos Jr., Edwin Alfonzo, Ziqi Li, Deirdre Hanley, Frances H. Arnold

Quinoline is a sp2-rich heteroaromatic compound that has gained significant recognition in the medicinal chemistry community because of its presence in numerous pharmacologically active substances, presenting a broad range of biological activities. Notably, the molecule's lack of unsaturated, chiral centers indicates potentially unexplored chemical space within its scaffold. Given precedent in the Arnold group, we posited that reduction and carbene transfer could significantly increase quinoline's sp3 centers. Here, we present a heme-enzyme, *Aeropyrum pernix* protoglobin (*Ap*Pgb), that can perform olefin cyclopropanation on products derived from quinoline reduction. This cyclopropanation occurs with significant chemo- and diastereoselectivity. Additionally, it is expected that the enzyme's chemo-, diastereo-, and enantioselectivity can be refined with directed evolution. These findings further elucidate the new-to-nature chemistry for heme enzymes and could lead to the discovery of new molecules with promising biological activity.

Unravelling the Structure of Seca: Analysis of WT- Vs. Y134S-Seca

Sabreen Alam

SecA and the integral membrane proteins SecYEG control almost all protein secretion across the bacterial membrane. A functional copy of SecA is essential for the survival of *E. coli*. Understanding the structure and function of SecA mutants can provide insight into the function of WT-SecA. Y134S-SecA is a crucial mutant because it can secrete preproteins with a modified signalling peptide more than WT-SecA. Understanding how this mutant affects SecA stability would begin the process of characterizing differences between WT and Y134S-SecA, and contribute to general knowledge of bacterial protein secretion. The objective of this project is to examine differences in structure and stability between WT and Y134S-SecA. The *WT-secA-* and *Y134S-secA-* gene was expressed in BL21 *E. coli*. The resulting protein was purified via Immobilized-Metal Affinity Chromatography, concentrated with a desalting column, and analyzed via tryptophan fluorescence (thermal unfolding from 310-400 nm) at temperatures ranging from 20-60°C. The mutation of tyrosine to serine at position 134 (WT to Y134S-SecA) resulted in a shift in the unfolding threshold of 2.6°C, indicating that tyrosine residues (amino acid 134 in WT-SecA) play a significant role in SecA stability. Human homologs of the SecA complex (Sec61α) can be compared against SecA to characterize human protein secretion. The position of amino acid 134 between domains of SecA involved in ATP hydrolysis suggests a possible link between protein stability and cell metabolism that can be explored in future research.

Relating Thermal/Colloidal Properties to Enzymatic Activity in the Intermediate and Native States of Amino Ester Hydrolase and Horseradish Peroxidase

Cameron Snyder, Kelly Badilla, Marcus T. Cicerone, Andreas Bommarius

While some proteins unfold according to the Lumry-Eyring model, many enzymes do not and instead exhibit more complex unfolding mechanisms. Enzymes that deviate from the Lumry-Eyring model may have intermediate states that exist between the native and unfolded state. Understanding the temperature dependence of the activity and stability of enzymes in their intermediate states may be crucial to optimizing their biotechnological and/or pharmaceutical applications. Horseradish peroxidase (HRP) and amino ester hydrolase (AEH) are enzymes observed to exhibit intermediate states in their unfolding. HRP has wide-ranging significance in biotechnology research for its utility as a reporter enzyme, in biocatalysis, and in bioremediation systems. AEH has garnered interest as a catalyst in the production of semi-synthetic beta-lactam antibiotics. A comprehensive characterization and comparison of either enzyme's native state to its intermediate state has yet to be undertaken. Thermal transition temperatures for each enzyme were identified from melting curves measured by nano-differential scanning fluorimetry (nanoDSF). Dynamic light scattering (DLS) experiments were performed at temperatures corresponding to the native states and intermediate states of each enzyme to obtain particle size distributions and diffusion interaction parameters k_D. In addition, HRP activity was measured in the native and intermediate state by utilizing a colorimetric assay. AEH activity in each state was measured using ampicillin hydrolysis. Preliminary results suggest thermal transition temperatures from the native state to the intermediate state of approximately 33°C and 42°C for AEH and HRP respectively. Additionally, HRP activity has been found to significantly decrease in the intermediate state compared to the native state.

High-Speed Magnetic Tweezers for Single-Molecule Manipulation

Caleb Maddry

Single-molecule force spectroscopy (SMFS) allows for the direct study of biological mechanisms such as protein dynamics and enzymatic activity by applying a force to a biomolecule of interest and studying the resulting effects. Thus, SMFS can unlock key insights into energy landscapes and fundamental biological processes. One common force spectroscopy instrument designed to manipulate single-molecules such as proteins and nucleic acids are magnetic tweezers (MTs).

MTs apply a magnetic field on microscopic magnetic beads to induce a force on a collection of biomolecules. This magnetic field can be produced by either a set of permanent magnets or an electromagnetic. In this configuration, we utilize a custom-made permanent magnet configuration designed to maximize the applied force. One main feature of MTs is the ability to rotationally manipulate biomolecules, as the intrinsic magnetic dipole of the magnetic beads allows for torque measurements to be made by rotating the magnetic field applied to these beads. This rotational ability can be used to measure properties such as torsional stiffness.

Traditionally, we have used optical tweezers (OT) and atomic force microscopy (AFM) to manipulate biomolecules. However, the rotational and constant-force capabilities of MTs make them a complementary addition to our current array of single-molecule techniques. While MTs are valuable for manipulation of single-molecules, they suffer from low spatial and temporal resolution compared to AFM and OT. This is because the tracking of a magnetic bead is done utilizing images of the bead at ~60Hz, however, this is often much longer than the characteristic time-scale of many biomolecules processes.

In this work, we implement various techniques to improve these resolutions and achieve a high-speed MT design. We use a high-speed complementary metal–oxide–semiconductor (CMOS) camera capable of capturing images at greater than 2000 frames per second in combination with GPU accelerated image processing. The accuracy of this design will then be analyzed by measuring a force calibration standard such as the elasticity of double stranded DNA.

Quantifying the Impact of Signaling Molecule Concentration on the Intercommunication of Sensory Protein Vesicles

Emma McDougal, Bornita Deb, Carl Denard, Yeongseon Jang

Bottom-up synthesis of synthetic cells holds the potential to enhance our comprehension of foundational biological principles and unlock biomedical applications, such as smart drug delivery systems. Globular protein vesicles (GPVs), formed through self-assembly using recombinant fusion proteins, stand as a promising platform for this purpose, due to their tunable properties and biocompatible production method. Herein, our work aims to develop and quantify sensory functions within the GPV platform, an important step towards constructing GPV-based synthetic cells. The designed sensory GPV system relies on the established protein binding interaction involving rapamycin, FKBP, and FRB. FKBP and FRB were incorporated into the GPV membranes, which were then exposed to specific concentrations of rapamycin, acting as a signaling molecule. A sensory response of FKBP- and FRB-decorated GPVs to rapamycin was indicated by the aggregation of the vesicles. It was found that the extent of intercommunication among sensory protein vesicles is impacted by the concentration of the signaling molecule. Quantitative analysis of fluorescent micrographs using ImageJ/FIJI with computing a coefficient of colocalization demonstrated that the variation in sensing extent across different working concentrations of the signaling molecule. Additionally, a range of optimal signaling molecule concentrations for maximal sensing, as well as upper and lower limits to sustain vesicle stability while inducing sensing. The same quantitative approach verified that the introduction of FKBP and FRB proteins to the GPV membrane is responsible for the sensory response to the presence of rapamycin. This work would provide a fundamental basis for integrating sensory protein functions into synthetic cells. This analysis method can also facilitate comparisons of intercommunication between vesicles with diverse characteristics.

Protease Activity Profiling Using Molecular Indexing of Proteins By Self Assembly (MIPSA)

Abhignya Panatula, Wayne Monteiro, H. Benjamin Larman

Background: Post-translational modifications (PTMs) are structural or chemical modifications to proteins that enable them to participate in various cell signaling processes. Their disregulation has been correlated with pathogenesis. For example, imbalances in proteolysis, a structural PTM, in the tumor microenvironment (TME) has been linked with many hallmarks of cancer like metastasis. Currently the substrates for most proteases are not known.

Methods: We propose to study the PTM proteolysis using Molecular Indexing of Proteins by Self Assembly (MIPSA). It is a display technology that produces libraries of full-length proteins covalently coupled to uniquely identifying DNA barcodes for analysis by next generation sequencing. Full-length proteins provide conformational structure that is important in many protein-protein interactions. The advantage of the covalent bond between the protein and barcode allows us to subject these proteins to denaturation assays without losing barcode information. To study proteolysis in cancer we aim to use a library of 16,700 human proteins that contain various classes of molecules like cytokines, chemokines, transcription factors, receptors, etc. We aim to incubate proteases of interest with the MIPSA library, separate or fractionate cleaved and un-cleaved proteins and subsequently sequence the fractions to identify non-canonical substrates. We propose two workflows to characterize protease activities: (1) An initial on-bead cleavage strategy to show proof-of-concept and (2) eventual in-solution cleavage strategy.

Results: Using the on-bead cleavage assay as a proof-of-concept, we show that MIPSA can be used to identify protease substrates from a mixture of proteins. But because the substrates are immobilized on beads, a certain degree of assay sensitivity might be lost. To overcome this, we plan to develop a version of the assay where the proteins are present in-solution and will have more access to the protease. We can fractionate the cleavage products based on their size, as cleavage products would experience a size shift thereby moving their attached DNA barcodes to other fractions. This could potentially increase the sensitivity of the assay owing to the many fractions we collect, each having better signal to noise ratios.

Conclusions: Protease activities can serve as diagnostic biomarkers and also as therapeutic agents for cancer, especially in the TME. MIPSA offers an unbiased, relatively low-cost and high-throughput way to profile protease activities. Identified substrates can be used to design diagnostic probes, therapeutic conjugates or further elucidate tumor pathway biology.

Understanding Improved Thermostability of Mutated BrCas12b Using Molecular Dynamics Simulations

Katelynn Horvath, Yinhao Jia, Janani Sampath

In the wake of the COVID-19 pandemic, the landscape of CRISPR-based diagnostics has undergone a major evolution, promising transformative advancements in the field of infectious disease detection. The innovative fusion of pre-amplification techniques, notably reverse transcription-loop-mediated isothermal amplification (RT-LAMP), with CRISPR-Cas reactions has helped create a seamless one-pot diagnostic assay. While this integrated approach holds the potential for rapid pathogen detection, it introduces challenges associated with the elevated operational temperatures necessitated by RT-LAMP. To surmount this challenge, CRISPR-Cas proteins have been engineered to confer enhanced thermostability. A recent example of this is BrCas12b, which, following strategic mutagenesis, has emerged as a robust contender capable of withstanding high temperatures with high efficacies in single-pot assays for infectious diseases is evident. However, the underlying mechanisms that bolster this stability remain unclear. Such insights are imperative, not only for understanding the inherent attributes of BrCas12b but also for guiding the precise tailoring of other resilient Cas proteins tailored explicitly for detection purposes.

To address this, we employ all-atom molecular dynamics (MD) simulations aimed at elucidating the unfolding dynamics of the wild-type CrCas12b protein at elevated temperatures. Comparative analysis is conducted with a mutated variant as proposed by Nguyen et al. (*Cell Rep. Med.*, 2023). Our assessment of unfolding dynamics employs a range of structural metrics, including root-mean-square deviation (RMSD), hydrogen bond count, and solvent accessible surface area (SASA). Notably, BrCas12b is divided into individual functional domains, to help capture the sequential order of their destabilization. Furthermore, we explore the thermal stability contributions of the N-terminus PEGylated BrCas12b and provide insights into its reinforcing effects. Looking ahead, we anticipate the correlation of our findings with experimental data to substantiate the predictive utility of MD simulations in devising effective strategies for enhancing stability. This study underscores the role of molecular dynamics simulations in unraveling the complexities of protein stability and offers a pathway for the rational design of thermally robust Cas proteins tailored for advanced diagnostics.

CRISPR-Ultimate: Leveraging Thermostable Cas12b and Tnpb Enzymes for a PAM-Less, Extraction-Free Detection Platform

Katelyn Meister, Santosh Rananaware, Long Nguyen, Noah Rakestraw, Jordan Lewis, August Bodin, Vedant Karalkar, Sarah Flannery, Piyush Jain

CRISPR based diagnostic platforms utilizing the collateral cleavage property of Cas12 and Cas13 enzymes have emerged as reliable tools for rapid, sensitive, and accurate nucleic acid detection. Given the potential for using these tools in point-of-care applications in resource-limited settings, there is a critical need to devise straightforward and efficient protocols for field use. Here we present CRISPR-ULTIMATE, a simple diagnostic platform based on the thermostable Cas12b and TnpB enzymes that can quickly detect nucleic acid targets from unextracted samples. Our platform combines nucleic acid extraction, RT-LAMP based amplification, and fluorescence readout in a single point reaction with a 5-minute heat inactivation step, thereby streamlining the assay preparation and reducing the risk of contamination. Importantly, along with a simplified assay, CRISPR ULTIMATE is also capable of detecting any target sequence without the constraint of a PAM or TAM sequence, thus making it a powerful tool for nucleic acid detection. We apply CRISPR-ULTIMATE for the direct detection of SARS-CoV-2 RNA from crude saliva samples providing a sample-to-result in less than 30 minutes. Thus, CRISPR-ULTIMATE promises to be a robust and powerful platform that can be used on the field or in a point-of-care setting.

Sigma Factor-Controlled Gene Expression in *E. coli* Using Different Spacer Lengths

Elizabeth Kornreich, Tharaka Alagiyawadu, Michael Anderson, Jason Boock

In synthetic biology, genetic circuits are desired to precisely control output expression while being orthogonal to native transcription. Non-native sigma factors, which initiate the transcription process through interaction of the RNA apoenzyme with a promoter, can be used to drive expression of heterologous genes while not influencing innate pathways. Sigma factors interact with the DNA upstream of a gene to be transcribed at the -35 and -10 binding regions. Spacer length and identity between sigma factor binding sites are known to affect transcription, but it is unclear how much either contributes to the level of transcriptional output. To decouple these, designed promoter regions recognized by sigma factor B (SigB) from *Bacillus subtilis* were created with spacer lengths ranging from 10 to 18 base pairs (bp), with the native promoter having a length of 12 bp. The designed promoters were located upstream of a gene encoding green fluorescent protein (*gfp*), which was used as a reporter of expression. We found that a 14 bp spacer had the highest fluorescence which matched the native distribution of the promoter. Libraries were created with random spacer identities were built to determine the contribution of spacer length and identity. Similar to the designed spacers, we found high fluorescence for 14 bp, medium for 11bp, and low for 17 bp libraries. We are currently working on designing a 15 bp spacer to be orthogonal to native sigma factors. The learning developed through this work will aid our understanding of bacterial transcription and design of gene circuits.

Evaluation of Decarboxylase Promiscuity and Suitability for Diverse Tryptamine Biosynthesis

Anna Wirsch, William Gibbons, J. Andrew Jones

Psilocybin, the key psychoactive molecule in magic mushrooms, has demonstrated psychotherapeutic properties in preclinical and clinical studies. The biosynthesis of psilocybin has been previously established using three enzymes of interest: PsiD, PsiK, and PsiM. PsiD is a decarboxylase responsible for converting tryptophan to tryptamine. PsiD is known to be a promiscuous enzyme, in fact, to enable psilocybin biosynthesis in *E. coli* it efficiently catalyzes the decarboxylation of the non-natural substrate, 4-hydroxytryptophan; however, it fails to act on bulkier and more reactive substrates. It is crucial to identify more promiscuous pathway enzymes as we push to enable the biosynthesis of a larger drug candidate pool. In this work, we evaluated the enzymes PsmH, RgnTDC, and CroTDC. PsmH is an enzyme within the biosynthetic gene cluster that produces physostigmine. PsmH natively acts on 5-hydroxytryptophan, suggesting it naturally has evolved a larger substrate binding pocket. CroTDC (*Catharanthus roseus* L-tryptophan decarboxylase) has been shown to effectively decarboxylate tryptophan in *S. cerevisiae*, more commonly known as brewer's yeast. RgnTDC (*Ruminococcus gnavus* L-tryptophan decarboxylase) has been shown to have approximately one hundred and thirty times more catalytic efficiency in comparison to its counterpart CroTDC. Here we evaluate the substrate specificity of these engineered decarboxylases, the impact of expression strength on their functional activity, and the ability of these enzymes to expand the synthesis of psilocybin derivatives in an *E. coli* host.

A Microfluidic Device to Compare Shear Flow Vs Chemokine Signaling Effects on Immune Cell Migration

Milan Patel, Sheetal Padhi, Seejal Padhi, Ashish Kokkula, Alexander Buffone Jr.

Microfluidics systems are widely used in various scientific fields for the analysis of components in micro/nano volumes of fluid. Analysis of concurrent signal influencing T-cell migration is best observed *in vivo*, but an effective *in vitro* model would be more beneficial for research applications. Specifically, studying cellular migration is important in order to better understand how chronic conditions occur, and to better evaluate treatments for these conditions. Neutrophils and T-cells can be co-cultured with endothelial human umbilical vein endothelial cells (HUVEC) or endothelial ligands to mimic immune cell migration on the vascular endothelial surface. There are systems on the market that allow for examination of cell migration for individual signals such as shear flow or chemokine gradient; however, there are no systems that allow for simultaneous examination of the effects of both cellular signals on cell migration. We aim to create a multiple-inlet microfluidic device to analyze the migration of T-cells in response to the chemokine gradient and shear stress flow. Our proposed design was created for a specific customer need: the ability to simultaneously view the effects of shear and chemokine flows. Polymethyl methacrylate (PMMA) will be used to build the microfluidic device. Endothelial cells will be seeded on a microscope slide that will be placed in the device, with T-cells seeded on top. A thermoplastic elastomer (TPE) strip will be used as a gasket on the microscope slide as a channel for the individual flows to follow. A peristaltic pump will be utilized to control flow rates of the shear flow and the chemokine flow. A prototype has been developed based on our design, and further modification to best suit clinical and research purposes is underway. Furthermore, we plan to conduct experiments that can determine which cellular migration signal will have a greater influence on T-cells using the microfluidics prototype. Based on the preference of the T-cells towards the chemokine gradient or the shear flow, the T-cells will migrate in the direction of specific outlets in our device, which will allow us to determine the cellular response to these signals. Through determining which cellular migration signal affects T-cells, we can understand the pathophysiology of chronic conditions and better target these biological mechanisms. We aim to have distinct detection of T-cells migration for creation of specific therapeutic interventions that can improve individuals' quality of living.

An Ionically Conductive Hydrogel Amplifies Cardiomyocyte Signaling *in Vitro*

Dominic Pizzarella, Ryan Koppes

(Please see image for the full Abstract; Figures could not be uploaded in the Text box)

An Ionically Conductive Hydrogel Amplifies Cardiomyocyte Signaling *In Vitro*

Authors: Dominic Pizzarella*[1], Katelyn Neuman[1], Anthony Zappala[1], Ryan Koppes, PhD[1]

[1]Northeastern University Department of Chemical Engineering, Boston MA

Introduction: According to the World Heart Federation (WHF), cardiovascular diseases are the world's leading cause of death, with 20.5 million deaths in 2021 alone [1]. The lack of treatment options is partly due to limitations in translatability of conventional screening platforms, 2D monolayer cultures and animal models, to human physiology [2]. Considering these limitations, hydrogels have shown promise as an alternative screening platform. Hydrogels are 3D crosslinked networks of hydrophilic, insoluble polymers [3]. These characteristics allow emulation of the native Extracellular Matrix (ECM) which results in a biocompatible material with tunable properties [3]. Despite these advances, it remains difficult to characterize electrophysiology in the hydrogel environment because of the intrinsic insulation properties of conventional hydrogel polymers [4]. This project investigates an ionically conductive hydrogel's ability to amplify action potentials (AP) from a culture of neonatal rat Cardiomyocyte (CM) cells. This work seeks to establish a biocompatible, ionically conductive, tunable scaffold that could create a physiologically accurate platform to screen for cardiovascular disease medications.

Materials and Methods: Ionic Liquid (IL) and Gelatin Methacrylate (GelMA) were synthesized based on previous methods [5,6]. Proton Nuclear Magnetic Resonance (^1H NMR) was conducted after synthesis to confirm the chemical structure. The hydrogels were fabricated with a photo-initiator, lithium phenyl-2,4,6-trimethylbenzoylphosphinate (LAP), using blue light (405 nm). To assess biocompatibility, a Live/Dead Viability assay and Immunocytochemistry were performed [6]. The elastic modulus was investigated following prior protocols [6] CM signaling was assessed by recording the AP and field potential with a microelectrode array (MEA). Before primary cell isolation, a chip was assembled onto a MEA (Figure 1A). Cells were encapsulated into a Gel-Amin (8% Gelatin Methacryloyl + 3.5% Choline Acrylate) or 9% GelMA control (n=2) in a cell density of 1e5 cells/μl. CM AP activity (i.e., Spikes) was monitored on day 7 post-seeding with a spike detection threshold of 20 μV. Spikes were analyzed using the MEAnalyzer tool [7].

(Images not able to be added)

Results: There were no significant differences in cell viability between the Gel-Amin and GelMA hydrogels (Figure 1 B). MEA testing revealed that incorporation of an IL significantly amplified action potentials (AP). This resulted in an interconnected, synchronous CM culture. The raster plot (Figure 1 C) shows a time stamp whenever a spike was recorded by the MEA. Based on the spike threshold, the Gel-Amin hydrogel was able to amplify CM signaling. This property allowed for a synchronous CM culture that was efficiently monitored by a MEA. Conversely, the GelMA hydrogel showed little activity which highlights the scaffolds limitations for excitable tissue studies.

Conclusion: The ionically conductive, Gel-Amin hydrogel overcame the insulating nature of the

GelMA polymer. This allowed for amplification of APs and an interconnected CM culture *in vitro*. Future work will be done to apply the Gel-Amin hydrogel to the development of a physiologically relevant heart-on-a-chip system.

References: [1] *Deaths from Cardiovascular Disease Surged 60% Globally over the Last 30 years: Report*. World Heart Federation. https://world-heart-federation.org/news/deaths-from-cardiovascular-disease-surged-60-globally-over-the-last-30-years-report/. [2] Marei I, et al. 3D Tissue-Engineered Vascular Drug Screening Platforms: Promise and Considerations. *Frontiers in Cardiovascular Medicine* **2022**, *9*, 847554. https://doi.org/10.3389/fcvm.2022.847554. [3] Sánchez-Cid, Pablo et al. "Novel Trends in Hydrogel Development for Biomedical Applications: A Review." *Polymers* vol. 14,15 3023. 26 Jul. 2022, doi:10.3390/ [4] Del Valle, L et al. (2017). Hydrogels for Biomedical Applications: Cellulose, Chitosan, and Protein/Peptide Derivatives. *Gels (Basel, Switzerland)*, *3*(3), 27. https://doi.org/10.3390/gels3030027 [5] I. Noshadi Sci. Rep., (2017) 4345 [6] J. Soucy. Tissue Eng. Part A., (2018) 1393-1405. [7] Dastgheyb, Raha M et al. "MEAnalyzer - a Spike Train Analysis Tool for Multi Electrode Arrays *Neuroinformatics* vol. 18,1 (2020): 163-179. doi:10.1007/s12021-019-09431-0

Development of a Point-of-Care Potassium Biosensor

Diya Godavarti, Alexandra Patterson, Mark Styczynski

Chronic kidney disease (CKD), the twelfth leading cause of death in the world, is difficult to diagnose due to a lack of symptoms in the early stage. CKD is treatable if detected early; therefore, concentrations of small molecules in the blood indicative of CKD must be frequently monitored to enable early diagnosis. For example, potassium levels above or below 3.5-5.0 mmol/L are correlated with CKD. However, current strategies to quantify potassium levels require expensive equipment and trained personnel, prohibiting use in low-resource nations where the burden of kidney disease is highest. To address this critical need, we are designing an inexpensive, user-friendly potassium assay by utilizing cell-free expression systems, which are affordable and transportable. Specifically, we are using a series of three enzymatic reactions to produce an easily detectable reporter protein, super folder Green Fluorescent Protein (sfGFP), in the presence of potassium. Here, we have shown that the latter ⅔ of the reaction pathway can be recapitulated by pre-expressing enzymes into cell-free lysates. Our findings show that we need higher levels of the enzymes than we could achieve with pre-expression. Thus, next steps will focus on combining the three enzymatic reactions in a cell-free lysate enriched with the three enzymes to create a reliable, robust sensor.

Impact of 3-D Confinement on the Migratory Behavior of Young and Aged Cells: Implications for Cellular Senescence and Age-Related Decline

Anya McDaniel, Farnaz Hemmati, Panagiotis Mistriotis

Cell migration is vital for development, wound healing, and tissue regeneration. Unchecked motility can trigger pathophysiological events, such as cancer metastasis. The prevailing notion is that cell motility reduces with aging. Most such studies use two-dimensional surfaces. Yet, in vivo, cells move through narrow, three-dimensional paths like micropores and channels. Confinement induces physical stress on cells, reshaping migration mechanisms. To date, the effects of confinement on aged and young cell migration are still unclear.

To compare young and aged cell motility in confinement, we fabricated PDMS-based microchannel devices using multilayer photolithography. The device had parallel 2D-like channels with perpendicular microchannels. Microchannels varied in width and height, keeping a fixed length ($L = 200\ \mu m$). Microchannel dimensions were approximately 10x10 μm^2, 10x3 μm^2, 3x10 μm^2, and 10x6 μm^2. For each trial, early (10-15 passages) and late (>25 passages) dermal fibroblast suspensions were added to a larger channel's inlet for pressure-driven flow. Devices incubated for 10 mins at 37°C, letting cells adhere near microchannel openings. Media was added, and devices placed under a Nikon Ti2 Inverted Microscope with a Tokai Stage-Top Incubator. Experiments lasted 16 hours, imaging every 10 mins. MTrackJ tracked cell movement, calculating cell entry, persistence, velocity, and speed in microchannels. Experiments were repeated thrice and significance was determined using appropriate statistical tests (e.g., Student's t-test, ANOVA, Tukey's test).

Cells with 10-15 passages were deemed young, while those with >25 passages were classified as old due to halted division and elevated senescence-associated beta-galactosidase levels. Experiments showed young cells had a significantly reduced speed and velocity with higher confinement, with no significant change in cell entry among devices. Adding a chemotactic gradient for young cells to boost confined migration affirmed speed and velocity decline with increased confinement.

In the weeks leading up to the conference, we will analyze the effects of confinement on the migration of old cells. After establishing baseline cell motility for young and old cells, we will delve into how key migration factors like myosin II, actin, focal adhesion, and nuclear hardening influence old vs. young cell motility in confinement. Our objective is to manipulate pathways for rejuvenating aged cells to emulate youthful behavior. This research targets alleviating cellular senescence's negative effects on function, offering potential solutions for age-related cellular decline.

Deciphering the Adhesion Mechanism of Segmented Filamentous Bacteria

Siddhartha Pinto, Radmila Janjusevic

Segmented Filamentous Bacteria (SFB) are gut microbes which are known to stimulate the host immune system by anchoring to intestinal epithelial cells. SFB have been shown to recruit vital Th17 and Th22 populations to protect the host from pathogenic and opportunistic bacteria. Except for the interaction through flagellum, the mechanism of adhesion of SFB is currently unknown. With more knowledge about the way in which SFB anchors to epithelial cells and induces immunity, this mechanism could be utilized for the efficient delivery of biologic drugs directly into the cytoplasm of patients' cells.

Bioinformatic analysis of predicted SFB adhesion proteins revealed P12260, a surface protein of the bacteria with a predicted carbohydrate binding domain. It is plausible that P12260 could bind to glycans in the intestinal mucin, the extracellular matrix, or on epithelial cell surfaces. In order to investigate the properties of this protein, His-tagged P12260 was recombinantly expressed in *E. coli* and purified using Nickel-Immobilized Metal Affinity chromatography. Prior work in our lab had used two-dimensional carbohydrate microarrays to probe carbohydrate binding, but results were inconclusive.

In order to immobilize sugars, monosaccharides D-glucose, L-fucose, L-mannose, L-rhamnose, and N-acetyl-D-glucosamine were bound by their hydroxyl groups to epoxy-activated agarose resin. FITC-tagged Concanavalin A, glucose binding lectin, and FITC-tagged Lotus tetragonolobus Lectin, known to bind fucose, were used as controls. All three proteins were loaded on individual columns made of agarose-immobilized specific sugar. After column washing and protein elution, we were able to determine each protein's binding affinity to each sugar. Western blot was used to determine the amount of protein in the elution. This method can be used to identify the principal monosaccharide residue which is involved in binding, but the full glycan structure is still unknown. In future research, epoxy-activated agarose resin can be used to immobilize P12260. This resin can be used in pull down experiments with gut epithelial cell lysate in order to identify the in vivo binding partner.

pH-Responsive Bioactive Polymer Plp-NDA: Computational Analysis on Membrane Interactions and Destabilization

David David, Thomas Nok Hin Cheng, Nicole Henry, Rongjun Chen

Poly(l-lysine iso-phthalamide) grafted with decylamine (PLP-NDA) is a bioactive polymer with potential for pH-responsive intracellular drug delivery. When drugs are internalized by cells through endocytosis, they are compartmentalized inside endosomes, which eventually merge with lysosomes causing degradation of the drug. This can be prevented by releasing the drug to the cytosol through the pH-triggered lysing of late endosomes by PLP-NDA. In past works, we have demonstrated that PLP-NDA exhibits membrane-lysis under acidic endosomal pH, but not at physiological pH, allowing for the lysis of endosomes without the destruction of cell membranes, preserving the cells. Because the strength of lysis depends on the degree of grafting with NDA, the polymer's activity can be fine-tuned for other biomedical applications.

Despite PLP-NDA's demonstrated activity, the underlying mechanisms of its interactions with membranes have remained relatively unexplored, especially in relation to its chemical structure, rendering rational design challenging. Past experimental works, observing the formation of ghost cells, have hypothesised a pore formation mechanism facilitated by the anchoring of hydrophobic NDA chains to the membrane. However, it remains unclear how a large polymer could efficiently anchor onto a membrane and only exhibit significant disruption upon protonation.

This study seeks to understand the biomolecular interactions of PLP-NDA and membranes, identifying the mechanisms of its pH-responsiveness through molecular dynamics simulations. Because the procedure is generalisable to other polymers, it allows for the computer-aided design of novel drug-delivery polymers through screening prior to experimentation. In this study, we investigate polymer binding and its effects on membrane deformation under different degrees of protonation and NDA grafting at intermediate polymer lengths.

With complete NDA grafting (100%), PLP-NDA does not bind persistently to the membrane, remaining unbound for 20% of the microsecond simulation time. In contrast, PLP-NDA with partial grafting (18%, experimentally optimal for lysis) or no grafting (0%) binds persistently for more than 90% of the simulation time, forming hydrogen bonds with the membrane. This result suggests the importance of lysine's hydrophilic carboxyl group in facilitating the initial binding and maintaining it. The bound PLP-NDA 18% and 0% are observed to cause time-averaged and localised membrane-thinning. PLP-NDA 100% did not cause membrane-thinning. Ionised PLP-NDA demonstrated weaker membrane-thinning, consistent with experimental findings of less membrane-lysis at higher pH. Although it is very challenging to observe membrane disruption in time-scales of atomistic simulations, these results may explain the early initiation of pores and demonstrates a quantitative criterion to benchmark novel polymers.

Development and Characterization of Synthetic and Biohybrid Hydrogel Enterosorbents for Removal of PFAS through the GI Tract

Kaitlyn Osborne, Maria Victoria Ximenes Klaus, J. Zach Hilt

PFAS (per- and polyfluoroalkyl substances) are a class of chemicals developed for various applications, but they typically do not break down in the environment for thousands of years making them nearly permanent fixtures once they are manufactured. In recent years, some PFAS have been found to cause adverse health effects, and thus, there exists a need to remove these chemicals from the human body to decrease the body burden of PFAS and prevent chronic diseases. Enterosorbents offer a non-invasive and promising future for the removal of these substances from the GI tract. In this research, poly(acrylamide) based hydrogels and hydrogel composites with the cationic co-monomers acrylamidopropyltrimethyl-ammonium chloride (DMAPA-Q) and diallyldimethylammonium chloride (DADMAC), the natural polymer gelatin, and the particulate calcium montmorillonite clay, were synthesized and characterized for their ability to perform as an enterosorbent. Synthesis was conducted via free radical polymerization and several characterization techniques were used including Fourier Transform Infrared Spectroscopy (FTIR), scanning electron microscopy (SEM), thermogravimetric analysis (TGA) and ultraviolet-visible spectroscopy (UV-vis) to determine material properties. The aim of this study was to confirm that these hydrogels could be synthesized and their cationic functionality persists through the polymerization, quantify the swelling capacity, and test their affinity for PFAS using model dye compounds. Successful synthesis of (poly)acrylamide hydrogels and hydrogel composites with cationic co-monomers DADMAC or DMAPA-Q or the gelatin with varying clay concentration was confirmed using FTIR and TGA, respectively. The sorption studies confirmed that cationic systems have high ability to remove model PFAS compounds from water. These results are promising for enterosorbent applications.

Pixcell: Novel Automated Software for Adipocyte and Lipid Tracing in Immunofluorescent Images

Tal Dassau, Elizabeth K. Johnston, Nickia A. Muraskin, Rosalyn D. Abbott

Adipose tissue is a highly dynamic endocrine organ that serves as the body's primary energy reservoir through the storage and mobilization of lipids. Adipocyte cellular size can be indicative of cellular status; for example, hypertrophic adipocytes are more prone to insulin resistance and the secretion of pro-inflammatory cytokines. Thus, accurate analysis of lipid size and distribution is vital, especially in biopsy samples, disease modeling, and regenerative tissue constructs. Presently, available software solutions are optimized for hematoxylin-eosin images where lipids are removed during the processing and appear as voids. Unfortunately, these softwares have difficulty with confocal derived z-stack images, struggling to capture cells both in the foreground and background of an image due to the interference of shadows and blemishes. This limitation has made manual analysis with ImageJ the most commonly utilized method for three-dimensional images. However as ImageJ is cumbersome, time consuming, and adds a degree of user-dependent subjectivity to analysis, we developed PixCell, an automated Matlab-based software, to reduce the subjectivity and time involved in adipocyte size analysis. PixCell sequentially thresholds the images and creates masking layers to capture the increased dimensionality of these images. The algorithm then cross-references cells captured across thresholding steps to prevent double counting. With its stepwise thresholding and masking steps, PixCell retains an average of >80% accuracy when tested on a variety of complex samples, including excised human adipose tissue, adipocyte laden collagen gels, and lipoaspirate seeded silk scaffold while reducing analysis time from hours to seconds. Additionally, PixCell has the added feature of tracking much smaller extracellular lipids, which are an item of interest in intercellular communication and disease states. This makes PixCell a powerful, user-friendly tool for use in both the clinical and preclinical setting, while greatly enhancing the efficiency and accuracy of lipid size analysis. While this tool was initially developed for adipose tissue analysis, it has broad applicability and future use in tracking cell proliferation and development, analyzing hepatocyte organoids, as well as investigating colloids.

Engineering Pan-Reactive VEGF Antagonists to Treat Neovascular Eye Diseases

Yuseong Oh, Paul Sargunas, Jamie Spangler

Ocular neovascularization drives the pathogenesis of neovascular age-related macular degeneration (NVAMD) and diabetic retinopathy (DR), two of the leading causes of blindness in adults. Vascular endothelial growth factor (VEGF) ligands are major stimulants of neovascularization, and blocking the binding of these ligands to their cognate receptors (VEGFRs) has shown clinical benefit in the treatment of both NVAMD and DR. Most VEGF-targeted therapies that have received FDA approval for use in treating NVAMD and DR block the activity of VEGF-A ligands, which are the most prominent regulators of blood vessel formation, by competitively inhibiting their signaling through VEGFRs on the surface of epithelial cells. However, VEGF-C is increasingly implicated in pathological neovascularization, and this ligand is found to be upregulated following anti-VEGF-A treatment, leading to therapeutic resistance. Moreover, VEGF-C is just as potent as VEGF-A in driving retinal neovascularization under hypoxic conditions. No current therapies bind both VEGF-A and VEGF-C, which would provide a more complete blockade of neovascularization. We designed an error-prone mutagenic DNA library templated on domains 2&3 of VEGFR-2 (VEGFR-2 D23), which binds both VEGF-A and VEGF-C. We then used the yeast surface display directed evolution platform to isolate and characterize high-affinity clones that bind both VEGF-A and VEGF-C. We further designed novel dual decoy receptor antagonists by fusing the binding domains of VEGFR-2 and VEGFR-1 to a human antibody Fc domain for a potent inhibition of both VEGF-A and VEGF-C activity. These promising proteins could act as superior therapeutics to treat neovascular eye diseases.

Supercas: A Supercharged CRISPR-Cas12a Enzyme for Enhanced Nucleic Acid Detection and Gene Editing

Jordan Lewis, Santosh R. Rananaware, Grace Shoemaker, Julio Ocana-Ortiz, Lilia Yang, Zoe Fang, Vedant Karalkar, August Bodin, Katelyn Meister, Ian Lange, Dylan Carman, Piyush Jain

In recent years, the Clustered Regularly Interspaced Short Palindromic Repeats (CRISPR) system has revolutionized genetic research, offering unparalleled ease and accuracy in gene editing and nucleic acid detection. Particularly, the CRISPR-Cas12a enzyme has shown significant promise in these domains. This paper introduces 'SupErCas', a novel variant of the CRISPR-Cas12a ortholog derived from Eubacterium Rectale (ErCas12a) that has been biochemically engineered to possess increased overall charge. We designed and characterized a library of ErCas12a enzymes with increasing positive and negative charges. We used structure-guided mutagenesis to systematically substitute neutral amino acids with those containing either positively charged (Lys, Arg, His) or negatively charged (Glu, Asp) side chains, thereby enriching the local charge at different domains of the enzyme. We observed that increasing the overall charge of the enzyme lead to an increase in its resistance to pH and temperature changes. Our charged Cas12a variants work at a broader temperature and pH range compared to the wild type Cas12a, thus making them ideal candidates for diverse genomic applications. We also demonstrate that one of the negatively charged variants characterized here possesses superior trans-cleavage activity that can be applied for highly sensitive detection of nucleic acid substrates. In summary, our work here explores the effect of charge engineering the CRSPR-Cas12a enzymes and turning these charged variants into tools for a broad range of gene-therapy and diagnostic applications.

Plastics to Fuel from an Unknown Sample

Eloise Thoreson, Jessica Ralph, Laura Osborne

Plastic is not infinitely recyclable, but little end of life processing has been developed outside of landfilling. The Plastics to Fuel project aims to change this standard and create an alternative use for plastic once it reaches the end of its life cycle. To do this, the Plastic to Fuel project can begin with an unknown mixed plastic material, characterize it through Differential Scanning Calorimetry (DSC), Thermogravimetric Analysis (TGA), and Fourier-Transform Infrared Spectroscopy (FTIR). Then, using a pyrolysis kiln reactor, a liquid carbon fuel, or diesel fuel, is produced. This is done slowly, in a single step, and without catalyst. The only required mechanical processing consists of shredding the feedstock, to increase packing density. By performing Flame-Ionization Gas Chromatography (GC-FID), and One-Dimensional Proton Nuclear Magnetic Resonance (1D 1H NMR) on the product, a comparison can be made to model systems of High-Density Polyethylene (HDPE), Low-Density Polyethylene (LDPE), and Polypropylene (PP), as well as with #2 diesel. Because of the variety in types and manufacturing techniques of plastic, the Plastics to Fuel project continues to test this methodology on many different feedstocks and mixtures to investigate the differences in the product. With further development, this project has the potential to reduce waste in landfills, prevent microplastics from leaching into the environment, while also reducing local cost for fuel in rural communities.

Regression-Based Machine Learning Approach for Predicting Viscosity and Gel-Strength of Drilling Fluids through Analysis of Drilling Mud Formulation Parameters

Kenneth Chong Yih Haur

The drilling process for onshore and offshore oil and gas wells requires drilling fluid, often known as drilling mud. Drilling fluid serves several major drilling functions, including cooling and lubricating the drill bit, carrying rock cuttings to the surface, and maintaining the formation pressure in the wellbore to prevent blowouts. The parameters of the drilling fluid, such as its viscosity, density, and pH, must be strictly controlled and modified during drilling operations. In the oil and gas sector, the use of machine learning in drilling fluid is becoming more and more crucial. Machine learning can be used to evaluate vast amounts of data from drilling fluid formulation operations, including data collected on the properties of the drilling fluid, such as its viscosity and rheological properties, as well as data on the geology of the wellbore. By using this data, regression machine learning algorithms can predict the behavior of drilling fluid under different conditions and formulations, optimize the characteristics of the fluid, and expected to improve drilling performance, save cost and time.

Cultivating Sustainability: Harnessing Plant and Animal Waste for Biofuel Production to Support Turbine Biomass Systems

Joshua He

The prevailing reliance on nonrenewable fossil fuels as our primary energy source poses significant economic and environmental challenges. This paper delves into the exploration of alternative energy sources derived from plant and animal waste. Specifically, it introduces an innovative approach to power generation, Turbine Biomass (TBS), which traditionally utilizes crop waste, coupled with an advanced biofuel production method employing transesterification and biofermentation to enhance its energy return on investment (EROI). Transesterification, a chemical reaction involving triglycerides commonly found in lipids and plant cell walls, yields alkyl esters with low flash points and high net energy yields. Biofermentation, on the other hand, harnesses animal manure and urine to produce methane and hydrogen sulfide. The combustion of these compounds generates substantial heat. Both processes are eco-friendly and economically viable due to their use of waste materials, low energy input, and minimal greenhouse gas emissions. TBS operates by burning such biomass to produce steam, which rotates a turbine connected to a generator. Particularly efficient on small-scale farms, TBS proves cost-effective by utilizing readily available crop and animal waste, all while boasting lower greenhouse gas emissions than fossil fuels. The excess heat can be channeled through a semiconductor for a thermoelectric effect or utilized in HVAC systems. Additionally, this process yields a nutrient-rich fertilizer from the residual organic matter. This paper advocates for a sustainable solution that employs biowaste as biofuel, presenting an environmentally friendly and economically viable energy supply system. As organic waste processing and energy generation technologies continue to advance, this method offers promising avenues for a more sustainable energy future.

Upgrading and Analysis of Red Oak Pyrolysis Oil to Develop Sustainable Aviation Fuels

Alayna Gilbert, Dan Cornett, Dr. Bret C. Windom

The aviation sector remains a large contributor to international carbon dioxide emissions. The global jet fuel market produces over 81 billion gallons of fuel per year [1] and demand is continually increasing with the 3.8 billion air travelers in 2016 expected to double by 2035 [2]. The expected increase in air travel provides challenges in corresponding emissions increases and increased demand for fossil fuels. This has led to an extensive interest in Sustainable Aviation Fuels (SAF(s)). SAFs are defined as non-petroleum-based aviation fuels that are produced from a variety of feedstocks including fats, greases, municipal solid waste, and biomass. This research focuses on SAF production using a biomass feedstock.

In the last several years, researchers have developed a process to create rubber from an arid woody shrub called guayule. This process has a high yield of plant waste resulting in opportunities for development of biproducts using the guayule biomass. Several university research groups have partnered to explore some of these opportunities including bug repellent, adhesives, and jet fuel.

This research targets the latter, focusing on the upgrading of pyrolyzed bio-oil waste into SAFs using a high-pressure/high-temperature hydrotreatment process and distillation. Due to the novelty of the rubber process, there is limited supply of guayule waste, thus red oak bio-oil has been used as a surrogate to develop the hydrotreatment and analytical methods that will be applied to guayule waste. This research discusses the experimental and analytical methods as well as the chemical processes involved through the upgrading process of red oak bio-oils. Preliminary results have been successful in yielding molecular components in the jet fuel range through hydrotreatment with distillation experiments and chemical property analysis ongoing.

References

[1] Holladay, Johnathan, Abdullah, Zia, & Heyne, Joshua. *Sustainable Aviation Fuel: Review of Technical Pathways*. United States. https://doi.org/10.2172/1660415

[2] International Air Transportation Association (IATA). 2016. "IATA Forecasts Passenger Demand to Double Over 20 Years." IATA Press Release No: 59, October 2016. https://www.iata.org/pressroom/pr/Pages/2016-10-18-02.aspx.

Quantifying Extraction Via Ultrasonic Cavitation of Bioproducts from Algae

Lorenzo Galang

Utilizing biofuels limits carbon emissions in the short term as the world works towards transitioning to clean energy sources. Biofuels can be used in existing fossil fuel infrastructure with minimal alteration, replacing natural gas with syngas and gasoline with ethanol. A particularly promising source of biofuels is algae, which only require ponds to grow and do not compete with food crops. Current extraction methods are energy intensive and expensive, counteracting the neutral carbon footprint of biofuels. Ultrasonic cavitation, the creation of small bubbles that pop violently using ultrasonic waves, is potentially cheaper and more energy efficient. Many phenomena associated with cavitation are conducive to extraction, including shearing forces from jetting, high temperatures, creation of free radicals, and collection by bubbling. To quantify the effect of these phenomena and assess the potential of this process for extraction of algal biofuel, a rapid lipid quantification method is necessary, especially when investigating process variables on the macroscopic scale. Current methods limit data collection, and fluorescent lipid dyes present an opportunity for rapid quantification through spectrophotometry. Several combinations of dyes and solvents were tested, as well as various infiltration times. Water, algal oil, live algae, and algae washed with an established method were compared to assess the correlation between the fluorescence and lysis of algae. Establishing a relationship proved difficult thanks to the lack of literature regarding the ability of the dye to indicate lipid concentration, rather than viability of algae.

Product Distribution at Different Conversions of Polypropylene (PP)

Emanuel Aponte

Plastics, derived from petrochemicals, have replaced metals and ceramics due to their low cost and lightweight nature. This has led to a surge in plastic production, resulting in increased plastic waste, with only 9% properly recycled. A Remade Project about chemical recycling, aims to address this issue. Our lab focuses on one of the four overall tasks: the development of mechanistic understanding of process components, where we work with the chemical degradation process known as pyrolysis. Pyrolysis decomposes organic materials into gases and oils useful for energy production. However, a lack of kinetic understanding of plastic pyrolysis restricts feedstock changes without compromising product quality. Therefore, we use a GCxGC system to analyze pyrolysis products, enhancing peak separation and yielding a three-dimensional chromatogram. We have previously observed that the total yield of polypropylene pyrolysis products increases with temperature. The current objective is identifying product distribution of non-catalytic PP pyrolysis at set temperature levels, in this case 400 degrees Celsius, but with different conversions. While GCxGC conditions don't affect pyrolysis, they do influence chromatogram quality. In the end, we made three conclusions. Cooling the reactor and using the cryotrap allowed this study to be possible. The major products identified were 2,4-Dimethylheptene, Pentane, and 2-methylPentene at observed conversion levels. And the information from the flame ionization detector and mass spectrometer allowed identification of these peaks in terms of carbon number and group (boiling point and polarity). Further steps involve quantifying these peaks for weight percentage and exploring different conditions and inputs to advance chemical recycling and reduce plastic waste.

Futile to Utile: Pyrolysis of Single-Use Plastics to Promote Circularity

Kate Williams, Uriel Perez, Jacob Walsh, Skip Rochefort

Single-use plastic waste is a significant problem today. Found in all industries, very little single-use plastic waste is recycled classically via mechanical recycling—grinding and re-forming into new plastic products—due to differences in design and composition. Chemical recycling via pyrolysis offers a way to introduce circularity to the single-use plastic industry. In pyrolysis, plastics are decomposed in the absence of oxygen at very high temperatures (>420°C) into diesel that can be used as fuel or further processed into monomers for plastic production. At Oregon State University, a 1-kg research kiln reactor has been used to successfully convert polypropylene (PP) syringes, polypropylene (PP) Dutch Brothers coffee cups, and mixed ocean plastics (typically a mixture of HDPE, LDPE, PP with some PVC and PS) into diesel. Each run, the reactor is loaded with up to one kilogram of shredded plastic waste and attached to a counter-current double pipe condenser. Electricity is used to power heating coils within the kiln. The polymer feedstock melts, then vaporizes, and is forced through a packed bed before exiting to the condenser.

A partnership with the Benton County Needle Exchange program provided used syringes, eliminating a portion of red bag waste the community must process. The Dutch Brothers Coffee shop on the OSU Corvallis campus donated used drink cups, keeping them out of the landfill. Ocean plastics were obtained from beach clean-ups in Kodiak, Alaska attended by OSU students in collaboration with the Ocean Plastics Recovery Project. Results will be presented on kiln reactor operation and complete characterization of input plastics (DSC, TGA, FTIR) and diesel characterization (GC, NMR).

Long term, the project goal is to produce larger versions of the OSU reactor (5-kg, 10-kg) for communities that lack recycling systems to promote circularity of plastic products and increase residents' access to diesel in their community.

Effects of a Star Styrene Butadiene Additive on the Thermodynamic Properties of a Group III Mineral Base Oil

Melanie Lindblom

Automotive transmission fluids are primarily composed of mineral base oils and often contain additives that help improve the performance of these oils. In this study, the effect of a star styrene-butadiene additive (MW=50,000 g/mol) on the thermodynamic properties of a petroleum-derived Ultra S4 mineral base oil was investigated. The interest has been on their properties under pressure.

The oils were evaluated at 1, 2, and 5 wt % addition levels of the styrene-butadiene modifier copolymer. Specific properties that have been evaluated include the density, isothermal compressibility, isobaric expansivity and the internal pressure.

Densities were determined at temperatures from 298 to 398 K over a pressure range from 10 to 35 MPa using a high-pressure variable-volume view-cell. The internal volume of the cell is changed with a movable piston, the position of which is altered with the aid of an automated pressure generator. The piston position and thus the internal volume of the cell is monitored with an LVDT (linear variable differential transformer) sensor at any given temperature and pressure. Densities are then determined form the initial mass loading which is known.

Density data are then correlated with the Sanchez-Lacombe Equation of State, which is a lattice fluid equation of state that is effective in accounting for and describing the pressure effects. The correlation equations are then used to evaluate the derived thermodynamics properties, namely, compressibility, expansivity and internal pressure.

The results showed that the inclusion of the additive increased the density of the base oil at a similar level for all these concentrations. Isothermal compressibility of the base oil also increased for the 1 and 2 wt % addition levels, but there was essentially no change for the case of 5% additive. The additive caused a decrease in isobaric expansivity of the base oil at all concentrations, with the 1% addition causing the greatest decrease. The internal pressure also showed a decrease at all concentrations, with the greatest decrease occurring at the 2% addition followed by the 1% addition.

Thermodynamic Modeling on Cloud Point of Fatty Acid Methyl Esters and Fatty Acid Ethyl Esters Biodiesel Mixtures with and without Additives.

Anthony Malshyti, Michael Senra

Biodiesel is a promising replacement for petroleum based diesel in a number of different applications. Although biodiesel offers numerous advantages over petrodiesel, one of its major disadvantages is its inferior cold flow properties. Previous work has shown that the challenge with the cold flow properites is strongly correlated to the composition of the biodiesel, which primarily contains fatty acid methyl esters (FAME). Taking advantage of this fact, researchers have used a wide variety of thermodynamic models of varying complexities to conduct solid/liquid equilibrium analysis to estimate the cloud point of the biodiesel. The major issues that can cause variations in these models can include such parameters as degree of saturation, presence of non-FAMEs and co-crystallizaton. This work comprehensively looks at cloud point data from a wide variety of sources and analyzes them using the simplest model, the ideal solution theory. Work has shown that with a few exceptions, ideal solution theory predicts the cloud point quite accurately, particularly when taking into account the variability of measurements such as cloud point and the heat of crystallization.

Investigating Sealing Methods for Three-Electrode Coin Cells

Virginia Parparcen, Nidhi Kapate, William Chueh

With rising CO_2 emissions, it is critical that we decarbonize the energy sector. Renewable energy resources offer a sustainable solution, but their intermittency reveals the need for better energy storage systems. Improving energy storage requires the study of the side reactions and transport limitations that happen inside of batteries. By isolating the anode and cathode potentials, three electrode-cells allow for a better understanding of these processes. However, they have low durability due to the opening at the junction of the third electrode, which allows moisture into the cell. In this study, we investigated three different epoxies (5-minute epoxy, Permabond 737, and Permabond 940) to determine the most effective sealing method. We used lithiated graphite inside of three-electrode cells as an indicator of the moisture content, and assembled four batches of cells to open at varying intervals. The resulting graphite was photographed to compare its color, since it is correlated to contact with water. The average RGB values of the images were extracted and compared statistically. Our study found that Permabond 940 allowed the most moisture inside the cell, while Permabond 737 and 5-minute epoxy improved cell longevity. There was not a significant difference between the Permabond 737 and the 5-minute epoxy. While this study provides valuable insights into sealing methods for three-electrode cells, further research with improved imaging and analytical methods is needed for conclusive results.

Developing Mechanistic Understanding of Zinc Plating Vs Zinc-Ion Intercalation in Chevrel-Phase Mo_6Se_8

Tricia M. Marchese, Sara Avraham, Brian Chen, Robert Messinger

Electrochemical ion intercalation is one charge-storage mechanism responsible for the operation of several battery chemistries such as rechargeable lithium (Li)-ion and zinc (Zn)-ion batteries. One possible failure mechanism for intercalation-based energy storage systems is uncontrolled metal electrodeposition (i.e., plating), where metal ions reduce on the ion-intercalation electrode surface instead of intercalating. Metal plating harms battery performance as the process consumes active material and can short-circuit the battery. The mechanistic interplay between electrochemical ion intercalation and metal plating is not yet well-understood. Elucidating a mechanism that describes the conditions necessary for metal to plate on ion-intercalation electrodes could facilitate development of suppression strategies.

In this work, we studied Zn plating in Zn-ion batteries wherein the interplay between electrochemical Zn^{2+} intercalation and Zn plating was measured in chevrel-phase Mo_6Se_8 cathodes. We demonstrate for the first time that Zn can plate on Mo_6Se_8 cathodes in Zn/Mo_6Se_8 cells. The occurrence of Zn plating on Mo_6Se_8 was studied through measurement of the nucleation overpotential, which appears as a minimum in the Zn/Mo_6Se_8 cell potential during galvanostatic discharge below 0 V. Variable-temperature (10 to 60 °C) and variable-current density (0.02 to 1 mA/cm^2) measurements of the nucleation overpotential were conducted to elucidate a mechanism that describes the competition between electrochemical Zn^{2+} intercalation in Mo_6Se_8 and Zn plating on the electrode surface. The variable-temperature galvanostatic measurements in Zn/Mo_6Se_8 cells show two unique electrochemical reaction mechanisms during Zn/Mo_6Se_8 discharge: (1) Zn^{2+} intercalation and Zn plating punctuated by the hydrogen evolution reaction (HER) in aqueous 1 M $ZnSO_4$ electrolyte, and (2) Zn^{2+} intercalation and Zn plating with mitigated HER in 1 M $ZnCl_2$ in H_2O:N-methyl-2-pyrrolidone (NMP) (1:17 mol/mol) electrolyte. Arrhenius analyses on the measured temperature-dependent nucleation overpotentials were conducted, revealing the critical role that ion adsorption plays prior to Zn plating or HER reactions. The presence of metallic Zn deposits was identified using optical microscopy and X-ray diffraction on galvanostatically discharged Mo_6Se_8 electrodes, confirming the occurrence of the Zn plating process on Mo_6Se_8.

It was verified that Zn can indeed plate on Mo_6Se_8 similar to how Li plating on graphite electrodes can occur in Li-ion batteries. This work enables the study of Zn plating as an analog to Li plating and establishes a connection between the two metal electroplating processes. Therefore, building a mechanistic understanding of both Zn and Li plating as failure modes could improve the practicality of each metal's respective battery chemistries in all conditions.

Synthesis of Desilicated Chabazite for Non-Oxidative Conversion of Methane (NOCM)

Jackson Boodry, Angel Santiago-Colón, Rajamani Gounder

Non-oxidative conversion of methane (NOCM) into value-added longer-chain hydrocarbons facilitates the transport of natural gas resources from remote drilling locations to centralized chemical plants and refineries. The NOCM reaction is hypothesized to be catalyzed by edge-defect sites on carbon-based supports at high (>1100 K) temperatures to overcome thermodynamic constraints. Desilicated chabazite (CHA) zeolite is one material proposed to form large densities of these edge defects through carbon deposition. The procedure to synthesize desilicated CHA involves preferentially removing silicon from the crystalline framework through an alkaline (NaOH) treatment while preserving the structure-directing agents (SDAs) occluded within microporous voids during crystallization to mitigate pore collapse. This selectively creates mesopores at crystallite exteriors to increase the accessible surface area for carbon formation during NOCM. The crystallinity and pore volume of the resultant materials were studied via X-ray diffraction and gas adsorption characterization. Broad applicability of desilicated CHA as a catalyst necessitates maintaining the number of Brønsted acid sites throughout the synthesis process, which is quantified through ammonia titration and temperature-programmed desorption (TPD). The structural stability of desilicated CHA depended strongly on the Si/Al ratio of the parent material and the temperature used for SDA removal. TPD results revealed an increase in the amount of ammonia desorbed at lower temperatures in mesoporous materials, possibly indicating the creation of weaker-binding, under-coordinated acid sites around regions of mesopore formation. These results collectively indicate a complex relationship between synthesis conditions and material stability, with application to NOCM necessitating stability analysis at conditions best suited for the reaction.

Electrochemical Kinetics and Performance Characterization of Graphene Oxide-Carbon Nanotube-Noble Metal Custer Electrodes for Energy Storage Applications

Michael J. Williams, Alexander Liesen, Avery Patel, Caspar Yi, Yash Joshi, F. John Burpo, Simuck Yuk, Yong Joo, Enoch Nagelli

As the consumer electronics, renewable energy, and electric vehicle markets have become integral to society, so have the energy store and conversion (ESC) technologies that enable them. The most versatile of these applications are lithium-ion batteries (LIBs) and hydrogen fuel cells (HFCs). The former of which have become ubiquitous due to their energy density, reliability, and cycle life while the latter deliver scalable and sustainable power with high energy density. Despite this, the need for further for high power and energy density drive demand for improvements in the materials chemistry in both technologies. However, the kinetics of the reactions that drive HFCs are limited by the chemistry and structure of the catalyst material. Moreover, LIBs suffer from dendrite formation in the solid-electrolyte interface layer and electrode materials with relatively low specific capacity (graphite anodes with specific capacity of 372 mAh g^{-}) in widespread usage. Structured carbon nanomaterials offer an alternative with ideal 3D properties for catalysis, improvements in battery performance, mechanical, chemical, and thermal stability. However, their synthesis is often lengthy, expensive and difficult to scale. Here, graphene oxide (GO) and oxidized Carbon nanotubes (ox-CNTs) are synthesized into a 3D carbon network. Then, mixed with 5 wt % polyacrylic acid (PAA) as a binder, they are air-controlled electrosprayed onto copper foil. High surface area noble metal (NM) nanoclusters are nucleated through the 3D carbon structure utilizing a spontaneous galvanic reduction reaction with NM salt solutions (K_2PtCl_4, and Na_2PdCl_4). The resulting 3D electrode structure of the GO/CNT-NM materials were characterized using Raman Spectroscopy, Scanning Electron Microscopy, and Energy Dispersive X-Ray Spectroscopy. The on-set potential of the material as a catalyst for the hydrogen evolution reaction (HER) and oxygen evolution reaction (OER) in hydrogen fuel cells were determined using linear sweep voltammetry (LSV). The electrochemical surface area and the redox potentials were verified using cyclic voltammetry (CV). Further, GO/CNT-NM electrodes were assembled into coin cell LIB half cells and characterized via electronic impedance spectroscopy (EIS). The GO/CNT-NM materials function as bifunctional catalytic electrodes, showcasing efficacy for catalysis in HFCs and improved specific capacity in LIBs. This improvement in kinetics and electrochemical performance is attributed to the coaction between the noble metal nanoclusters and the 3D GO/CNT network.

MgO As a Water Scavenger in Lithium-Ion Batteries' Separators

Adel Fadhul, Fei Hu, Wyatt Tenhaeff

In lithium-Ion batteries, multifunctional separators are able to improve performance and energy storage depending on the additives used to synthesize the separator layers. This study explores the possibilities of adding Magnesium Oxide (MgO) to lithium-ion batteries' separators to act as a water scavenger. It was hypothesized that MgO, a ceramic nanoparticle, would bind to water molecules which prevents them from corroding the battery by reacting with the lithium-based electrolytes. A one-step fabrication method, namely UV-initiated polymerization-induced phase separation (PIPS), was used to fabricate microporous separator layers. As a base, Polyethylene Glycol Diacrylate (PEGADA 400), Ethylene Carbonate (EC), and a choice of photoinitiator were mixed, then varying amounts of MgO were added to the solution. MgO had shown that it does not interfere with the curing process of the layers, as well as an ability to remain trapped within the layers following the washing process. It had also shown to increase the porosity of the layer by 8.4% when 11.11% MgO is in the layer, however, lumping was observed in samples containing 20% MgO, decreasing porosity by -4.87%. Upon cycling a battery with a layer sample containing 11.11% MgO, the coulombic efficiency of the battery drops to 93.5% after 50 cycles, only 3.3% lower than a Celgard commercially sold separator layer sample. The voltage profiles, however, show that the capacity of the batteries were stable across the 50 cycles for both samples.

Computational and Experimental Studies to Analyze Pilot-Scale Petroleum Packaging Facility

Steven Roth, Sean Curtis, David Theuma, Michael Fracchiolla, Emma Padros, Swapana Jerpoth, Barnabas Gao, Robert Hesketh, Kirti Yenkie, C. Stewart Slater, Mariano J. Savelski

Petroleum packaging facilities process multiple lubricant oils per day to meet the required market demands. As the number of packaging lines are limited, they need to clean the lines between product change-overs. During the product change-over procedure, the lines are cleared to remove the previous product. However, due to the highly viscous nature of most petroleum products, there remains a nontrivial amount of residual oil in the packaging line, which can contaminate the new product. The solution is to flush the line with new product, which mixes with the residual product to generate commingled oil. Further flushing with the new product will clear the line of commingled oil and completely fill it with new product that meets the desired specifications. While this is a successful solution to the issue, the commingled oil created during this procedure is a downgraded product that does not meet company standards with little economic value.

The goal of this work is to accurately model the industrial packaging plant and apply a well-thought scale-down strategy for the construction of a pilot scale plant that predicts the mixing properties observed at the real plant. To address this, we started with a computer model that is designed to predict how different factors affect the flushing operation and whether the new flushes meet industry specifications. Utilizing the historical data and machine learning approach, our model was able to correctly predict 110 out of 116 flushes. Furthermore, the scale-down evaluations to pilot scale were accomplished using the Reynolds number, and the pipe diameter to length ratio. With the completion of the pilot plant of the packaging line, further experimental studies were conducted to better understand the actual mixing properties for products of varying specifications in the packaging line. While maintaining the same residual and flush product, the factors that were analyzed were flowrate, filter casing volume, and filter type. Residence time distribution curves were created for both plants. Dimensionless analysis was conducted to directly compare the pilot plant experiments to the industrial plant. Through this, we gained insights on how to alter the pilot plant to better mimic the packaging plant. We concluded that the filter volume is an important factor in modeling the packaging line, whereas the type of filter was not as important. Further tests are underway for other factors such as variable flowrates for laminar to turbulent regimes and the type of change-overs based on viscosity gradients.

Synthesis of $CoMoO_4/NiMoO_4$ Composites As Supercapacitor Electrodes

Teja Patrice, Kunli Yang

The energy crisis and climate change have emerged as global concerns due to the unsustainable reliance on fossil fuels for energy production. This has spurred increased research with the intention of satisfying the global energy demand through renewable sources. Novel and more efficient methods for energy generation and storage are necessary. One promising solution lies in the exploration and application of electrochemical systems, such as supercapacitors, which can store electrical energy through fast surface redox reactions. Supercapacitors offer a range of advantages, including high capacitance, exceptional power density, and long lifecycle. They act as a bridging technology between traditional capacitors and batteries, capable of storing more energy than capacitors while delivering and accepting charges faster than batteries.

This research focuses on the investigation of transition metal molybdates, which have attracted significant attention as promising materials for supercapacitor electrodes due to their exceptional electrochemical characteristics. Our approach employs a straightforward hydrothermal synthesis method to synthesize cobalt-doped $NiMoO_4$, which is deposited onto a nickel foam substrate without binders. Subsequently, the as-synthesized materials undergo annealing at different temperatures, resulting in the formation of α and β phases at distinct temperature ranges. Despite sharing a similar nanoneedle morphology, these two crystalline phases exhibit contrasting electrochemical behaviors. Specifically, the β phase has a higher specific capacitance while exhibiting inferior long-term stability compared to the α phase. This investigation underscores the significant influence of the crystalline phase of transition metal molybdates on their electrochemical performance.

Refining the Isothermal Cell Technique to Measure the Temperature Coefficient of Redox Active, Aqueous, Organic Compounds

Kantapat Boonme, Samuel Kopfinger, Shrayesh Patel

Thermo-electrochemical cells (thermocells) are devices used in the conversion of waste heat into sustainable energy. The performance of a thermocell is dictated, in part, by the temperature coefficient. The temperature coefficient (also referred to as Seebeck coefficient) is defined by the equation, where the change in voltage with temperature, is affected by a contribution via thermal gradient induced ion migration, S^*_M, and a contribution via entropy of reaction, S^*_E. Traditionally, the temperature coefficient is measured with a non-isothermal cell, which consists of two cells at different temperatures connected by a salt bridge. Recently, a new experiment called the isothermal cell, consisting of only one electrochemical cell, has been proposed as a viable technique to measure the temperature coefficient of redox compounds. Due to its ease of use and lack of a thermal gradient, the isothermal cell is an accessible alternative to a non-isothermal cell while also reporting a temperature coefficient more intrinsic to a compound. In this work, we verified the technique in multiple experimental conditions and environments by studying the effects of redox-active concentration, supporting electrolyte concentration, and CV scan rates on the reported temperature coefficient of the common aqueous ferri/ferrocyanide redox couple ($Fe(CN)_6^{3-/4-}$) with sodium sulfate (Na_2SO_4) as the supporting electrolyte. Next, we used the isothermal cell to report the temperature coefficient of 4-hydroxy-2,2,6,6-tetramethylpiperidine-1-oxyl-4-yl ($TEMPO^{1+/0}$), ethyl viologen dibromide ($EtVioBr^{2+/1+}$), and 2,6-dihydroxyanthraquinone ($DHAQ^{2-/4-}$) to be -0.08, -0.11, and -1.25 mV/k respectively. This result aligns with the classical model that suggests the temperature coefficient increases in magnitude with the oxidative/reductive charge of the species. We further looked into effect of 2,6-DHAQ concentration and KOH concentration on the temperature coefficient and see trends similar to that of the $Fe(CN)_6^{3-/4-}$ survey, implying that the effect of varying concentration on the temperature coefficient is similar for different chemical species. Finally, we show that the location of the hydroxide groups on anthraquinone also affect the temperature coefficient by reporting the temperature coefficient of 1,2-DHAQ and 1,8 DHAQ to be -0.6 mV/k and -0.8 mV/k respectively. This result is especially interesting because it suggests that the temperature coefficient can be controlled not only by redox center, but by the type and location of charged pendants connected to the organic molecule.

Optimization of the Transesterification Process and Analyses of Resulting Cold Flow Properties of Neat and Blended Biodiesels

Ryan Pignotti, Michael Senra

Transesterification reactions produce biodiesel, an important alternative to petrodiesel; however, poor cold flow properties (CFPs) are a major drawback of biodiesel. The present work varied the transesterification process in order to optimize the procedure for laboratory analysis. Smaller-scale reactions were beneficial, as saponification was mitigated. The work also performed cold flow analyses on blended biodiesels, finding that interaction and dilution effects are present, altering CFPs.

Renewable Natural Gas Feasibility Assessment for Hard-to-Decarbonize Sectors in New York State

Celina Yeung, Eric Mitchko, Vatsal Bhatt, Devinder Mahajan

Several manufacturing industries are significant contributors to the emission of greenhouse gases into the atmosphere as a result of the production process. Particularly, industries such as metals, minerals, and pulp/paper are known to be highly carbon-intensive due to high processing temperatures and reliance on fossil fuels. As a result, these three industries are classified as hard-to-decarbonize (HTD) sectors, indicating resistance to efforts made to reduce carbon emissions and transitions to more sustainable methods. Renewable natural gas (RNG) refers to "upgraded" biogas interchangeable with conventional natural gas, a fossil fuel, due to similar chemical composition. RNG is unique in that it is derived from organic resources such as manure, food waste, and landfill gas and is produced through anaerobic digestion. The process of anaerobic digestion recycles carbon dioxide from the atmosphere, which can be used as a strategy to achieve carbon-negative emissions. This case study evaluates RNG feasibility as an alternative to traditional natural gas used by New York's facilities in HTD sectors. This work extracts all active facilities categorized under each HTD sector (metals, minerals, pulp/paper) and their annual reported direct emissions (natural gas, coal, process emissions) from the United States Environmental Protection Agency (US EPA) Flight Tool. From this data, the quantity of natural gas and coal consumed per sector can be computed by converting the reported metric tons of carbon dioxide equivalent (MTCO2e) emitted to billion British thermal units (BBTU) consumed. This conversion is utilized to gauge the current demand for natural gas by assessing how much each sector consumes based on their reported emissions. Using the Organic Resource Locator tool by the Rochester Institute of Technology (RIT), the annual availability of RNG is determined from the identification of sites and the approximate yield of RNG derived from organic resources. This includes existing ADG projects, manure from confined animal feed operations (CAFOs), and food waste. With this data, RNG resource availability from identified sources can be compared to the natural gas demand based on the location of HTD facilities. The results comparing only estimated RNG produced from resources from CAFOs to natural gas demand vary by location where only some clusters had a high percentage of demand met. In future works, the same analysis will be conducted with different organic resources such as landfill gas and wastewater. With this, the demand for natural gas may be fulfilled or even be exceeded.

Star in a Jar: The Transformation of Sound into Light

Conner Campbell

Sonoluminescence is the conversion of sound energy into light. This study involves the impact of the fluid medium and the temperature of the fluid on the resulting light intensity. In addition, an analysis of the thermodynamic principles to theorize the driving force behind single bubble sonoluminescence is explored.

Phase Equilibria for Tetra-Aryl-Phosphonium Based Ionic Liquids: Enabling More Efficient Solar Thermal Energy Production and Storage

Allan Wilson, James H. Davis Jr., Brooks Rabideau, Kevin West

Climate change is one of the most pressing issues facing humanity in the 21st century. A major contributing factor to this problem is the continued use of fossil-fuel based sources for energy. Traditional renewable solar energy poses itself as a potential solution to providing low-cost energy but is unable to fully replace fossil-fuel based sources due to difficulties storing energy during off-peak production times. Concentrated solar power (CST) with thermal energy storage (TES) provides a potential opportunity to support this transition away from fossil fuels but suffers from high costs and extended downtimes. Current heat transfer fluids used in these processes often consist of a mixture of alkali metal nitrates with melting points above 200°C. For CST and TES to be viable, a thermally stable, low melting fluid is needed. Ionic liquids (ILs) provide a very feasible form of TES due to their high heat capacities and low melting points. Thermally robust ILs can be synthesized by limiting the structural diversity of the species to thermally stable moieties on the cation and coupling these cations with a thermally stable anion. However, these synthetic restrictions result in salts with melting points at or above 100°C. Mixtures of these salts, with each other or with structurally similar molecular species, can yield fluids with melting temperatures low enough to be effective thermal storage fluids. In this project, the phase equilibrium of a binary mixture of tetraphenyl phosphonium bistriflimide (TPP) and benzanilide was examined to provide valuable information on eutectic behavior for organic ionic liquids with aromatic compounds. The use of digital scanning calorimetry (DSC) (Solid/Liquid Equilibrium, SLE) along with cloud-point analysis (Liquid/Liquid Equilibrium, LLE) was used to study this behavior. The phase transitions, including eutectic behavior, are expressed on a binary T-x diagram and show how the phase behavior varies with temperature and composition. Solid/liquid equilibrium behavior is compared to the ideal solution model to aid in understanding molecular-level interactions in the mixture. The study showed a minimum eutectic melting temperature of below 100 °C and liquid-liquid interactions in compositions having high amounts of TPP compared to benzanilide. The findings of this study will be used to influence selection of future materials to combine with TPP.

Ionic and Electronic Conductivity of Graphene Oxide Thin Films and Flowable Suspensions for Electrochemical Flow Capacitors and Redox Flow Batteries

Edward Chen, Jorden Corpuz, Blake Smith, Woodson Squier, Jeffrey Chin, Matthew Armstrong, Ryan P. Murphy, F. John Burpo, Simuck Yuk, Enoch Nagelli

Graphene oxide in aqueous suspension has multiple promising applications as novel nanocomposite electrodes in free-standing film and flowable electrodes for energy storage applications such as redox flow batteries. From the nanomaterial level, interactions between suspended aqueous graphene oxide particles can be used to assemble thin film electrodes or as flowable particle suspension electrodes. Using layer-by-layer self-assembly, 3D electrodes of graphene oxide films are created from all solution based aqueous spin coating techniques. Noble metal salt solutions were spin coated in between sequential graphene oxide layers. The electrostatic interaction of the noble metal complexes and the oxygenated functional groups at the edge and basal plane on graphene oxide help control 3D spacing of graphene films preventing aggregation from self-stacking and loss of surface area. This platform synthesis procedure can be utilized to produce scalable free-standing film graphene oxide electrodes with controlled 3D spacing and with enhanced conductivity from the presence of the noble metals. In addition to free-standing film electrodes, graphene oxide suspensions were electrochemically characterized at different particle loadings as flowable electrodes. Materials characterization techniques such as Scanning Electron Microscopy (SEM) and Energy Dispersive X-Ray (EDX) are used to study the particle size, morphology, and surface elemental composition. Electrochemical techniques such as impedance spectroscopy and cyclic voltammetry are used to study electronic and ionic conductivity and electrochemical surface area. Both the SEM imaging and electrochemical testing confirmed the fabrication of thin films with similar resistivity and conductivity to industry produced chemical vapor deposition films. The free-standing graphene oxide-noble metal cluster thin films were characterized for their potential application as stationary porous electrodes used in electrochemical flow capacitors or batteries. In addition to free-standing film testing, analysis of EIS plots for aqueous graphene oxide solutions demonstrates the potential for suspensions as flowable electrodes. These diverse applications in static and flow from aqueous graphene oxide suspensions show promising application before integration as electrodes into energy storage applications such as flow batteries and flow capacitors.

*Contact: Dr. Enoch Nagelli, **Email:** enoch.nagelli@westpoint.edu

Production of Ammonia from Hydrogen Obtained in an Electrolysis Pilot Plant Powered By Energy from Geothermal Sources

Valeria Arias Corzo, Adrian Paredes, Jesus Mamani

Ammonia has been industrially synthesized since 1908 and has valuable applications in industries such as agribusiness, pharmaceuticals, textiles, refrigerants, plastics, and explosives. However, traditional production methods, such as methane steam reforming, are highly polluting due to greenhouse gasses emissions (CO and CO2). Therefore, this study evaluates the feasibility of environmentally low-impact ammonia production through the Haber-Bosch synthesis, as illustrated in Figure 1. This proposal integrates geothermal energy to supply the required power for 1) the ammonia synthesis stage and 2) the production and purification of synthesis stage reactants: N2 and H2. The aforementioned energy source has been chosen for its sustainability and constant availability. The identified geothermal reservoir, located in the Paucarani District, Tacna, Peru, has a theoretical power supply of 61.1 MW for 37 years, calculated using the volumetric method. Based on its characteristics, such as temperature and flow rate of the manifestations, the implementation of a simple flash power plant is proposed. Regarding H2 production, an electrolysis process with minimal carbon emissions has been proposed.

Consequently, to generate the results, material and energy balances were conducted in DWSIM and Python. An analysis of scenarios was performed in the mentioned software to obtain the amount of ammonia produced from a specific water consumption and the number of required wells, each with a power output of 1.5 MW. It is noteworthy that under the proposed design, 1 well is required to obtain 2.57 kmol/h of NH3 at 99.55% mol, based on a feedstock of 350 kg/h of water and 1050 kg/h of air. As a preliminary result, it was found that it is possible to power the processes involved in ammonia synthesis using geothermal energy, considering the streams described for the design of a pilot plant. Finally, for future research, it is recommended to consider the alternative of including a reverse osmosis plant to obtain the water entering electrolysis, as well as to assess the feasibility of scaling up the process.

Procedural Enhancements to Industrial Pipeline Flushing Operations

David Theuma, Sean Curtis, Michael Fracchiolla, Steven Roth, Emma Padros, Swapana Jerpoth, Barnabas Gao, Kirti Yenkie, Robert Hesketh, C. Stewart Slater

Every year in a single petroleum packaging facility, approximately 1,400 pipeline flushing operations occur between 1,000 petroleum products. Worldwide, this totals to hundreds of thousands of pipeline flushes annually. Pipeline flushing occurs during a product changeover, where the same pipeline must be used to process different types of petroleum products while maintaining the required product specifications. In typical flushing operations the existing product in the pipeline is addressed as the residual oil and the new product is known as the flushing oil. The new product (or flushing oil) is passed through the pipelines to remove any residual oil within the pipeline system. This results in mixing of the residual and flushing oils, thus generating large quantities of commingled oil. Due to the specifications for each of the pure petroleum lube oils, the commingled oil is non-usable for intended applications and must be sold as a downgraded product. This leads to economic losses and environmental concerns. Therefore, there is a need to optimize the pipeline flushing operations at an industrial scale to minimize the production of commingled oils. The goal of this work is to optimize this flushing process using creative solution strategies such as procedural enhancements at both pilot and industrial scales to minimize commingled oil production.

Procedural enhancements have been the focus of this research, involving little to no capital cost with no system reconfigurations required. We first analyzed the current pipeline flushing operations at an industrial partner facility to determine ways to improve current operations, such as the incorporation of air-blowing to better clear the lines of residual oil prior to flushing. The existing and suggested procedures were then tested on a newly developed benchtop pilot plant to determine whether the air-blowing technique was an improvement to the existing flushing procedure. After quantifying the improvement in flushing operations, this procedure was presented to the industrial partners, who accepted these changes. Our team then implemented this improved procedure on their industrial drum filling lines and trained their operators on this air-blowing technique. As a result of this improved procedure, there has been a significant reduction in flushing failures at their facility from approximately 10% to 1.4%. The cost-benefit analysis further revealed that on wider implementation of this new procedure, a single facility can save up to 40,000 to 70,000 gallons of lubricant per year, which is approximately 1.2 million dollars annually, and a reduction in waste by 30-50%.

Metallic Nanoparticles for Green Hydrogen Production and Energy Application

Ghadeer Al-Ajmi

Hydrogen's potential in the contemporary energy sector is substantial, driven by its clean energy attributes, adaptability, and storage capabilities. Moreover, it offers a means to decrease reliance on imported fuels, benefiting from continual technological progress. Particularly noteworthy is its role in decarbonizing challenging sectors. As a result, hydrogen is pivotal in the worldwide shift toward a more sustainable, environmentally-conscious energy landscape. Recent research shows that the energy sources used to produce hydrogen from water are not clean, and it is generated from carbon-based fuel, which increases the global warming effect, is expensive, and is not sustainable. Therefore, it is necessary to design cheap, flexible, and sustainable innovation technology. The project focuses on utilizing Metallic Nanoparticles as a photocatalyst to produce green hydrogen, presenting a groundbreaking advancement in sustainable energy technology. Metallic nanoparticles combined with cellulose fibers have interesting characteristics, and new combinations of both materials may lead to promising functional nanocomposites with unique properties. Here, multi-metallic nanoparticles are synthesized in the presence of cellulose fibers to form cellulose–metallic nanoparticles composite materials for hydrogen production under LED light irradiation. The nanocomposite fabrication methods and hydrogen production application have been discussed. The samples were systematically examined by Transmission electron microscopy (TEM), X-ray diffraction (XRD), and UV-vis spectroscopy

Key words: Metallic nanoparticles, photocatalyst, cellulose fibers

Understanding Resolution Requirements for Sensing Technologies and Emission Scenarios: Toward Effective Emission Detection and Mitigation in the Permian Basin

Sandhya Sethuraman, Arash Fathi, Grant Seastream, Felipe Cardoso-Saldaña, Anantha Sundaram

Localizing methane emissions from oil and gas facilities is a complex and challenging task, especially when those emissions are intermittent, vary in intensity, and are highly contingent on wind and weather patterns.

In order to effectively detect emissions when multiple oil and gas operators are active and present — as is true in the Permian basin — multiple sensing systems, like ground sensor networks, aircrafts, and satellites, should effectively work together. However, these technologies each have their own advantages, limitations, and requirement for optimal operation.
In this study, we strive to understand the impact of wind and model resolution, emission scenarios, sensing technologies, and detection and response time. Due to the complexity of our models, halving the resolution can increase the computational run time by an order of magnitude, making it imperative that we identify a critical time period for our work given our limited resources. Following an extensive literature review, we choose to find a representative week for each quarter of the year by constructing a joint frequency distribution (in space, through time) for each component of our high-resolution wind data. Ultimately, we verify our methods with divergence metrics and simple averaging as a means for comparison, and robust visualizations that demonstrate the similarity between our chosen week and the overall distribution for the quarter.

Next, we run our simulations on this week of wind for each quarter, and model a series of emission scenarios (10, 100, and 1000 kg/hr), which are representative of medium to super emitters. We draw conclusions on the optimal grid resolution to model various emission scenarios, and observe the impact of height, which roughly corresponds to the nature and precision of different sensing technologies, from our simulations. We demonstrate when grid resolution is between 500 m to 1.5 km, the evolution of the plume at the ground level is well-captured. Further work is needed to understand how satellites and aircrafts "see" methane, and to accurately model their sensing profiles.

Modeling methane dispersion from oil and gas sources using the appropriate resolution and timescale is crucial for accurate, reliable, and timely detection of abnormal emissions, leading to effective mitigation strategies.

Spray Ignition of n-Alkanes in Diesel and Jet Fuels: Effects of Temperature, Pressure, and N2 and NO Addition.

Feras Almadani, S. Mani Sarathy

This study investigates the effects of NO-addition and N2-dilution on the autoignition process of n-C10–n-C14 alkanes, which are important components of diesel and kerosene-based jet fuels. The study is motivated by the need for a comprehensive understanding of the effects of nitrogen oxides (NOx) and N2 on the fuel ignition process. This understanding is essential to enable the effective employment of advanced combustion engine technologies, such as exhaust gas recirculation (EGR) and Rich-Quench-Lean (RQL) gas turbine combustors. Ignition delay time (IDT) measurements were performed under various combustion conditions relevant to practical engine operation and low-temperature oxidation chemistry in a constant-volume combustion spray chamber device. The experimental results show that N2-dilution consistently results in longer IDTs for the fuels tested. On the other hand, NO-addition has a temperature-dependent effect, shortening the IDTs for $T > 692$ K and vice versa. The magnitude of reactivity promotion or inhibition depends on NO concentration, temperature, pressure, and equivalence ratio. Based on the tested NO-addition levels, the inhibition of reactivity is approxi- mately ten times greater than the promotion of reactivity caused by NO. The study fills a gap in the literature by providing comprehensive fundamental combustion data for n-C10–n-C14 fuels. This data, along with the corre- sponding analysis, can be used by engine researchers in developing more efficient and cleaner combustion engine technologies.

Fluid Dynamics of Jet Mixing Reactors for Building Simulation Models

J. Garrett Jackson, Jessica Winter, Faiz Khan, Khalil Tran, Nicholas Brunelli

Continuous flow microreactors are used in the synthesis of nanoparticles, critical materials for the biomedical and agricultural industries. Their value results from their unique, size-dependent properties that differ from those of larger, bulk materials. At the lab-scale, nanoparticle synthesis is achieved primarily in batch. However, these processes become uneconomical at the industrial scale up because of poorer mixing that results in inhomogeneous concentration gradients and polydisperse particles. Continuous flow configurations, such as the jet mixing reactor (JMR), would improve industrial synthesis of these materials by improving mixing.

The JMR consists of three input streams that meet at a central mixing point and exit through the output line. The three streams consist of two 0.02" jet lines that flow antiparallel to each other and one 0.04" main line entering perpendicular to the jet lines. The main and jet lines have different compositions and mix in the central region to produce the product line. The primary goal of this project is to obtain experimental mixing time data to be used to validate and construct COMSOL fluid dynamic models.

We compared experimental measurements and COMSOL simulations of mixing time. In well-mixed situations, a kinetically-limited regime prevails that results in monodisperse particles. Small mixing times of 0.5-100 milliseconds directly correlate with a kinetically-limited regime. Mixing time was measured through the commonly used Villermaux-Dushman reaction, which consists of a competing reaction set. Poorer mixing results from a stoichiometric defect of acid that produces triiodide anions, whose yellow color can be measured through UV-Vis Spectrophotometry. Different flow rates in symmetric flow conditions from 8 to 14 mL/min were tested, and the mixing time was determined to be <100 milliseconds for all conditions, which compared favorably to previous studies of the JMR.

Asymmetric flow configurations are also being studied because limiting the flow of one line could provide economic, product, or safety benefits. We are testing the effect of main to jet line flow ratio on mixing time. Our long-term goal is to examine different JMR geometries to further improve the mixing time, guided by machine learning-directed COMSOL simulations.

Real-Time Time-Dependent Density Functional Theory Calculations for the Study of the Effects of High Laser Intensity on Liquid Crystals

Julia Kubes, Kenneth L. Marshall

Liquid crystals (LC's) are useful materials in laser applications in devices such as wave plates and circular polarizers. Despite their remarkable resistance to incident laser energy, under certain conditions, these materials can still experience laser induced damage. The mechanism of the laser-induced damage in LC's is not well understood, which makes it necessary to study the materials on the molecular level to explain the nature of laser damage. Density functional theory (DFT) is a computational chemistry method that uses a simplified version of the Schrödinger equation to calculate a molecule's electron density at its ground state. Real-time time-dependent density functional theory (RT-TDDFT) builds off the ground state calculation to model electron density as a function of time, including its response to an external electric field or a laser pulse. The work described here utilized RT-TDDFT to model the effects of increased laser intensity on the LC molecule 4-cyano-4'-pentylbiphenyl (5CB). Each simulation included a single 5CB molecule in the gas-phase exposed to individual, 1053 nm, 22 fs Gaussian laser pulses of increasing incident energy each time. It was hypothesized that as the incident laser intensity was increased a threshold would be found in which the molecule's electron density would be irreversibly altered after the pulse had passed through the LC material. The results of this experiment show that as incident laser intensity increased the peak magnitude of the induced dipole moment also increased which corresponds to the relationship between intensity and electric field strength. Increased differences in the electron density as intensity increased, shown by the initial and final electron density maps, also align with the hypothesis and may be an indication of the molecule reaching its damage threshold.

PAL 2.0: A Physics-Driven Machine Learning Algorithm for Material Discovery

Yiran Wang

The lack of efficient discovery tools for advanced functional materials remains a major bottleneck to enabling advances in the next-generation of energy, health and sustainability technologies. A major factor contributing to discovery difficulties is the large combinatorial space of materials (with respect to material compositions and processing conditions) that is typically redolent of such materials-centric applications. Searches of this large combinatorial space are often influenced by expert knowledge and clustered close to material configurations that are known to perform well, leaving undiscovered potentially high-performing candidates in unanticipated regions of the composition-space or processing protocol. Experimental characterization or first principles quantum mechanical calculations of all possible material candidates can be prohibitively expensive, making it infeasible to conduct an exhaustive approach to determine the best candidates. As a result, there remains a need to develop computational algorithms that can efficiently search a large parameter space for a given application.

This poster introduces PAL 2.0, a machine learning method that combines a physics-based surrogate model with Bayesian optimization. The key contributing factor of our proposed framework is the ability to create a physics-based hypothesis using XGBoost and Neural Networks. This hypothesis provides a physics-based 'prior' (or initial beliefs) to a Gaussian process model, which is then used to perform a search of the material design space. We demonstrate the usefulness of our approach on three test cases: (1) discovery of metal halide perovskites with desired photovoltaic properties, (2) design of metal halide perovskite-solvent pairs that produce the best solution-processed films and (3) design of organic thermoelectric semiconductors.

The two most compelling results of PAL 2.0 are that we:

1. Demonstrate superior optimization performance, finding the optimal target within the lowest number of iterations as compared to state-of-the-art models such as Genetic Algorithms, an off-the-shelf Bayesian optimization package, SMAC, as well as one-hot-encoded Gaussian Process models for material discovery; and

2. Provide a *predictive* physics-informed model for the material space, offering valuable chemical insight.

To the best of our knowledge, PAL 2.0 is the first computational materials discovery framework that utilizes predictive, physics-based surrogate models within a Bayesian optimization framework by combining feature engineering and material space modeling.

Surface Halide Vacancies Influence the Phase Transition in $CsPbI_3$

Samuel Smith, Rory Campagna, Jonathan Outen, Zachery R. Wylie, Jeffrey Christians

Halide Perovskites are an emerging technology with promising potential for photovoltaic applications. The improvement of halide perovskite stability and efficiency in recent years has begun to make halide perovskites a potential competitor to traditional Silicon based photovoltaic technologies. However, despite these continued improvements, halide perovskites continue to lack long-term stability limiting their viability in large markets. This research was conducted using Cesium Lead Iodide ($CsPbI_3$). Cesium Lead Iodide is known to have severe stability issues, one of which is that this perovskite will undergo a change in state which can be visibly observed as the perovskite film will transition from a dark colored phase to a light, yellow phase impacting the materials ability to effectively act in a photovoltaic system. The Cesium atoms in this perovskite's crystal structure are too small, ultimately allowing the structure to tip over causing this change in state. Additionally, when manufacturing halide perovskites, vacancies within the crystal structure. These vacancies can then act as a nucleation site for the phase change. One proposed method of limiting these crystal vacancies is through doping the perovskite with additional material This research had the primary focus of exploring the effects of doping halide perovskite films by doping Cesium Lead Iodide perovskite with Iodine using a solution of either Cadmium Iodide (CdI2) or Cesium Iodide (CsI) dissolved in Isopropyl alcohol. This research consisted of doping Cesium Lead Iodide perovskites with Iodine using the previously mentioned solution at varying concentrations. Trials with Cesium Iodide and Cadmium Iodide were done at concentrations ranging from 0.001 mmol/ml to 0.01 mmol/L. The effects of doping these perovskites were observed by measuring the change in phase fraction of the doped perovskites over time. Absorbance data was collected via spectroscopy and converted into phase fraction through the use of the JMAK model, $x(t) = \exp(-kt^n)$ where n is the growth coefficient, k is the rate constant and t are representative of time. Graphing the calculated phase change data against time displayed decreasing rates of change with increasing concentrations of doping material. It was then concluded that doping these perovskites slowed increased the materials stability suggesting that halogen doping increases halide perovskite stability.

Dielectric/Impedance Spectroscopy for in-Line Measurement of Moisture Content

Sophie Zhou

Pharmaceutical production requires strict moisture content control in order to ensure quality, stability, and efficacy of medicinal products. Moisture content can be manually tested but a rapid and reliable solution for real-time moisture content monitoring was necessary to improve overall efficiency in pharmaceutical manufacturing. This research delved into the practicality of using electrochemical impedance spectroscopy (EIS) as a dependable technique to determine moisture content. By investigating the electrical properties of materials, we established a robust connection between shifts in electrical properties with variations in moisture levels. We employed an equivalent circuit model to unveil the underlying mechanism which provided valuable insight into the sensitivity of impedance spectroscopy to changes in moisture content. Furthermore, this study incorporates AI techniques, employing a 1D Convolutional Neural Network (1DCNN) model to effectively process the complex spectroscopy data. The proposed model integrates frequency spectrum information, equivalent circuit indices, and material characteristics to estimate moisture content with an average predictive accuracy of 1.06%. Our work represents a significant advancement in the field, promising to enhance process control, ensure product quality, and drive overall efficiency improvements within pharmaceutical manufacturing.

Hexagonal Boron Nitride Modulates Crystallinity and Charge Mobility in PEO–NaNO3 Electrolytes

Colby Snyder, Shreyas Pathreeker, Russell J. Composto

Composite polymer electrolytes (CPEs) hold great promise for the development of safe and sustainable batteries. In this study, we find that 2D hexagonal boron nitride (h-BN) as a filler has a non-monotonic effect on polymer crystallinity and total ionic conductivity in PEO-$NaNO_3$ electrolytes. The dual Lewis acidity and basicity of hexagonal boron nitride (h-BN) allows it to interact with dissociated salt ions and the polymer matrix itself. PEO crystallinity was quantified using differential scanning calorimetry (DSC) and X-ray diffraction (XRD), and complex formation between $NaNO_3$, PEO, and h-BN was studied using ATR-Fourier Transform IR (FTIR) spectroscopy. Total ionic conductivity was determined using electrochemical impedance spectroscopy (EIS) as a function of temperature. We find that h-BN has two competing effects in CPEs: 1) nucleation-enhanced crystallization of PEO on h-BN surfaces at low h-BN loading, and 2) spherulitic confinement of PEO at higher h-BN weight loading. These competing effects reveal that h-BN modulates both polymer crystallinity and charge mobility in our CPEs. Density Functional Theory (DFT) calculations confirm strong attractive interactions between h-BN and both free ions (Na^+ and NO_3^-), and we also find lesser attractive interactions between h-BN and PEO. These new findings for Na-polymer electrolytes support our experimental results. Our findings highlight the importance of filler geometry and chemical characteristics in designing future CPEs for Na-ion batteries.

Silver Coated Ionic Liquids for Liquid Phase Space Optics

Titus Szobody, Vivek Dwivedi, Michael Dickey, Meixiang Wang

Origami inspired deployment of space telescope mirrors is a complex approach to increasing the aperture size of these orbital science assets with potential multi single point failures, see the James Webb Space Telescope (JWST). The size, mass, and general handling issues of these mirrors have been a cost, time, and energy-intensive challenge to space agencies like NASA. To address these challenges, a novel approach is needed. Herein we report the development of an alternative multiwavelength reflecting-based mirror that in theory is geometrically independent from these orbital assets.

Our research demonstrated the successful coating of an ionic liquid with silver nanoparticles under UV light exposure to form a reflective, mirroring surface. Ionic liquids are salts with extremely low melting points and negligible vapor pressure. These properties render them ideal candidates for liquid deployment in space without mass loss in vacuum or freezing in ultra-low temperatures. The general procedure includes using candidate ionic liquids such as 1-ethyl-3-methylimidazolium ethyl sulfate ([EMIM][ESO_4]) and dissolving a silver containing salt, silver tetrafluoroborate ($AgBF_4$), with a photoinitiator diphenyl(2,4,6-trimethylbenzoyl)phosphine oxide (TPO) followed by UV light exposure. This resulted in the self-assembly of silver nanoparticles at the surface of the ionic liquid caused by the photoreduction of Ag^+ cations by free radicals from the homolytic cleavage of the photoinitiator, forming a smooth and reflective surface. The liquid mirror exhibits reflective and self-healing properties while maintaining stability in vacuum conditions. Additionally, there is potential for the introduction of polymeric chemistry to enable the tunability of the mirror viscosity. The outcome of this research demonstrates alternatives to conventionally used space telescope mirrors applicable to the next generation observatory.

Gravitational Effects of Electrodeposition on Zinc Anodes

Elena Medina, Ngoc Nguyen, Christopher Fetrow, Shuya Wei

Zinc is abundant, safe to handle, non-toxic, and has a high theoretical capacity of 820mAh/g. These benefits cause zinc metal to be a promising anode material and alternative to lithium-ion batteries. Zinc's prolific dendritic growth is the main challenge that prevents the widespread and commercial use of zinc as an anode material. Dendrite growth increases the surface area which can cause the anode to corrode faster. In addition to the increased surface area, the dendrites can connect the working and counter electrodes, causing the battery to short and fail. Zn||Zn parallel visualization cells were set up to further understand the dendritic growth of zinc anodes. Three visualization cell setups were tested with a 2M ZnSO4 electrolyte and were used to explore the growth behavior of the dendrites. After constant current charge or discharge, the anodes were examined under an optical microscope to analyze the dendrites' morphology. The optical microscope images demonstrated that dendritic growth on zinc anodes is influenced both electrochemically and structurally by gravity. Zinc deposition with ions moving against the gravitational field showed more even surface coverage, a more compact structure, and a more stable plating and voltage potentials.

Simple Preparation of Metal-Impregnated FDM 3D-Printed Structures

Diana Flores, Jose Noboa, Mickaela Tarapues, Karla Vizuete, Alexis Debut, Lorena Bejarano, Daniela Almeida Streitwieser, Sebastián Ponce

Three-dimensional (3D) printing has gained widespread acceptance within the engineering field due to its cost-efficiency and versatility in design. Nevertheless, certain applications need surface modification of 3D-printed materials to improve its compatibility. This study is primarily focused on the impregnation of Fused Deposition Modeling (FDM) 3D-printed Polylactic Acid (PLA) samples with silver nanoparticles and an iron salt. The impregnation method involves cleaning, activating, and etching before impregnating the material to create a porous surface on the PLA. Silver nanoparticles, both spherical and non-uniform in shape, were successfully synthesized and securely affixed to the surfaces of the PLA samples using a microfluidic reactor. Simultaneously, iron salt was robustly anchored to the 3D-printed stirrer caps, serving as a catalyst for dye degradation processes. To confirm particle synthesis and surface adhesion, several analysis methods were run such as UV/Vis spectrometry, transmission and scanning electron microscope, energy dispersive spectroscopy and X-ray diffraction. Reaction and adhesion tests were employed to assess the strength of impregnation and the chemical performance of the new material. These include a flow test on the silver nanoparticle milli-fluidic reactor and a kinetic analysis for dye degradation with the iron salt stirrer caps. This research contributes to the expanding spectrum of 3D printing applications in engineering, showcasing its adaptability in accommodating diverse material modifications for a wide range of functional applications.

Effects of Geometry on the Locomotion of Active Particles Driven By Self-Diffusiophoresis and Induced-Charge Electrophoresis

Zoe Cruse, Arkava Ganguly, Kendra Kreienbrink, C. Wyatt Shields IV, Ankur Gupta

The discovery of self-propelled particles has spurred interest in creating synthetic microswimmers that mimic biological entities for autonomous propulsion with medical and environmental applications. Phoretic phenomena demonstrate how colloidal particles can be driven by external fields generated by a variety of gradients such as concentration (diffusiophoresis), temperature (thermophoresis), or electrostatic potentials (electrophoresis). This work examines the geometric effects of patch shape on the motion of patchy spherical colloids for two mechanisms: self-diffusiophoresis and induced-charge electrophoresis (ICEP).

Our work on self-diffusiophoretic particles focuses on achieving tunable propulsion by modifying phoretic interaction between the catalytic particle surface and the solute gradient. We determine the resultant solute concentration by numerically decomposing the patch effects into a combination of spherical harmonic modes through an optimization scheme. To calculate the propulsion velocity, we obtain a continuum-scale slip velocity from the tangential gradient of the surface solute concentration and employ the Lorentz reciprocal theorem. We validate our methodology using well-known results for model problems, such as: (i) a uniformly active particle (no motion); (ii) a Janus particle (translates). Our results demonstrate how catalytic patch and the phoretic interaction can be fine-tuned to control both translational and rotational motion.

Additionally, we have examined how patch orientation affects the motion of particles when they are propelled via ICEP. To theoretically describe this range of ICEP propulsion, we first solve for the electric potential in the fluid bulk arising out of an imposed electric field and a given surface potential profile. From a given polarizability asymmetry, we can obtain an expression for the induced electrophoretic slip via the Einstein-Smoluchowski equation. Finally, using the obtained slip velocity, we derive expressions for the translation and rotation of the particle. Through this theoretical analysis, we found qualitative agreement between the predicted and experimentally observed cluster trajectories[1].

Finally, we have begun investigating geometric effects of patch shape for particles driven *via* ICEP. By fabricating arbitrary patches on spherical particles using a two-photon lithography technique, our preliminary results indicate that the particle trajectory depends on patch geometry and can result in linear, helical and trochoidal trajectories. In the future, we plan to conduct further computations and experiments to fully describe the autonomous motion of these particles.

Relevant Publications:

1. Lee, Jin Gyun, et al. "Magnetically locked Janus particle clusters with orientation-dependent motion in AC electric fields." *Nanoscale*, 2023, p. 9.

Electrochemical and Rheological Properties of Poly(vinyl alcohol) and Carbon Nanotube Composite Binders for Lithium Sulfur Battery Applications

Matthew Moellering, Michael J. Williams, Avery Patel, Caspar Yi, Patrick Bowers, Sam Cowart, F. John Burpo, Simuck Yuk, Yong Joo, Enoch Nagelli

This study develops a method for the air-controlled electrospraying of poly (vinyl alcohol) (PVA)/carbon nanotube (CNT) nanocomposite binders for enhanced thermal stability, conductivity and electrochemical performance in in lithium-sulfur (Li-S) batteries and electrochemical supercapacitors. CNTs are comprised of carbon atoms in hexagonal formations with a backbone of alternating carbon-carbon double bonds that form linear high aspect ratio walled tubes. Chemically oxidized CNTs have carboxylic acid, carbonyl, and epoxide functional groups on the side walls and tips of the carbon hexagons along the CNT structure. In this method, CNTs were chemically oxidized in strong acids at high temperature, PVA was hydrolyzed and mixed with the oxidized carbon nanotubes, and the resulting PVA/CNT mixture was air-controlled electrosprayed to form nanocomposite PVA/CNT binder films through hydrogen bonding and physical crosslinking. The PVA/CNT binder films are characterized using Fourier transform-infrared spectroscopy, thermogravimetric analysis, rheometry, Raman spectroscopy, scanning electron microscopy (SEM), and with electrochemical techniques such as cyclic voltammetry and impedance spectroscopy to show evidence of chemical functionalization, thermal and mechanical stability, enhanced conductivity, low charge transfer resistance, and lithium-sulfur (Li-S) and electrochemical charge-discharge performance, respectively. Performance showed promising cycling, high capacitance, and conductivity. Ultimately, our analysis allows for the scalable application of the thermally stable and conductive binder in high-performance batteries and supercapacitors.

Exploring the Cobalt–Carbon System at High Pressures

Zeynep Alptekin

Metallic carbides are ideal in heavy-duty applications for hypersonic flight and aerospace engineering. In renewable energy research, there is a need for a sustainable permanent magnet that can support technological innovation without being as environmentally taxing as lanthanide rare earth permanent magnets. It has been shown that carbon may have a magnetic pinning effect on carbon, inducing permanent magnetic effects. However, cobalt–carbon only exists as a nanomaterial or thin film, where surface effects can stabilize the metastable compound. We investigated the cobalt–carbon system computationally via random structure searching done with Ab Initio Random Structure Search (AIRSS) interfaced with Castep to perform DFT calculations that generate random structures of cobalt–carbon and adjust them into their local energy minima. We randomly sampled phases of cobalt–carbon at incremental pressures of 5 GPa between 0 and 25 GPa and found a system of 16-atom *Pnma*-Co3C that became thermodynamically accessible above 7 GPa. We assembled four diamond anvil cells (DACs) containing carbon and cobalt to perform laser heated synthesis at the Argonne National Laboratory.

Equivalent Circuit Model Analysis for Electrochemical Characterizations of Textile-Based Supercapacitors

Laurel Hilger, Alyssa Grube, Mona Bavarian

Textile-based supercapacitors (TSCs) hold untold potential benefits as lightweight, wearable, safe, and convenient source of energy for commercial and personal electronics. TSCs are fabricated with highly conductive yarns knitted together into electrodes. Herein, $Ti_3C_2T_x$ MXene, a two-dimensional transition-metal carbide with a conductivity as high as 10 kS/cm, is used to fabricate conductive yarns. In this study, cotton and wool yarns are coated with MXene flakes to build TSCs. Subsequently, electrochemical characterizations including cyclic voltammetry (CV) and electrochemical impedance spectroscopy (EIS) were conducted. Equivalent circuit models (ECMs) were constructed based on the physics of the device and fitted to the EIS measured data to determine the theoretical resistance and capacitance. EIS analysis reveals that the TSCs stray from an ideal Randles circuit model and act as a constant phase element (CPE), resembling an imperfect double-layer capacitor. The selected model adds a Warburg impedance in parallel to adjust for further mass diffusion resistance. These models will be used in future fabrication and analysis of improved TSCs with lower resistance and more pseudocapacitive behavior. From electrochemical characterization, cotton TSCs achieved an areal capacitance of 823.9 mF/cm^2 while wool TSCs achieved 284 mF/cm^2 at a scan rate of 5 mV/s.

Optimizing Spin Coating Parameters for Uniform Photoresist Deposition on Silicon Wafers in Diffraction Grating Development

Coenradt Taylor

This poster presents a study focused on achieving uniform deposition of photoresist on silicon wafers for use in diffraction grating development. Spin coating, a common technique for thin-layer deposition, was utilized to achieve uniform photoresist layers on the wafers. Through a comprehensive Design-of-Experiments (DOE) approach, various spin coating parameters were systematically examined to enhance the deposition process. The variables explored included spin speed, spin time, ramp speed, ramp steps, exhaust percentage, wafer bake time, and use of a primer agent. These elements offer insights into the factors governing uniform photoresist deposition on silicon and fused silica substrates. It was observed that shorter bake times, lower exhaust percentages, and slower ramp speeds resulted in the desired photoresist layer of 800 to 1000 nanometers. In addition, a sufficiently large first ramp step in the spin-up process resulted in more uniformity of deposition. The resulting coated wafers from this process are later used in photolithography and reactive ion etching processes to improve the manufacturing process for optical diffraction gratings.

Enhancing Protonic Conductivity in $Ba(Zr_{0.4}Ce_{0.4}Y_{0.1}Yb_{0.1})O_3$ Electrolyte for Solid Oxide Electrolysis Cells through Sintering Aids and Single-Grain Layer Design

Ha Tran, Xuemei Li, Shaoshuai Chen, Wenyuan Li

One of the formidable challenges is to reduce the reliance on fossil fuels and the greenhouse effect. Therefore, the development of storable and transportable chemical fuels by employing electricity and water vapor to drive an endothermic chemical reaction is the key. The electrolyte in solid oxide electrolysis cells serves a critical role in facilitating ionic conductivity. Among various investigated compositions, this project aims to develop a $Ba(Zr_{0.4}Ce_{0.4}Y_{0.1}Yb_{0.1})O_3$ electrolyte with high protonic conductivity for proton-conducting solid oxide electrolysis cells. It is known that at 1500°C, the high-temperature densification process leads to Ba evaporation, Ni diffusion from other components, and phase segregation of Y_2O_3 and Yb_2O_3, all of which contribute to a reduction in the proton conductivity of this material. Sintering aids along with other additives have been investigated to achieve a single-grain layer electrolyte structure, thus avoiding the issues associated with high-temperature densification and high resistance from grain boundaries. Appropriate amounts of ZnO in addition to Ba, Y, and Yb oxides are introduced as sintering aids to improve the ionic conductivity and lower the electronic conductivity of the material with better Ni electrode performance. The effect of Ni diffusion from the substrate to the electrolyte and its coarsening is reduced due to lower electrolyte densification temperature. The electrolyte's phase purity, microstructure, and proton conductivity are further measured and analyzed.

Synthesis and Characterization of Combustion-Generated Particles for Nanostructured Materials

Cassandra Volpe

Biogas, a gas produced through the anaerobic digestion of organic waste, offers a sustainable alternative to natural gas. However, the usability of biogas is hindered by excess gasses, including carbon dioxide and hydrogen sulfide. Developing a method to remove this excess gas is necessary to utilize biogas as a sustainable natural gas alternative.

The overarching goal of this project is to utilize flame spray pyrolysis (FSP) for synthesizing an efficient adsorbent for carbon dioxide (CO2). FSP, a technique that sprays a liquid precursor mixture into a flame to synthesize small particles, was chosen for its ability to produce products with fewer impurities than wet chemistry methods. It also can synthesize only the desired product, reducing the need for costly separation processes.

The project focused on optimizing an updated spray nozzle system within the burner and evaluating its performance compared to an earlier design. The nozzle configuration was modified using mechanical adjustments, and operational settings, including liquid flow rate and air pressure, were examined to determine the optimal conditions for droplet formation. The similarity between the two designs was quantified by analyzing the adsorbents from the updated system and comparing them to those from the previous design using a Scanning Mobility Particle Sizer (SMPS).

Cellulose Nanofiber-Carbon Nanotube Biotemplated Composite Nickel/Nickel Oxide Aerogels for Pseudocapacitor Electrodes

Grant Lee, Paul Trackey, Rosemary L. Calabro, Enoch Nagelli, Galen Mandes, F. John Burpo, Stephen F. Bartolucci, Joshua A. Maurer

Electrochemical pseudocapacitors made from metal oxides are a critical technology for energy storage. While there has been a considerable body of research into pseudocapacitor synthesis, there are several challenges to its widespread implementation, namely, mechanical durability and active surface area. To address this need and to develop low-cost, transition porous metal nanomaterials for energy storage applications, one strategy is to use naturally occurring biopolymer-based templates to better tune material properties such as feature size, pore structure, conductivity, material phase, and mechanical strength. Examples of this approach to control nanostructure and functionality include biotemplated gelatin, silk fibroin, and cellulose nanofiber composite metal aerogels. Where biopolymers confer nanostructure shape control and chemical functional groups to mediate metal phase deposition, the addition of carbon allotropes such as graphene and carbon nanotubes allows for strength and electrical and thermal conductivity. In this work, we demonstrate the synthesis a porous, mechanically strong material with tunable pore size, nanowire length, diameter, and material phase which offers the possibility of electro-mechanical materials that enable device fabrication at multiple length scales with an overall decrease of systems mass and corresponding increase in performance metrics. This is achieved using a hybrid carboxymethyl cellulose nanofiber-carbon nanotube (CNF-CNT) composite hydrogel. Hydrogels equilibrated in a nickel salt solution are chemically reduced, rinsed, and then supercritically dried. Subsequent pyrolysis, and thermal annealing results in a porous NiO psuedocapacitor aerogel. Aerogels were characterized using scanning electron microscopy, energy dispersive x-ray spectroscopy, x-ray diffractometry, nitrogen gas adsorption-desorption with and Brunauer-Emmett-Teller surface area analysis, vibrating-sample magnetometry, and compressive stress. Electrochemical performance was determined with electrochemical impedance spectroscopy and cyclic voltammetry. Due to the low-cost synthesis and tunable properties, these composite CNF-CNT-metal/metal oxide aerogels are expected to provide material solutions to a wide range of energy storage, catalytic, and fuel-cell applications.

Rapid Synthesis of Metal-Organic Frameworks Onto Polyester for the Filtration of Particulate Matter

Lauren Hernon, Ankit Dhakal, Gaurav Giri

In light of the COVID-19 pandemic and an increasing amount of air pollution, there is a current need for face coverings that are more effective at filtering particulate matter than disposable medical masks and fabric masks. Metal-organic frameworks (MOFs) are highly porous, crystalline compounds that have been used to filter particulate matter. The scalability issues that arise from prior work are that the MOF formation can take hours to days and requires harsh organic solvents, such as DMF. In this work, we synthesized a zirconium (Zr)-based MOF (UiO-66) on the surface of polyester fabric using sequential dip coating (SQD) within minutes using water as the solvent.

To begin, small swatches of MOF fabric (2-inch diameter circles) were made and tested for particulate matter filtration. Then, the process was scaled up using a roll-to-roll coating process to make full sized face masks (6" by 9" rectangles). The full-sized MOF-fabric masks showed enhanced filtration capacity and passed the ASTM F3502-21 Level 1 standard for face coverings, which is filtration greater than 20% efficacy and a pressure difference fewer than 15 mmH2O when tested at an air flowrate of 85 liters per minute. In this study, we show that we can synthesize large area MOF-fabrics in aqueous medium for better particulate matter filtration within minutes.

Fundamental Structural Study of Boron Nitride (BN)-Based Materials at Low and High Temperatures

Jae-Won Seo, Amol Pophali, Seongwoo An, Chi Seng Lee Liang, Sihan Li, Henry Liu, Eunkwang Park, Jaewoo Kim, Tae Jin Kim

In the recent past, boron nitride (BN)-based materials have raised interests due to their physical and chemical properties (i.e., wide band gap (5-6 eV), high oxidation resistance, high thermal conductivity, high mechanical strength, and high surface area). Amongst these BN-based materials, boron nitride nanotubes (BNNTs) and hexagonal-boron nitride (h-BN) have acquired great attention due to unique structures. The BNNTs have a tubular structure with boron and nitrogen atoms in hexagonal lattice, whereas the h-BN showed 2D geometry in a hexagonal configuration. This work aims to study the stability of BNNTs and h-BN materials under broad temperature range (-50 to 800 oC) in terms of weight and molecular structure. The BN-based materials were tested using thermogravimetric analysis (TGA), Fourier Transform Infrared spectroscopy (FTIR), and ex-situ/in-situ Raman spectroscopic techniques to understand the thermal and structural properties under harsh conditions. Raman spectroscopy results reveal that BNNTs responded well only to the UV (325 nm) laser, whereas structure of h-BN was well layered and detected by all (UV, visible and near IR) lasers. Low temperature (room temperature to -50 oC) in-situ Raman studies were performed for BNNTs and h-BN revealing their stable molecular structure in these conditions. On the other hand, h-BN showed a red-shift in Raman spectra when treated at 800 oC under inert (N2) and oxidative (O2) conditions confirming expansion in the hexagonal layers. This was further confirmed by FTIR analysis which showed structure modification at higher temperatures for h-BN due to structural expansion, while no changes were observed for BNNTs. The collected characterization data confirm the stability of BNNTs under high and low temperature with inert/oxidative conditions. Future scope of this work is the use of stable BNNTs as an effective support material for catalysts synthesis and use for various gas phase chemical reactions.

Interpretation of Platinum Microelectrodes Impedance Response Using Measurement Model

Taskina Zaman Jui, Cynthia Ezeh, Mark E. Orazem

Microelectrodes are used for cellular and neurophysiological applications. This study is focused on identifying the electrical properties of platinum electrodes using electrochemical impedance spectroscopy. Impedance spectroscopy is a powerful tool to study the non-ideal electrical circuits by considering frequency-dependent potential as input and frequency-dependent current as output. The impedance data can be fitted by a process model to predict the material and transport properties of a system.

The impedance data was collected with the Gamry Reference 3000 Potentiostat at three potentials: open circuit (OC), OC+10mV, and OC-10mV. A sinusoidal perturbation of 25 rms was applied for a frequency range of 1MHz to 0.1Hz. The electrolyte was phosphate-buffered saline, and the working electrode was connected to a platinum-mesh counter electrode and a silver/silver chloride (Ag/AgCl) reference to form a 3-electrode cell configuration. The measurements were taken under aerated conditions and controlled at a temperature of 37°C.

The measurement model program by Watson and Orazem [1] was used to analyze impedance data. The program fits the Voigt element series to impedance spectra within the 95.4 % confidence interval. The stability and causality of the impedance response were inspected by analyzing the stochastic error structure. The linearity was confirmed by studying Lissajous plots. The process model indicated that the electrochemical system included reduction of oxygen, which was influenced by diffusion, oxidation of water, and the constant-phase-element behavior of the electrode.

Acknowledgements:

The authors thank Prof. Won Tae Choi and his student Seokjun Han for providing the electrodes used in this study.

References:

[1] W. Watson and M. E. Orazem, EIS: Measurement Model Program, Version 1.8, ECSArXiv, 2023, 10.1149/osf.io/g2fjm."

High Temperature Dilatometry on Tungsten

Michael Zaza, Davis Conklin, Hailey Loehde-Woolard, Alan Weimer

Refractory metals such as tungsten belong to a small class of metals that have a high melting point, density, tensile strength, electrical conductivity, and a low thermal expansion coefficient. Tungsten (W) is used in nuclear control rods and rocket nozzles for its high melting point, and is used in other fields for its chemical and mechanical properties. Tungsten is fabricated through the process of powder metallurgy- where metal powders are compacted and then heated, in a process known as sintering, to achieve a higher density. A master sintering curve (MSC) is a versatile model that is used to optimize the sintering process. MSCs are used to determine ideal sintering conditions based on the density desired; they combine the effects of temperature and time into a single parameter that relates temperature profiles to a predicted final density and provide an activation energy of sintering. These curves are typically generated using dilatometry as the density of the sample is measured at every point in the temperature program. However, this approach has not been widely utilized due to the high temperature tungsten needs for sintering. Here we show the results of high-temperature dilatometry on a tungsten pellet to produce a master sintering curve from one experiment. A green tungsten pellet was produced by pressing 1.5g of 1-5m powder at 1 metric ton with no binders or lubricant in a 6mm die. The green density was 63% of the theoretical density of W. We performed dilatometry using a Netzsch 402c push-rod dilatometer equipped with a graphite furnace capable of testing at 2000°C. 3% hydrogen with balanced argon was used to purge the sample holder. Shrinkage measurements are shown below in Figure 1. We found that heating the tungsten pellet at a ramp rate of 2°C/min, reaching 1950°C, then cooling the pellet at a ramp rate of 20°C/min resulted in tungsten having a higher final density of 77.74%. Based on the shrinkage data and using the activation energy for W sintering from literature, a MSC will be calculated for this experiment. The MSC is modified by substituting the densification ratio and densification parameter to identify the regions where shrinkage occurs. Furthermore, a MSC will be produced to analyze the shrinkage resulting from different ramp rates of tungsten.

Synthesis of h-BN/Nanosilica Hybrid Reinforced Room Temperature Vulcanizing Silicone Rubber (RTV SiR) Nanocomposites for High Voltage (HV) Insulators

Muhammed Zeeshan Ahmed

High voltage (HV) insulators play a crucial role in power transmission, microelectronic components, transformer systems, and other energy applications. However, they often face challenges related to the surface accumulation of moisture and solid contaminants, which can lead to thermal runaways and flashovers. Such problems ultimately reduce the insulator's durability and overall efficiency. To address these concerns and minimize maintenance costs, polymeric composite-based coatings have become a popular solution.

This study focuses on the fabrication of room temperature vulcanizing silicone rubber (RTV SiR) based nanocomposites as external coatings for HV insulators. The primary objective is to enhance the thermomechanical and degradation-resistant properties of RTV SiR using hexagonal boron nitride (h-BN) and nanosilica reinforced hybrid filler. Using perflouro-octyl-triethoxysilane (FTS) functionalized h-BN (h-BNNS), improved filler dispersion is achieved within the SiR matrix, which offers increased mechanical strength and hydrophobicity. Different loadings of h-BNNS, h-BNNS/SiO2, and SiO2 are prepared using a facile synthesis method and dispersed homogeneously into the RTV SiR matrix. The resulting nanocomposite is spray-coated onto ceramic tiles to evaluate the impact of filler type and loading amount on thermomechanical stability, surface wettability, and resistance to surface tracking.

Through characterization, h-BN/SiO2 reinforced RTV SiR nanocomposites exhibited a manifold increase in electrical breakdown strength in the Inclined Plane Test (IPT) in comparison to pristine RTV SiR, while also showing uniform filler dispersion using Scanning Electron Microscopy (SEM). It also outperformed pristine RTV SiR in Thermogravimetric Analysis (TGA), Water Contact Angle (WCA), and Dynamic Mechanical Analysis (DMA), indicating the nanocomposite's superior thermal stability, dielectric properties, interfacial interactions, and minimal agglomeration, rendering it ideal for energy applications in harsh environments.

Mitigating the Effects of Road Salt on Concrete

Annamarie Lechleidner

Concrete is 10% cement, 30% sand, 40% gravel, and 20% air and water. Cement is made up primarily of lime (60-65%), silica (17-25%), and alumina (3-8%).

Concrete can be damaged by the freeze/thaw cycle of water seeping into pores in the concrete which causes hydraulic pressure build-up and cracking.

Chloride salts such as calcium chloride, magnesium chloride, and sodium chloride are often used to minimize the freeze/thaw cycle by lowering the freezing point of the water.

Unfortunately, these salts can lower the alkaline pH (usually around 11) of the concrete in addition to leaching the calcium hydroxide from the cement, weakening the structure of the concrete.

Multipurpose Lubricants from Composite Nanomaterials

Benjamin Jackson, Nicholas Migaldi, Halle McGuire, Meagan B. Elinski

Emerging technologies have shown increasing dependence on lubrication and wear protection. Composite systems composed of nanoparticles and two-dimensional (2D) materials are well-positioned for potential use as multipurpose liquid lubricants for applications such as electric vehicles or renewable energy systems. Nanoparticle additives, like zirconium dioxide, form thin tribofilms between sliding surfaces that protect the material from mechanical wear. 2D materials, such as phosphorene and MoS_2, have well-controlled chemical composition, nanoparticles showcase low friction through a nanoscrolling mechanism where the 2D material wraps around the nanoparticles. The induced strain further selectively controls the bandgap in 2D materials, useful for electronic applications. This project seeks to investigate the potential of a composite system with tribofilm forming nanoparticles, a reduced friction environment as a result of nanoscrolling, and adjustable electronic and optical properties from 2D material deformation.

Additive Manufacturing Porous Carbon with Complex Geometries

Nathan Smith

Porous Carbon is important for applications such as filters, energy storage, and catalysts. Current manufacturing methods for producing these porous carbon structures are time and energy intensive, and use harmful chemicals. The popular Template method of porous carbon production is limited by the difficulty, inefficiency, and wastefulness of the template removal process. Alternatively, 3D printing aqueous inks loaded with polymeric carbon precursors offers a material-, cost-, and time-efficient route to mass producing carbon-based prints with complex geometries which yield carbon with tunable porosity upon heat treatment. Polyvinyl alcohol (PVA) is added for steric stabilization of the carbon precursors, in which the particles are coated with hydrophilic tails that prevent flocculation through weakening of the Van der Waals attractive forces between particles. This addition of PVA increases solids loading, which aids in mass manufacturing because it which reduces shrinkage upon drying, allowing for easy standardization amongst shape. A water solvation post-processing step removes the PVA to leave behind a geometry of carbon precursors. Upon this step we characterize the pore structure via micro-CT and SEM, and analyze the geometry with high precision 3D scanning, giving optimization parameters for finding the ideal PVA weight percent. In the future carbon doped with specific derivatives such as carbon black or CNTs can be added easily, allowing for other properties such as electrical conductivity to be altered.

Additive Manufacturing of Energetic Materials

Matthew Ellsworth, Patrick Bowers, Justin Hunter, Faris Wald

Abstract:

This abstract describes the methodology and objectives aimed at investigating the compatibility of ammonium perchlorate and aluminum (AP/Al) with Viton (VAPAL) for use in additive manufacturing. The project is comprised of two phases. Phase 1 involves a compatibility analysis of the materials, followed by the formulation and rheological analysis of the VAPAL and a mock mixture (substituting ammonium chloride for AP) to predict optimal printing parameters. Phase 2 uses a Filabot extruder to form filaments of both materials for use in a Fused Deposition Modeling (FDM) printer. The project also aims to plot the energetic performance of the material in relation to solids loading and/or binder selection.

Results: The procedure for suspending AP/Al in Viton was effective, with a theoretical maximum density of 99.01%. Furthermore, the results from both sensitivity testing and analysis showed that Viton is compatible with AP/Al as the stability of the product increased when compared to pure ammonium perchlorate. Additional results from a differential scanning calorimeter show that the decomposition of the product occurs at the same temperature as ammonium perchlorate indicating that no new compound was created, and our AP was intact in Viton. Lastly, samples taken along an extruded sample show density of AP/Al similar to the measured density of 99.01%, validating our filament extruding process. Further research will be done exploring the feasibility of the VAPAL (Viton-Al/AP product) within 3D printing. Challenges we expect with this next phase of research include how Viton will behave in the extruder head of a printer, how high definition we can print with this filament, safe operating protocols when printing with VAPAL, and the variance in burn rate and deflagration rate of printed VAPAL.

PDMS Stamping Device for Double-Sided Janus Particles

Luna Silva

[1]Luna Silva Esquinas, [1]Eric McPherson, and [1]Prof. Ilona Kretzschmar

[1]The City College of New York, Department of Chemical Engineering

Janus particles have at least two different materials on their surface and are studied for potential applications in targeted drug delivery, optical sensing, and more. Janus particles can also be coated with materials on both sides, i.e., double-sided Janus particles, which can have unique properties such as active magnetic fluids and magnetic assembly. The process of placing two coatings requires a complicated stamping approach that is difficult to control, leading to reduced reproducibility of the stamping process and low quality of the double-sided Janus particles.

Here we report on the design of a device that has the capacity to simultaneously stamp up to seven slides with Janus particles onto PDMS-coated microscope slides. This approach makes the stamping of the particles easier, faster, and more reproducible. It also scales up the process currently used in the laboratory. With the help of the PDMS stamping device, we are able to process more than one slide at a time and control the force applied during the stamping process. The device's performance has been tested using sensors on different spots where the slides are placed, collecting data to determine the best time and weight for the stamping, and using the device to make double-sided particles, which have been characterized to check the quality of the resulting particles. Double-sided particles have been coated using physical vapor deposition (PVD) and the coatings have been checked using a scanning electron microscope (SEM). Moreover, the behavior of the double-sided Janus particles that have been made with this device have been studied, which have magnetic properties and can self-assemble.

The Effect of Slurry Composition on the Surface Quality of Potassium Dihydrogen Phosphate Crystals Finished By Fluid Jet Polishing.

Dylan Rykert

Potassium dihydrogen phosphate (KDP) is an indispensable optical crystal utilized in fusion-class laser systems for frequency conversion and polarization control. Fabrication of a freeform KDP optic in which the exit surface possesses areas of varying thickness is currently under exploration as a method to spatially alter polarization states of the laser beam and improve beam irradiation on the targets. Due to the properties of the crystal such as water solubility, softness, brittleness, and sensitivity to thermal shock, however, there are limited methods of fabricating freeform surfaces with acceptable laser-induced–damage thresholds (LIDTs) and optical quality.

Fluid jet polishing (FJP) is one such method that can meet this challenge, yet improvements to the process are needed to produce higher-quality surfaces with more robust LIDTs. The FJP method consists of pumping a polishing slurry (abrasive particles suspended in a liquid medium) through a nozzle toward a surface to generate a sub-aperture removal spot. The simplicity allows for flexibility of the slurry, which can be specialized for the surface to be polished, such as KDP. The current slurry employs alumina abrasive particles suspended in mineral oil. Although, the method produces surface defects, such as cracks, scratches, and deep pits. Additionally, the high viscosity of the slurry is a limiting factor for the pumping pressures and nozzle diameters that are desired. In this work, we studied the FJP performance of new slurry candidates, employing alumina particles of varying sizes and hardness dispersed in a low-viscosity mineral oil. We found a clear correlation of particle size and hardness on surface quality, which can provide guidance for further slurry optimization.

Fabrication of a freeform, potassium dihydrogen phosphate (KDP) wave plate by fluid jet polishing (FJP) is under active investigation at the University of Rochester's Laboratory for Laser Energetics (LLE). Implementation of such a freeform optic on the OMEGA Laser System would provide dramatically improved laser-beam uniformity for fusion experiments. The current FJP process produces a variety of defects such as cracks and pits that reduce optical surface quality and laser-damage resistance. This work investigates alternative slurry compositions for improving the FJP method. Slurries with varying abrasive particles and surfactants were used in the FJP process and generated different surface finishes. The surfaces were characterized by microscopy and interferometry to measure surface roughness and defects. The results, which demonstrated a clear correlation of particle size and hardness to surface quality, will provide guidance for further optimization work.

Optimizing Silver Selenide Vertical Devices for High Sensitivity Near Infrared Imaging

Letian Li

Near infrared imaging has promising potential applications including but not limited to detecting crop conditions, enabling night vision, and studying changing weather patterns. In the technology, colloidal quantum dots, which are semiconductor particles a few nanometers in size, are used as photodetectors, converting absorbed light in the range of 700 nm to 1600 nm into electrical current. In our work, we build silver selenide (Ag2Se) vertical devices, for its potential as a near infrared photodiode. By manipulating particle sizes, ligands, and particle deposition methods, we can alter the electrical properties of silver selenide. Photodiodes, which consist of hole and electron blocking layers interposed between the colloidal quantum dots layers, are built and measured. By evaluating the impact of each factor on the rise of conductivity and responsivity of the silver selenide photodiode, we aim to optimize Ag2Se vertical devices for high sensitivity near infrared imaging.

Mxene Derived Electrolytes: Advancements in Solid-State Battery Technology

Jacob Voigt, Sahand Serajian, Syed Ibrahim Gnani Peer Mohamed, Mona Bavarian

In recent years an increased demand for solid-state batteries (SSBs) has arisen due to concerns with current liquid electrolyte batteries including battery thermal runaway, dendrite formation, and electrolyte leakage. SSBs have gained significant traction due to their improved safety and increased lifespan compared to liquid electrolyte containing batteries. In SSBs, the solid electrolyte (SE) is used instead of flammable liquid electrolytes, contributing to the mentioned benefits. However, limitations in SE ionic conductivity when compared to liquid electrolyte counterparts has slowed the implementation of SEs. Thus, our research focuses on how SE performance can be improved through the addition of $Ti_3C_2T_x$ MXene into polymer electrolytes. $Ti_3C_2T_x$ MXene is a two-dimensional transition metal carbide and nitride which has been shown to improve the ionic conductivity of SEs for use in SSBs. By integrating MXene into a polymer matrix of PEO and LiTFSI, we created MXene-containing Composite Polymer Electrolyte (MCPE). Our MCPEs are prepared with differing weight percentages of MXene and the resulting MCPEs are analyzed through methods such as X-ray diffraction (XRD), scanning electron microscopy (SEM), thermal gravimetric analysis (TGA), and electrochemical impedance spectroscopy (EIS). Current and future research will extend beyond our current understanding of MCPEs' composition and its impact on ionic conductivity, and will provide insight on the role of chemical nature of MXene on the MCPEs properties.

ZIF-Based Novel Green Nanomaterial for Cr(VI) Removal

Bhoomika Chegu Vinay, Swathi M. Wadavi, Prathima B., Krishnamurthy Sainath

Cobalt (Co)-based zeolitic imidazolate frameworks (ZIF)-L is a porous material having a high surface area, and high thermal, chemical, and hydrothermal stability. ZIF-L (Co) finds various applications in catalytic reactions, drug delivery, gas separation, adsorption of heavy metals, and wastewater treatment. Hexavalent chromium [Cr(VI)] is a heavy metal that is extremely carcinogenic in nature. According to WHO, the permissible limit for chromium in drinking water is 50 $\mu g/L^{-1}$, confirming its removal to be paramount. The objective of this work includes optimization of ZIF –L (Co) nanoparticles (NPs) synthesis by varying the addition rate of 2-methyl imidazole solution to the metal precursor solution (Cobalt salt) and studying the removal efficiency of chromium by carrying out adsorption studies using the prepared NPs. The synthesized NPs were characterized for evaluation of morphology, crystal structure, and functional groups using scanning electron microscopy (SEM), X-ray diffraction (XRD), and Fourier-transform infrared spectroscopy (FTIR), respectively. The leaf-like structure revealed by SEM images confirmed the formation of ZIF-L (Co) NPs, and the average size of NPs was found to be 250 ± 88 nm. The XRD diffractogram confirmed all the peaks pertaining to ZIF-L, whereas the FTIR spectra revealed the functional groups on ZIF-L (Co) NPs bearing important peaks at 427, 754, 1143 and 1386 cm^{-1} corresponding to vibrations of Co-N, imidazole ring, C-H, and C=N, respectively. Batch adsorption studies were conducted to evaluate the Cr(VI) removal efficiency by varying the contact time, adsorbent dosage, and concentration of Cr(VI) ions. The results depicted that equilibrium was reached within 20 min for an initial concentration of 20 mg/L of Cr(VI) when the adsorbent dosage was 0.6 g/L. This study investigates the use of ZIF-L (Co) NPs for the adsorption of Cr(VI), and further studies are required to use this material for the remediation of the Cr(VI) contaminated water.

Aqueous Synthesis of $Li_3InCl_{6-X}Br_X$ for Low-Cost Electrolyte Processing

Rebecca Choi

Rare-earth alkali halides of the form Li_3MX_6 are known to have high lithium-ion conductivity, some approaching superionic conductivity (≥1 mS/cm) at room temperature. These solid-state materials are candidates for replacing typical lithium-ion battery liquid electrolytes, which are flammable and pose a combustion risk, and could significantly decrease safety concerns associated with lithium-ion batteries. Li_3InCl_6 (LIC) is one such solid electrolyte. It has been shown that LIC can be synthesized in an aqueous solution with controlled dehydration. However, high pressure must be maintained on devices made with LIC for them to perform as well as devices with liquid electrolytes. Replacing chlorine with bromine can soften halide sublattice dynamics which may lead to more facile Li transport and more favorable mechanical softness. This project explored $Li_3InCl_{6-x}Br_x$ synthesis from aqueous solutions and whether these derivatives yield better conductivities. The x=0 and x=0.5 syntheses yielded pure products, according to XRD analysis. If x≥1, the resulting powders contained residual lithium chloride that did not react in synthesis. Additionally, if x≥4, the resulting product required an extra annealing step to produce $Li_3InCl_{6-x}Br_x$. In all cases, adding bromine did not increase the room temperature conductivity compared to LIC.

Mechanistically Studying the Degradation of Cesium Lead Iodide Perovskite Phases Due to Temperature and Humidity

Rory Campagna, Samuel Smith, Riley Nelson, Jeffrey Christians

Halide perovskites offer exciting potential as photovoltaic materials and simply as semiconductors. Perovskites' ever-increasing efficiency, stability and lowering manufacturing costs makes them attractive as a competitor to silicon solar cells. The focus of this research was on cesium lead iodide ($CsPbI_3$). A major drawback of $CsPbI_3$ is its limited stability in the perovskite crystal phase, the colored phase with desirable optoelectronic properties. Cesium is somewhat too small of a cation to stabilize the perovskite structure and in the presence of outside factors, such as increased humidity and temperature, the metastable perovskite structure changes to the thermodynamically favored orthorhombic phase. It is suspected water acts as a catalyst to lower the activation energy of the metastable perovskite phase, thus accelerating the phase change. In this study, the absorbance of a $CsPbI_3$ film was analyzed over time while controlling humidity and temperature. Absorbance data collected was converted to phase fraction data and modeled by the JMAK model, $x(t) = \exp(-kt^n)$, where n is the growth coefficient and k is the rate constant. It is found that relative humidity increases the phase transformation rate exponentially, indicating a first order process; however, increasing temperature leads to a monotonic decrease in the rate constant due to increased surface water desorption. Experiments are consistent with the phase transformation rate being nucleation rate limited where individual grains transform very rapidly once a non-perovskite phase nucleates. SEM imaging shows growth across domain boundaries. The growth coefficient n was found to have little variance with an average value of 2.54. Holding this value constant, a global fit was applied to the phase fraction data to determine the activation energy of the phase transformation as well as the energy of adsorption for water to the perovskite surface. Work continues to model the adsorption and desorption mechanism of water onto the surface of $CsPbI_3$ to obtain more accurate activation energy and adsorption values. Further description of this process will help researchers design more durable materials to increase commercial use of halide perovskites.

A New Methodology in Quantifying Surfactant Interactions

Chetan Chilkunda

This study rapidly quantifies binary and ternary surfactant interactions in the context of formulation development. This work utilizes model-driven design of experiments paired with ML-based approaches to determine the critical micelle concentration (CMC) of surfactant mixtures in one 96 well plate and one week. The high-throughput, novel methodology efficiently and accurately reproduced the literature-reported mixture CMC for the Sodium Dodecyl Sulfate (SDS) and Tetraethylene Glycol Monooctyl Ether (C8E4) as well as the mixture CMC for SDS / C8E4 in 0.5 M NaCl. Additionally, this methodology determined that Non-ideal Solution Theory (developed by Hua and Rosen) fails to adequately model ternary mixtures. Current work is related to expanding this high-throughput approach to ternary surfactant models and looking at the cosolvent and cosurfactant effects of hexanol.

Understanding Zinc Metal Electrodeposition for Aqueous Batteries

Claire Ely, Lacey Roberts, Michael F. Toney

Aqueous zinc metal batteries (ZMBs) could enable safe and cost-effective electrochemical energy storage needed to transition to intermittent renewable energy sources. Zinc has an electrochemical potential of -0.76 V vs. SHE, allowing non-flammable aqueous electrolytes. Zinc also has a high volumetric specific capacity of 5854 mA per cubic centimeter which makes ZMBs a strong candidate for grid-scale energy storage applications. While promising, ZMBs face reversibility issues which prevent long-term rechargeability. This irreversibility is caused by 1) undesirable side reactions of zinc with the electrolyte, including hydrogen evolution and 2) non-homogeneous plating morphology. Non-uniform plating morphologies can lead to electrochemically inactive zinc or dendrites which short-circuit the battery. Understanding the mechanisms that influence homogeneity of zinc electrodeposition is of interest to design highly reversible ZMBs.

In this work, we characterize electrode morphology under different electroplating conditions in 1 molal zinc sulfate ($ZnSO4$). We also investigate the effects of the electrolyte additive dimethylformamidium triflurosulfonate (DOTf), which is designed to suppress the hydrogen evolution reaction (HER). To understand morphology, heterogeneity of zinc plating, and electrode surface composition, we used characterization techniques such as scanning electron microscopy (SEM), energy dispersive spectroscopy (EDS), and X-ray photoelectron spectroscopy (XPS). Electrochemical techniques including cyclic voltammetry and galvanostatic cycling are performed to support morphological findings and understand the conditions under which side reactions occur. Long term galvanostatic cycling is used to determine performance metrics of different electrolyte systems.

We find that certain electroplating parameters, such as increased current density at low capacity, allow for more compact and homogeneous zinc deposits. Low current densities show electrochemical behavior above the zinc plating potential that could be attributed to side reactions, alloying with the copper electrode, or underpotential deposition. This electrochemical behavior impacts morphology throughout cycling. DOTf as an additive leads to more compact zinc plating and shows reduced side products at the surface. However, NMR shows the DOTf additive electrolyte is not chemically stable and degrades over short periods of time, impacting reversibility. High concentrations of DOTf in the electrolyte also worsen long-term cycling. More stable additives that lead to similarly compact deposition should be considered. This work investigates the path towards smooth zinc electrodeposition for viable ZMBs.

Ruthenium(II) Polypyridyl Complex with Phosphoric Acid As Anchoring Group: Synthesis, Characterization and Application

Abbey Blue, Kadarkaraisamy Mariappan

Ruthenium complexes containing phosphoric acid as anchoring group are well known as dyes for water splitting reactions when combined with TiO_2. A three-pot synthesis using 2,6-diacetylpyridine, *n*-octyloxybenzaldehyde, phenylhydrazine and lead (IV) tetraacetate yielded a Pyridine dipyrazolinel tridentate ligand (**L3**). The ligand **L3** is well characterized by ^1H NMR. Reaction of ligand with $RuCl_3.3H_2O$ yielded [$RuCl_3$.**L3**] (**2**), which was characterized by single-HR mass spec. The ruthenium (III) complex (**2**) is further reacted with tetraethyl[2,2′-bipyridine]-4,4′-diylbis(phosphonate) to make [RuCl(**L3**).tetraethyl[2,2′-bipyridine]-4,4′-diylbis(phosphonate)]Cl (**3**). The black dye [RuCl(**L3**).tetraethyl[2,2′-bipyridine]-4,4′-diylbis(phosphonatc.NCS]ClO_4 (**4**) is synthesized by reacting compound **3** with an excess of potassium thiocyanate followed by a metathesis reaction with $LiClO_4$ (aq). This new black dye was characterized by ESI MS, UV and emission spectroscopy and it needs to be hydrolyzed prior to test for water splitting reactions after thorough analytical characterization. Long alkyl chains adorn this new pyrazolyl ligand to limit solubility and promote stability of the ruthenium dye complexes in dye-sensitized solar cell applications.

Novel n-Doping and Characterization of Pbttt-C14 for Applications in Organic Diodes

Colin Brown, Antoine Kahn

The use of organic semiconductors (OSCs), i.e., conductive polymers, has risen greatly over the past few decades, finding applications in organic light-emitting diodes (OLEDs), solar cells, and wearable technologies. Developing OSCs requires effective n-type and p-type doping to enhance their electrical conductivity, but it is challenging to develop n-doped polymers that reach the same conductivity of p-doped OSCs. Given that many electronic devices such as thermoelectrics and OLEDs require both n- and p-doped components, this stark difference raises the need for further research into better-performing n-doped polymers to allow for further expansion of OSC use.

This work contributes to the pool of available n-doped polymers by demonstrating the feasibility of a novel combination of polymer and n-type dopant, poly[2,5-bis(3-tetradecylthiophen-2-yl) thieno[3,2-b]thiophene] (PBTTT-C14) and (pentamethyl cyclopentadienyl) (1,3,5-trimethylbenzene) ruthenium dimer (RuCp*Mes)2, respectively. PBTTT-C14 is commonly p-doped due to its high morphological order, which is advantageous for electron mobility in OSCs, and (RuCp*Mes)2 is a powerful n-dopant that has previously been used in the Kahn Lab. We hypothesized that PBTTT-C14's relatively high electron affinity would allow for this n-doping reaction.

First, UV-Vis spectroscopy was used to identify the formation of polarons, which are characteristic of electron localizations, in the n-doped PBTTT-C14 to confirm the doping reaction. Next, this work characterized how the conductivity of this new material changes as a function of different preparation and measurement conditions. Ultra-high vacuum current-voltage measurements revealed positive correlations between conductivity and dopant concentration, temperature, and UV light exposure, reaching a maximum value of 1.05 x 10-4 S/m. In general, the conductivity can be precisely tuned and heightened through all of these conditions.

At the same time, this work revealed that conventional spin coating techniques prove ineffective for n-doped PBTTT-C14, and alternative film preparation methods create a more disordered polymer morphology. Additionally, evaporation doping, which previous literature showed to be effective for raising conductivity in OSCs, did not show any improvement over mixed solution doping. Despite these concerns, n-doped PBTTT-C14 still reaches within an order of magnitude to the conductivity of p-doped PBTTT-C14 prepared in a comparable manner in literature. Therefore, this comparison suggests that, if developed, synthesis methods which produce more ordered morphology can yield highly-conductive n-doped PBTTT-C14. Paired with current p-doped OSCs in literature and industry, this new material could be deployed in a wide variety of fully organic diode devices.

A Quantitative Figure of Merit for the Conductive Properties of Battery SEI

Katelyn Lyle, Bo Liu, Yuzhang Li

Solid electrolyte interphase (SEI), the passivation film found on the face of battery electrodes, has been extensively studied since its discovery in 1979. However, despite its impact on battery performance, many properties of the SEI, including its conductivity, remain unquantified. In this paper, we use a new cell structure in which the SEI serves in place of a separator in symmetric Li|Li cells to illustrate the ionically conductive properties of SEI and numerically quantify these values. Thus, the conductive properties of SEI are much more clearly and quantifiably understood as it is used as a separatory cell component.

Developing a Novel Continuous Flow Process Using Hydrazine to Synthesize VO2 Nanoparticles for Thermochromic Films

Emhyr Subramanian, Owen Wostoupal, Jie Li

Thermochromic materials can selectively block solar infrared (IR) radiation, allowing for "smart" window films to improve insulation and reduce energy usage in buildings. Vanadium dioxide (VO_2) nanoparticles have been the primary focus of thermochromic research for several decades, yet three fundamental issues are major barriers to widespread adoption:

1. Current processes require batch reactors that take several hours to produce particles.

2. VO_2 particles must be in a pure "M-phase" crystalline structure to be thermochromic, which can be difficult to achieve with faster processes.

3. VO_2 particles should be as small as possible (<30 nm) to produce effective films.

This project aims to solve these scalability and film effectiveness issues by developing a novel continuous flow process for synthesizing ultra-small M-phase VO_2 particles using hydrazine monohydrate ($N_2H_4 \cdot H_2O$). Vanadium oxide (V_2O_5) was reduced by $N_2H_4 \cdot H_2O$ to form a precursor solution of V^{4+} ions, which was subsequently nucleated via a continuous flow hydrothermal reactor. By varying several parameters—such as component ratios, precursor flow rates, and reactor temperatures—and using DLS, TEM, SEM, and XRD instruments to determine particle size, morphology, and phase, optimal parameters were obtained. Precursor synthesis time was reduced 20-fold, pure M-phase VO_2 particles were consistently produced, and particles under 60 nm were synthesized. Future optimization with capping agents and sodium hydroxide will likely yield smaller particles. This project has introduced a novel methodology for VO_2 nanoparticle synthesis that demonstrates great promise in mass-producing thermochromic materials.

Optimization of Integrated Electrodes Using Laser-Made Bimetallic Nanocatalysts

Lydia Schultz, Madeleine K. Wilsey, Astrid M. Müller

The research aimed to immobilize laser-made [NiFe]-layered double hydroxide nanocatalysts on hydrophilic carbon fiber paper with uniform nanoparticle distribution. These bimetallic nanomaterials are excellent electrooxidation catalysts, 1 and we chose laser-made [NiFe]-layered double hydroxide nanocatalysts 2 because the laser synthesis is surfactant-free. 3 We used hydrophilic carbon fiber paper as an electrode support due to its high surface area, conductivity, cost-effectiveness, durability, and scalability. 4 To maintain hydrophilicity, we employed a rapid green chemistry process developed by the Müller group. 4 Uniform nanoparticle distribution on the carbon fiber paper was achieved using aqueous citrate-capped gold colloid immersion. 5 However, laser-made [NiFe]-layered double hydroxide nanoparticles aggregate in aqueous suspension, posing challenges for uniform distribution in integrated electrodes. To address this, we systematically explored various surfactants, a dispersant, and sonication methods to load these nanoparticles onto the carbon fiber paper with minimal aggregation while retaining adhesion of nanoparticles on the carbon. We kept catalyst mass constant and tested citrate, CTAB, SDS, and Triton-X with different concentrations as surfactants, applied different volumes of isopropanol as a dispersant to surfactant-free suspensions, and used a sonication bath or sonication horn with varying times. We analyzed nanoparticle distribution by scanning electron microscopy (SEM) and energy-dispersive X-ray spectroscopy (EDX) mapping. SEM provided insights into nanoparticle aggregation, while EDX mapping provided relative catalyst amounts on carbon surfaces. In summary, our study demonstrates the successful adhesion of nanocatalysts to integrated electrodes with minimized aggregation using a combination of methods.

References:

1) B.M. Hunter, H.B. Gray, A.M. Müller, Earth-Abundant Heterogeneous Water Oxidation Catalysts. Chem. Rev. 2016, 116, 14120-14136.
2) B.M. Hunter, J.D. Blakemore, M. Deimund, H.B. Gray, J.R. Winkler, A.M. Müller, Highly Active Mixed-Metal Nanosheet Water Oxidation Catalysts Made by Pulsed-Laser Ablation in Liquids. J. Am. Chem. Soc. 2014, 136, 13118-13121.
3) R.C. Forsythe, C.P. Cox, M.K. Wilsey, A.M. Müller, Pulsed Laser in Liquids Made Nanomaterials for Catalysis. Chem. Rev. 2021, 121, 7568-7637.
4) M.K. Wilsey, K.R. Watson, O.C. Fasusi, B.P. Yegela, C.P. Cox, P.R. Raffaelle, L. Cai, A.M. Müller, Selective Hydroxylation of Carbon Fiber Paper for Long-Lasting Hydrophilicity by a Green Chemistry Process. Adv. Mater. Interfaces 2023, 10, 2201684.
5) R.C. Forsythe, C.P. Cox, M.K. Wilsey, W. Yu, A.M. Müller, High surface area assemblies of gold nanoparticles on hydrophilic carbon fiber paper with ionomer overlayers for aqueous CO2 reduction electrocatalysis to clean syngas. Top. Catal. 2023, 1-19.

Process Intensification through Data-Rich Approach: Dynamic Exploration of Continuous Flow Quantum Dot Parameter Space

Hannah Dickerson, Fernando Delgado, Milad Abolhasani

A quantum dot (QD) is a high performing semiconductor material, exhibiting more than one tunable optoelectronic property through precise control of size, morphology, and composition. Recently, colloidal synthesis of II-VI and III-V Quantum Dot (QD) nanomaterial has been successfully achieved through the hot-injection and heat up synthesis batch process technique, resulting in near unity photoluminescence quantum yield and enhanced photostability. To broaden QD integration in the renewable energy, display, and chemical industry, the search for more efficient QD nanomaterial and the corresponding sustainable manufacturing route has received increased attention in both academia and industry. This exploration through the QD synthesis space can be drastically accelerated by combining continuous flow chemistry and autonomous experimentation with reaction and data intensification. Specifically, applying process intensification in the form of microscale flow chemistry has demonstrated the ability to enhance heat and mass transfer through an increased surface to volume ratio, shortening startup and shutdown time, lowering reaction time, and decreasing precursor consumption, when compared to traditional batch chemistry. Moreover, continuous flow chemistry is capable of facile integration with data-rich experimental methodology, generating accurate *Big Experimental Data* for any precise physical and chemical condition. However, most analyses performed in continuous flow chemistry discard the transient information obtained during experiment startup and shutdown, as well as between each experiment, consequently limiting data acquisition to only the steady state condition. Recuperating the discarded, transient information through data-rich experimental methodology can not only utilize data more effectively, but also significantly expedite the pace of research acceleration. In this work, we demonstrate the importance of analyzing transient phenomena for intensification of the experimental design space associated with nanoparticle synthesis through the systemic evaluation and optimization of the single stage continuous flow synthesis of an exemplary QD, cadmium selenide (CdSe). Additionally, we present reduced raw material consumption and lowered reaction timescales, relative to traditional steady state flow experimentation, thereby achieving considerable intensification of the development and discovery of colloidal functional material. By embracing flow chemistry's compatibility with data-rich experimental methodology and extensively studying dynamic behavior, the pathway for accelerating experimentation is notably fortified, providing greater precision and efficacy for navigation of the complexity associated with QD discovery and development.

Functionalization and Controlled Porosity within 3D Covalent Organic Frameworks

Maddy Davis

Maddy Davis, Jade Greg, Lucy Park, Elena Marchetti, Brody Mistrot, Brian Smith

Functionalization and Controlled Porosity within 3D Covalent Organic Frameworks

3D Covalent Organic Frameworks (COFs) are porous, crystalline polymers with unique interpenetrated structures and large surface areas. Their structures allow for potential implementation in a variety of fields, including energy storage, catalysis, drug delivery and water purification. Specifically, 3D COFs' interpenetrated structure provides a means to control pore size, which has been demonstrated with both porous and collapsed crystalline forms of the same network. In addition, the modularity of COFs provides the opportunity to vary functional groups through the systematic selection of starting materials. The addition of functional linkers to COF networks is vital for applications and optimizing the use of internal surface area, specifically in the case of purification through targeted molecule capture. However, common functionalization approaches for 2D COFs and metal organic frameworks (MOFs) are generally unsuccessful when applied to imine-linked 3D COFs. Here we explore the mechanism of 3D COF polymerization and crystallization to pursue a general approach for modular functionalization and controlled porosity. In particular, the different solubility of functionalized monomers has central importance for both optimizing conditions for COF crystallization as well as understanding the single crystal growth of small molecule analogs. We intend to ultimately develop a library of functional linkers and understand the impact of that functionality on both network growth and final porosity.

The Impact of Interlayer Spacing on Alkali Metal Insertion in Layered Electrode Materials

James Aravena, Lauren Marbella, Susie Park

The Marbella Lab's research focuses on understanding the electrochemical behavior of lithium, sodium, and potassium ion batteries, specifically the insertion mechanisms of monovalent alkali metals (Li and K) into layered electrode materials. Despite their potential cost benefits, the primary challenge with K batteries is the shorter cycle life and lower energy density, likely attributable to the larger atomic size of potassium. With tunable synthetic properties, the Re6Se8X2 two-dimensional van der Waals material offers an avenue to investigate these mechanisms. Comparative studies were conducted using K and Li half cells with Re6Se8Cl2 and LiNi0.33Mn0.33Co0.33O2 (NMC111) cathodes, respectively. Key methods included cyclic voltammetry, galvanostatic cycling, X-ray photoelectron spectroscopy (XPS) to study the solid electrolyte interphase, and X-ray diffraction (XRD) to evaluate crystalline structures and interlayer spacing changes. Results from the cyclic voltammetry showed apparent differences between K and Li in Re6Se8Cl2: while Li had two discernible intercalation events, K exhibited just one. This suggests that potassium's larger atomic size complicates its intercalation process, producing broader peaks and eliminating a second intercalation event.

Further, the discharge cycle analysis showed Li undergoing two stable intercalation plateaus compared to K's single event. Screening effects amplified by K's ionic size result in enhanced ion-ion repulsion, impairing its transportation in the cathode. The slope data established less screening in the K cell compared to the Li cell, hinting that repulsions significantly hindered K's reversibility. An intriguing observation was the difference in interlayer spacing in Re6Se8(S-Ph)2 and Re6Se8Cl2. The former's larger spacing allowed for dual K ion intercalation without subsequent deintercalation. K's insertions are size-dependent. Our study underscores the size of potassium ions as a primary factor affecting their behavior in Re6Se8Cl2. Enhancing our understanding of these mechanisms is paramount to advancing the potential use of K batteries in the industry.

Enhanced Heat Transfer Via Electrostatic Resonance for Thermal Management

Hunter Black, Thomas Corbin, Jason Livesay, William O'Brien, Ranga Narayanan

With many fields requiring forms of thermal management, new devices can have groundbreaking changes across a variety of industries. This especially applies when working in reduced gravity environments where buoyancy driven convection is reduced, and actively pumping fluids can cause cavitation. Currently, thermal management devices in space can be categorized into either active or passive systems. Active systems typically rely on moving parts while passive devices control temperature without external motion.

Here we propose a system that increases heat transfer through an electrostatic resonance instability in a bilayer fluid interface. In this system, we generate flow via resonance by parametrically forcing the system to its natural frequency. The natural frequency of the system is the frequency at which the interface oscillates in the absence of any external forcing. When parametrically forced to the natural frequency of the system, the flow becomes unstable, and waveforms are generated on the interface of the fluid bilayer. The generation of these waveforms increases convection within the test cell thus increasing the overall heat transfer.

In our experimental trials, we have proved that heat transfer is enhanced due to these unstable waves within our system. An enhancement of heat transfer of up to 55% is obtained when resonant onset occurs. This large increase in heat transfer is normally observed at lower frequencies such as 0.5 or 1 Hz. While this is a substantial increase, optimization of the system by changing fluid layer heights and operating fluids is needed and is part of our future work in addition to testing under microgravity conditions.

Funding is acknowledged from NSF 2025117, NASA 80NSSC 21K0352, and FSGC 80NSSC20M0093.

Impact of Sample Volume on Magnetic Particle Imaging Quantification Accuracy

Sebastian Montero

Magnetic Particle Imaging (MPI) is a noninvasive and radiation-free imaging technique with the ability to directly quantify superparamagnetic iron oxide nanoparticle (SPION) tracers. MPI uses nonlinear re-magnetization behavior found in SPIONs to generate a signal directly related to SPION mass. This allows for negligible background signal and tissue attenuation making MPI an increasingly attractive modality for medical applications such as cancer diagnosis and tumor sizing. Despite this, the quantification accuracy of MPI may be hindered by its finite resolution limits which are seen in other medical imaging devices, such as MRI, PET, and CT. With resolution-based errors being a normality in novel imaging modalities, this study aims to detect the presence of these errors arising from changing volume sizes in MPI.

To assess quantification error, 3D-printed disc shaped phantom models were designed to have volumetric cavity sizes ranging from 16-512μL while maintaining a constant depth. Models were then loaded with Feruocarbotran, a commercially available SPION tracer, using a constant mass dilution series, generating a set of models with constant mass, constant depth relative to scan projection, and varying volumes. Disc models were then imaged to produce 2D projections of tracer distributions in different MPI built-in scan modes: High Sensitivity (HS), High Resolution (HR), and High Sensitivity/High Resolution (HSHR). A set of spherical 3D-printed models were loaded using the same procedure but imaged to produce 3D projections of tracer distributions in HR scan mode. The mass of each model was estimated by using three constant volume calibration curves which constructed a scalar relationship between the total signal and tracer mass. The difference between estimated mass and ground truth defines the error.

Results show significant quantification error in tracer mass estimation. As the sample volume size deviated from that of the calibration curve, the percent error increased indicating a relationship between calibration curve size, model size, and quantification accuracy. Quantification error showed a scan mode dependency with the High Sensitivity scan mode having the least amount of error. 3D scans showed more accurate quantification likely due to spatial resolution improvement from numerous scan angles which allowed for overlaid projections, possibly reducing the effects of resolution-based errors in one direction. Although some scan parameters showed more accurate results, there were still consistent quantification errors present throughout the trials. Because of this, addressing the quantification inaccuracies associated with volume variation is the next pivotal step to increase MPI's suitability for clinical applications.

MOF-Derived Fe, N Co-Doped Porous Carbon Polyhedrons for High-Performance Lithium-Ion Battery Anode Materials

Kaiyi Qin

The lithium storage performance and structural stability of anode materials have a significant influence on the capacity performance and cycling stability of lithium-ion batteries. Despite the cost, conductivity, non-toxicity, and non-polluting advantages of graphitic carbon materials, their cycle stability is compromised by volume changes during lithium storage. Additionally, the theoretical capacity of 372 mAh/g is no longer sufficient to meet the growing demand for high-capacity LIBs. Metal-organic frameworks (MOFs) and their derivatives have emerged as promising anode materials for LIBs due to their tunable structure, high surface area, and high porosity. In this study, we successfully synthesized Fe, N co-doped carbon materials (Fe-N-C) with exceptional properties by calcination and co-doping Fe and N elements into a ZIF-8 precursor. Scanning electron microscopy (SEM) results revealed that the morphology of the synthesized samples was identical to that of ZIF-8. X-ray diffraction (XRD) and X-ray photoelectron spectroscopy (XPS) analyses confirmed the successful co-doping of Fe and N into the carbon material. The porous structure provided a buffer for volume changes and improved cycling performance. After 300 cycles at a current density of 500 mA/g, the Fe-N-C anode exhibited charge and discharge specific capacities of 538 mAh/g and 543 mAh/g, respectively. The co-doping of Fe and N enhanced the conductivity and rate performance of Fe-N-C, resulting in excellent cycling stability. This study demonstrates the exceptional electrochemical activity and potential of Fe, N co-doped carbon polyhedrons for high-performance lithium-ion batteries. Carbonization of metal-organic frameworks and co-doping with Fe and N elements represents a promising strategy to enhance the electrochemical performance of graphitic carbon anode materials.

Promising Thermoelectric Properties for Barium Pnictide Material

Victoria Kyveryga, Philip Yox, Arka Sarkar, Gayatri Viswanathan, Kirill Kovnir

Thermoelectric materials are a critical tool in energy conservation because they are capable of regenerating electricity from thermal energy, i.e. from waste heat. The thermoelectric power of a material is quantified by the dimensionless thermoelectric figure of merit, ZT ($ZT{=}S^2T/\rho\kappa$).

The ideal thermoelectric material with maximized ZT will have low electrical resistivity ρ, low thermal conductivity κ, and high Seebeck coefficient S. However, designing a material which meets each of these conditions is non-trivial, as these parameters are interdependent. Because the thermoelectric power is proportional to the square of the Seebeck coefficient, one strategy to enhance ZT is to focus on enhancing the Seebeck, which can be achieved by lowering the carrier concentration in the material. Here, we report the thermoelectric properties of $Ba_4AgGa_5As_8$, a semiconducting bulk powder with a very high room-temperature Seebeck coefficient.

In its original report, $Ba_4AgGa_5As_8$ has been explored only in terms of its nonlinear optical potential. Here, we optimize the synthesis such that phase-pure powder is consistently obtained and study thermoelectric properies on sintered $Ba_4AgGa_5As_8$ pellets. Seebeck coefficient measurements on densified pellets reveal a high room-temperature Seebeck coefficient of 290 μV/K, which is expected to further increase with temperature. Coupled with its low thermal conductivity, $Ba_4AgGa_5As_8$ remains extremely promising for thermoelectric applications.

Identifying the Efficacy of Mutants of the *De Novo* Protein S824 As Capping Ligands for the Biosynthesis of CdS Quantum Dots

Vivian Payne, Leah Spangler

Quantum dots are ideal for biomedical applications such as fluorescent cell labeling, targeted drug delivery, and precision imaging therapy. However, the use of quantum dots is currently limited due to the high toxicity of chemicals used in synthesis. To make quantum dots biocompatible, several biological synthesis routes, such as biomineralization, have been used to produce quantum dots using proteins. While effective, these routes are currently limited to specific metal sulfide materials because only a few proteins have been identified and applied to control quantum dot synthesis. Here, we aim to identify new mutated proteins to expand and improve upon the quality and materials palette of quantum dots that can be produced using biomineralization. We utilized site-directed mutagenesis, transformation, and recombinant protein expression to synthesize mutated proteins of interest. Specifically, we identified and tested mutants of the *de novo* protein S824. Multiple mutants of interest were observed to stabilize CdS quantum dots and maintain quantum confinement as shown by UV-Vis spectroscopy. We screened proteins of interest with Na2S and Cd2+ to identify quantum confined CdS complexes using spectra data. We assessed binding strength of the proteins using centrifuging exclusion and biolayer interferometry. Ultimately, we hope to identify amino acid sequence motifs that correlate to a strong metal binding affinity of naturally evolved proteins to apply them in later research involving 'designer' engineered proteins for quantum dot capping.

Supercapacitor Electrodes Based on Reduced Graphene Oxide Composited with Manganese Metal Oxide

Mariana Milano-Benitez

The growing global demand for energy has increased the need for sustainable and green energy storage solutions. This project proposes a rapid, inexpensive, yet facile technique, called Laser Ablation Synthesis in Solution (LASiS) to synthesize hybrid nanocomposites (HNCs) composed of manganese oxides (MnO_x/Mn_3O_4) and reduced graphene oxide (rGO) as active electrode material in energy storage devices. During LASiS, the HNCs were synthesized under extreme conditions of a liquid-confined plasma plume, resulting in a structure consisting of MnO_x/Mn_3O_4 nanorods and nanoparticles embedded on rGO nanosheets. The HNCs, alongside graphene and carbon nanofibers (CNFs) were then 3D printed via inkjet printing method for electrode fabrication. Electrochemical testing demonstrated that the LASiS-synthesized MnO_x/Mn_3O_4-rGO electrode had a significantly higher specific capacitance than electrodes created with commercially available Mn_3O_4-graphene nanocomposites. The result from this study highlights the potential of LASiS-synthesized functional materials and additive manufacturing techniques for creating all-printed supercapacitor devices with superior performance.

Improving the Performance of Lead-Acid Batteries through Guitar Coated Fibers

Katelyn Shadley

Preliminary results from University of Idaho research show, that GUITAR coated materials improve the performance of lead-acid batteries as a battery paste additive. Theoretical modeling has also shown greater improvement in performance than observed in experimental results. However, GUITAR coated materials tend to be hydrophobic which is detrimental to liquid contact during the mixing of battery plate paste and to the battery electrolyte solution during operation. By utilizing a heat-treat treatment on GUITAR coated materials the hydrophobic property can be reduced, and the materials will become more hydrophilic. This project is comparing standard factory plates with standard hand pasted plates and heat-treated GUITAR coated fibers as an additive for lead-acid battery test cells and cycled to assess if the additives improve battery performance.

The batteries were tested using a sequence of discharge rate cycles that would be of interest to battery manufacturers for industrial applications. Done by doing a schedule starting with a one hour, fifteen minutes, thirty minutes, one hour, ten hours, and ending with a twenty-hour discharge rate with five cycles for each rate. The preliminary results of the test cells with standard factory plates are within range of what was expected and will serve as a basis for the test cells with heat-treated GUITAR additives. Standard hand pasted and additive plates have been fabricated and cycle testing is being initiated.

Investigating Ca-Ion Conduction in PVDF-Based Electrolytes

Edward Fluker

Beyond Li–ion energy storage is gaining considerable attention owing to several challenges encountered with Lithium metal, such as scarcity, availability, high cost, and safety. Among the best potential alternatives is Calcium, which possesses many properties desirable to Lithium, and is more Earth–abundant, sustainable, and cost–effective [1]. Recent research in battery electrolytes has been directed toward alternatives for liquid electrolytes due to their risk of flammability and leakage. Solid polymer electrolytes (SPEs), gel polymer electrolytes (GPEs), and poly-ionic liquids are a few examples. However, many of these alternatives still lack the ionic conductivity necessary for commercial use. The objective of this work is to develop Ca-ion conducting PVDF-based gel polymer electrolytes (GPEs), wherein N–methylpyrrolidone (NMP) is used as a liquid solvent and $Ca(TFSI)_2$ as the salt. Specifically, we investigate ionic conductivity in these GPEs as a function of drying temperature and find that at a fixed salt concentration of 0.5 M, ionic conductivity decreases with increase in drying temperature due to decrease in NMP content. The highest ionic conductivity obtained in this study is in the range of 10^{-4} S/cm at room temperature and up to 10^{-3} S/cm at elevated temperatures [2]. These results are promising, and some areas that could be further explored are its transference number, mechanical strength, and the relationship between salt concentration and ionic conductivity.

[1] Hosein, I. D., *ACS Energy Letters*. (2021)

[2] Edward C. Fluker, Shreyas Pathreeker, and Ian D. Hosein, *The Journal of Physical Chemistry C*. (2023)

Impact of Phase Separation on Marangoni Synergism in Binary Surfactant Mixtures

Linda Liu, Robert Tilton

Impact of Phase Separation on Marangoni Synergism in Binary Surfactant Mixtures

Linda T. Liu[1,2], Stephen Garoff[3], Robert D. Tilton[1,2], Department of Chemical Engineering[1], Department of Biomedical Engineering[2], Department of Physics[3]

Carnegie Mellon University, Pittsburgh, PA 15213

Surfactants are molecules that reduce the surface tension of a liquid by adsorbing to the liquid surface. A non-uniform distribution of surfactants on a liquid surface causes a surface tension gradient that drives Marangoni spreading from low surface tension to high surface tension regions. This flow plays an important role in a variety of surfactant applications such as pulmonary aerosol drug delivery, coatings, and agrochemical sprays. Many of these products are formulated with mixtures of surfactants to take advantage of synergistic surface tension reduction effects. A spreading phenomenon recently discovered in cationic and anionic surfactant mixtures is Marangoni synergism. This occurs when a binary surfactant mixture prompts a greater Marangoni spreading velocity than the individual surfactants. It remains to be determined how widespread Marangoni synergism may be.

Many mixtures of cationic and anionic exhibit phase separation. This could impact Marangoni spreading. This project will focus on Marangoni spreading in binary surfactant mixtures that exhibit phase separation. Spreading experiments were performed with binary surfactant samples at various ratios and total concentrations, where each mixture exhibited a different extent of phase separation. The spreading velocities were tracked for each sample and allowed for evidence about the role of phase separation in Marangoni spreading. This work showed that systems that exhibited the most extensive phase separation were not synergistic, but in fact antagonistic. The Marangoni spreading velocity was lower than either of the individual surfactants by themselves. This can be attributed to the inadequate mass transfer of surfactants from the aggregate phase to the air/water interface to maintain the necessary surface tension gradient. This work may improve the design of complex fluid products that rely on spreading at fluid interfaces.

Controlling the Network Structure of a Novel Hydrogel: Pentenoate Modified Dextran (PDEX)

Sienna Sun

Controlling the Network Structure of a Novel Hydrogel: Pentenoate Modified Dextran (PDEX)

Sienna K. Sun, Jeayoung Jeong, Joseph M. Scalet, and Stevin H. Gehrke

[1]*Department of Chemical and Petroleum Engineering, University of Kansas*

Certain ocular diseases, including aged-related macular degeneration, require repeated drug deliveries via intravitreal injections to the posterior cavity of the eye, an uncomfortable and expensive treatment. This exploration investigates a specific hydrogel's viability as an alternative to frequent injections by attempting to control the network structure of the hydrogel, *pentanoate modified dextran (PDEX).* As a biocompatible, degradable, and viscoelastic material, this novel gel was studied to justify whether the thiol-ene "click" chemistry reaction creates more efficient and uniform network structures, allowing better control of properties such as diffusion rates in drug delivery systems. Diffusion tests were performed with vitamin B-12 to quantify the diffusion coefficients of various thiol-ene ratio PDEX gels. Another studied drug-delivery hydrogel, divinyl sulfone dextran (DVS), was synthesized with different DVS crosslink ratios and used to compare network properties with PDEX hydrogels, hypothesizing better network control for the thiol-ene reaction versus the chemically crosslinked DVS-dextran gels. DVS gels were utilized specifically for comparing diffusion coefficients of gels with similar swelling degrees. Additional property tests performed include uniaxial compression to measure mechanical properties and swell testing to measure the water sorption capacity. With these measurable properties, crosslink density was found to quantify the average network mesh size. PDEX hydrogels were successfully synthesized and primarily displayed diffusion trends as predicted, with the optimal controlled thiol-ene crosslinking ratio as 1:1. The structure relationship of the PDEX and DVS gels should be investigated further for potential PDEX application in drug delivery systems.

Hydrogel Technologies to Prevent Vaginal Toxicity and Promote Healing after Pelvic Radiotherapy

Emma Cohen, Jennifer Weiser, Woojin Han, Deborah Marshall

Hydrogel Technologies to Prevent Vaginal Toxicity and Promote Healing after Pelvic Radiotherapy

In 2022, one million women in the U.S. were diagnosed with pelvic cancers[1] , and roughly half required radiation therapy. About half of patients undergoing pelvic radiation experience vaginal stenosis, which adversely affects vaginal health, sexual function, and examinations. To mitigate this, an injectable bioadhesive hydrogel was synthesized to address post-radiotherapy damage to the vaginal canal by positioning damaged tissue apart. To promote re-epithelialization of vaginal mucosal tissue, polylactic-co-glycolic acid microspheres (PLGA-MS) loaded with estradiol (E2) were synthesized and will be embedded in the bioadhesive. Embedding E2 loaded PLGA-MS will help promote the normal healing process.

8-arm polyethylene glycol vinylsulfone (PEG-VS) was crosslinked with VPM, a dithiol protease-cleavable peptide, to generate bioadhesive hydrogels[2-3]. PEG-VS reacts with thiol-rich tissue and offers tunable gelation kinetics, ensuring easy injecatbility[4]. To address vaginal toxicity and promote healing post-pelvic radiotherapy effectively, the hydrogels must be mechanically optimized, easily injectable, bioadhesive, and compatible with a controlled release drug delivery system.

Mechanical characterization of 20kDa PEG-VS (8%, 10%, and 12% w/v) bioadhesives was done using ASTM F2150-19 to characterize compressive moduli. The bioadhesives (n=3 per formulation) were placed in phosphate buffered saline (PBS) to swell overnight before testing. Increasing PEG-VS concentration saw an increase in bioadhesive stiffness. To optimize injectability, bioadhesives were formulated at different pH values (3.0, 6.5, and 7.5), and rheometric analysis was done using ASTM D2556 to determine gelation kinetics. The bioadhesive solutions (n=4 per formulation) were casted onto the rheometer to polymerize while the test was run. pH 6.5 was found to provide an appropriate working time of 15 minutes for ease of injection. Preliminary bioadhesion testing was done using ex-cadaveric vaginal tissue. The bioadhesives were cast onto the tissue using a 5 mm biopsy punch that remained in place for 20 minutes. The tissues were washed in 1X PBS. Visual inspection showed strong adherence of the bioadhesive to tissue. To test feasibility of using PLGA-MS in our bioadhesives, PLGA-MS were prepared using a single-emulsion technique. PLGA-MS were imaged on a microscope and analyzed using ImageJ for size and distribution analysis (n=150). The average particle size was 45-50 microns, which led to an appropriate size for embedding into bioadhesives during polymerization[5]. E2-loaded PLGA-MS embedded bioadhesives will be characterized the same as those without the embedded PLGA-MS. Future studies aim to optimize the stiffness of the bioadhesive to match that of vaginal tissue and study the controlled release of E2-loaded PLGA-MS embedded in the bioadhesives.

References

1. SEER*Explorer: An interactive website for SEER cancer statistics [Internet]. Surveillance Research Program, National Cancer Institute. [Cited 2022 Apr 15]. Available from https://seer.cancer.gov/explorer/.

2. Han WM, Anderson SE, Mohiuddin M, et al. Synthetic matrix enhances transplanted satellite cell

engraftment in dystrophic and aged skeletal muscle with comorbid trauma. Sci Adv. 2018;4: eaar4008.

3. Patterson, J., & Hubbell, J. A. (2010). Enhanced proteolytic degradation of molecularly engineered peg hydrogels in response to MMP-1 and MMP-2. Biomaterials, 31(30), 7836–7845. https://doi.org/10.1016/j.biomaterials.2010.06.061

4. Wang, J., Youngblood, R., Cassinotti, L., Skoumal, M., Corfas, G., & Shea, L. (2020). An injectable peg hydrogel controlling neurotrophin-3 release by affinity peptides. Journal of Controlled Release, 330, 575–586. https://doi.org/10.1016/j.jconrel.2020.12.045

5. DiStefano TJ, Vaso K, Panebianco CJ, Danias G, Chionuma HN, Kunnath K, Karoulias SZ, Wang M, Xu P, Davé RN, Sahoo S, Weiser JR, Iatridis JC (2022). Hydrogel-Embedded Poly(Lactic-co-Glycolic Acid) Microspheres for the Delivery of hMSC-Derived Exosomes to Promote Bioactive Annulus Fibrosus Repair. Cartilage, 13(3). PMID: 36040157.

Effect of Thermal and Mechanical Rejuvenation on a Model Thermoreversible Colloidal Gel

Charlotte Gottilla, Khushboo Suman, Norman J. Wagner

The aging behavior of adhesive hard sphere (AHS) colloidal gels is of interest due to the role of these gels in common pharmaceutical and consumer products. In this work we examine the possibility of shear rejuvenation in an octadecyl-coated silica nanoparticle system, which has previously been shown to exhibit anomalous aging behavior following a thermal quench (*Suman and Wagner, The Journal of Chemical Physics 157 (2), 024901 2022).* This AHS system is observed in comparison to Pluronic F-127, which is expected to show shear rejuvenation following a thermal quench. We observe the phase transition behavior of Pluronic-F127 and the AHS system, then investigate the behavior of each system following thermal versus shear rejuvenation. Our comparative analysis shows differences in the aging behavior between systems and demonstrates a lack of shear rejuvenation in the AHS system. Our rheological characterization at different amplitudes, temperature ramp rates and pre-shear history shows intriguing effects of non-equilibrium parameters on the kinetics of aging behavior of colloidal gels. This study of phenomenon of aging in model colloidal gel system is of both fundamental importance and industrial relevance in determining the shelf-life of products based on colloidal gels.

Structural Biphasic Electrolyte Embedded Carbon Fiber for Aerospace Applications

Alexander Albrecht, Jodie Lutkenhaus, Tasya Nasoetion

Energy storage in aerospace is of significant interest, driven by the renewed focus on space travel. The industry requires batteries that provide high energy density, structural integrity, and can withstand extreme low temperatures. This research investigates the potential of a structural battery, capable of serving as both an energy storage and load-bearing element in an aircraft. We assessed the suitability of selected multifunctional materials and found the importance of desizing the carbon fiber and maintaining low levels of electrolyte in increasing performance. However, further improvements are necessary to optimize their performance.

Synthesis of Model Virus Nanoparticles for Charge-Driven Separations

Yasamin Hashemi, Barbara Knutson, Malgorzata (Gosia) Chwatko, Stephen Rankin

Silica nanoparticles (SNPs) can be employed as model systems in the design of separation of biologics (vaccines, cells, therapeutic proteins). The use of SNPs addresses the need to safely develop and design advanced separation techniques outside biological labs as part of multidisciplinary collaborative teams. This research tests the hypothesis that the mobility of dispersed SNP (model viruses) in an electric field can be tuned with surface functionalization and used to demonstrate charge-driven separation in a membrane process. To mimic virus functionality, SNPs are surface functionalized. Bare silica nanoparticles have a negative surface functionality. By controlling the extent of functionalization to the SNP, positive surface charge can be added to mimic the virus surface. Another approach to SNP surface functionalization is to add lipids on the outer surface of the particles to give the SNP additional cell-like or virus-like surface properties. Incorporating a dye on the SNPs aids in the quantification of the SNPs.

Nonporous silica particles of around 60-70 nm were synthesized and characterized using a Scanning Electron Microscope (SEM) and Image J software. Dynamic Light Scattering (DLS) was used to confirm the size of suspended particle solution, which was determined to be 60-80 nm. Next, the SNPs were amine-functionalized using 3-aminopropyl)triethoxysilane. To quantify the extent of amine functionalization, acid orange II was bound to the particles at low pH and recovered at high pH. The amount of bound dye was determined from UV-vis spectroscopy. The SNPs were modified by attaching a lipid bilayer containing a dye on the outside surface of the particles for further analysis of charge separations. The surface charge of the SNPs was measured across different batches of the particles by performing zeta potential using DLS. The movement of suspended dye-loaded particles across various membranes in the presence of an applied voltage was tracked. Future work will consist of additional electrodialysis experiments that relate the applied voltage and SNP surface properties to the effectiveness of the charge-based separations.

Temperature Dependence of Viscosity for Mineral Oils Used in Drug Patches

Anthony Orlando, Kenneth Mineart, Ridwana Bashar

One application of mineral oil-based gels is for use as transdermal drug patches that diffuse a drug into a patient through their skin. The rate of the drug diffusion out of the patch is vital to effectiveness and safety of drug patches. Mineral oils of varying viscosities can be used in these drug patches and the viscosity of the oil used in the gel influences the diffusion rate of the drug out of the patch. Additionally, the temperature of the gels could vary depending on which part of the body the patch is applied to and the climate that the patient is living in. Intuitively, it makes sense that higher temperatures and lower viscosity oils will result in gels with higher diffusion rates, but quantifying the relationship between the temperature, viscosity, and diffusion rates of the drug through patches is important for designing the desired release from drug patches. As a first step toward this overarching goal, we examine and compare the temperature-dependent viscosity of several aliphatic mineral oils (squalane, Hydrobrite® 200, 380, 550, and 1000). We observe lower viscosities at higher temperatures, as would be expected, but the strength of this correlation depends on the specific mineral oil. The correlation strength between viscosity and temperature is quantified by the Arrhenius Equation, specifically the activation energy associated with flow of the mineral oils. Ultimately, an Arrhenius Equation model can be made to predict viscosity of an oil at different temperatures which can then be used to predict drug diffusion rate for patches made with different oils.

Diffusion in Macroscopically Layered Polymer Gels

Nick DeVita, Kenneth Mineart

Transdermal drug delivery is a vital mechanism for skincare, hormone replacement, and other biomedical applications. Organic polymer gels have been recently identified as candidates for this drug delivery mechanism. Our present work focuses on controlling model drug release rate in a layered polymer gel system consisting of multiple organogels and a polystyrene backing. Typically, one organogel contains a diffusion probe [AOT (sodium bis(2-ethylhexyl)sulfosuccinate)], tri-block copolymer, and an organic solvent whereas the other contains only tri-block copolymer and organic solvent. The tri-block copolymer forms a physically crosslinked network within the gels that consists of spherical polystyrene domains and a plasticized rubbery matrix consisting of ethylene-co-butylene and aliphatic mineral oil (organic solvent). The matrix phase is fluid-like and amenable to mass transport, which allows for probe diffusion. Using Fourier Transform Infrared (FTIR) spectroscopy, the overall probe release rate, which stems from this diffusion, can be tracked. In particular, we are interested in the comparison between the aforementioned layered systems and a 'traditional' system with a single organogel containing the AOT diffusion probe. We also seek to investigate how changing the tri-block copolymer concentration of the barrier organogel layer - that which contains no AOT - changes the overall diffusion probe release rate. Our measurements show that the release rate of layered systems exhibits a time-delay phenomenon. Furthermore, systems with higher tri-block copolymer concentration in the barrier layer exhibit a longer time delay.

Impact of Degradation PEO-b-PCL on Aqueous pH

Tara Whipkey, Joanna White, Kayla Jarski, Ryan Van Horn

Plastics are made up of long chains of repeated structures, called polymers. These polymers can have unique characteristics especially when combined into copolymers which have characteristics of two or more polymers. Polyethylene-block-Polycaprolactone (PEO-b-PCL) is one such copolyester that demonstrates both hydrophilic and hydrophobic characteristics respectively. The amphiphilic nature of PEO-b-PCL makes it an ideal candidate for sustained drug delivery, especially with hydrophobic drugs. The polycaprolactone (within PEO-b-PCL) is a class of polymers called polyesters, that are known to release carboxylic acid when they degrade in water. Through this research, it has been found that the carboxylic acid, specifically caproic acid, produced by PCL degradation in water decreases the overall pH of its environment. Previously, degradation studies were conducted with clean DI water after each time trial with the previous DI water being thrown away. The main objective for past degradation studies were to examine the speed of degradation instead of degradation effects on its environment. This study showed how the PEO interacted with the caproic acid production of PCL. While there was an initial decrease in pH, the PEO seemed to create micelles that surrounded the caproic acid either increasing the pH or maintaining it, until the micelles degraded. The plateau shows a brief stability before pH continues to decrease compared to just PCL. Different areas of the body maintain constant pHs for maximum efficiency, thus maintaining that pH is important. This research gave insight to the safety of using the copolymer in the body or a marine environment.

Synthesis of Magnetically Organized Ionic Domains of Sulfonated Poly(styrene-isobutylene-styrene) with Potassium Ferricyanide Nanoparticles for Fuel Cell Applications

Sebastian Sánchez

This research explores the incorporation of ferrocyanide nanoparticles into sulfonated poly(styrene-isobutylene-styrene) (S-SIBS) through ultrasonication as a strategy to augment proton conductivity and overall performance of proton exchange membranes (PEMs) for fuel cell applications. SIBS was sulfonated to a high ion exchange capacity, followed by membrane fabrication with potassium ferricyanide (KFeCN6) nanoparticles previously ultrasonicated and dispersed into the polymer matrix. Several characterization techniques were performed, including infrared spectroscopy, thermogravimetric analysis, proton conductivity measurements, water uptake analysis, swelling ratio assessments, and methanol permeability testing. The membranes incorporating nanoparticles exhibited a remarkable 32% increase in proton conductivity when compared to their pristine counterparts, along with reduced methanol permeability and improved thermal stability.

Sustainable Production of Wool and Cotton Textile - Based Supercapacitors

Kaitlin McKenzie

Textile-based supercapacitors (TSCs) are being developed to become convenient and safe energy sources to power a variety of electronic devices. TSCs are unique in that they combine the flexibility and lightweight nature of textiles with the energy storage capabilities of traditional supercapacitors. The TSCs are prepared by coating wool and cotton yarn with $Ti_3C_2T_x$ MXenes, a 2D conductive material. In this study, an auto-coater was used to pull yarn through a MXene bath and then dried the yarn. This one-step process was repeated twelve times to increase the mass loading of the conductive material, increase the conductivity, and ultimately enhance the overall capacitance of TSCs. In addition to TSC device testing and characterization, we developed a method to test a two-electrode system by using individual yarns to classify their intrinsic capacitive behavior. Electrochemical characterization was performed on MXene-coated wool and cotton to determine the charge storage and resistive behavior of yarns. Cyclic voltammetry (CV) curves were conducted to represent the charge storage and discharge behavior. Electrical Impedance Spectroscopy (EIS) was performed to characterize the impedance behavior of the yarns and build an equivalent circuit model to determine the system parameters. Wool yarn has an intrinsic specific capacitance of 0.49 mF/cm while cotton yarn has an intrinsic specific capacitance of 0.0935 mF/cm. These tests represent the potential of using wool based TSCs versus cotton based TSCs.

Evaluating the Mechanical Properties of Silk Fibroin Scaffolds through Tensile Testing for Use As an in Vitro Culture Platform

Asha Rao, Emily Fussell, Elizabeth Aikman, Whitney Stoppel

One goal for engineered *in vitro* skeletal muscle models is to answer fundamental questions[1] about rare diseases, especially when animal models fail to recapitulate human disease.[2] One successful method by which simple skeletal muscle cell-based tissues are engineered is via *in vitro* addition of human myoblasts into an anisotropic biomaterial scaffold.[3] To apply these methods to study the progression of rare muscular dystrophies, our scaffold material must withstand over 8 weeks of mechanical stimulation, to enable time for muscle myoblast growth and maturation (4 weeks) and time for a hypothesis driven experiment. To explore the feasibility of a long-term *in vitro* culture platform, we quantified tensile mechanical properties of anisotropic silk fibroin biomaterial scaffolds[4] as a function of fabrication parameters via static and dynamic tensile testing. We evaluated polymer concentration (3% or 5% wt/v silk), freezing rate (50% or 100% ethanol and dry ice[4]), and crystallinity (water annealing or autoclaving). We performed analyses of static and fatigue tests as well as frequency and amplitude sweeps. Young's modulus and ultimate tensile strength (UTS) was determined by hydrated static testing through stretching different scaffold formations from rest to 100% strain. The different fabrication parameters resulted in tunable Young's modulus values of 600 kPa to 2800 kPa, relevant to skeletal muscle. Stress-strain curves show that the elastic and plastic response differ between autoclaved and water annealed samples. The fatigue tests were performed via continuous mechanical stimulation at 10% strain and 1 Hz for over 6000 minutes, and results show minimal change in the storage and loss modulus. Lastly, dynamic frequency and amplitude sweeps were performed. Frequency sweeps analyze the impact of the rate of reoccurring 10% strain on the material, a requirement for our *in vitro* model. The results of the frequency sweeps show that for all formulations of the anisotropic silk scaffold is not frequency dependent below 2 Hz, which is critical for use of these materials for skeletal muscle tissue engineering. Amplitude sweeps analyze the impact of stretching the material from a strain from 0% to complete failure while maintaining frequency value at 1 Hz. The results of the amplitude sweep show that these materials are highly amplitude dependent above 5% strain. Future work focuses on assessing human skeletal muscle myoblasts (HSMMs) within the 3% anisotropic silk scaffolds, 100% ethanol freezing solution, and water annealing, with the addition of decellularized extracellular matrix to improve adhesion and aid in cell maturation.

Probing Remodeling of Responsive Synthetic Extracellular Matrices with Nanoscale Characterization Techniques to Expand the 3D Cell Culture Toolbox

Qi Zhang, Breanna Huntington, Eric Furst, April Kloxin

Engineered hydrogels are powerful tools capable of recapitulating biophysical and biochemical properties of the extracellular matrix (ECM) in different cellular microenvironments. Currently, two-dimensional (2D) cell culture on synthetic hydrogels is employed to study cellular response to environmental stimuli. However, such culture systems have limitations when simulating specific native microenvironments and related multidimensional cellular responses, for example, cell migration and proliferation. In this project, we utilize an on-demand externally and cellularly responsive hydrogel to probe cellular responses to dynamic stimuli events, with a focus on the application of nanoscale tools for examining changes in the mechanical properties of the cellular microenvironment. We apply atomic force microscopy (AFM) in conjunction with bulk rheometry to understand hydrogel mechanics from a local, nano level to a bulk level. The modulus of non-degraded hydrogels free of cells, hydrogels with cells encapsulated, and hydrogels that have been enzymatically degraded have been evaluated using both bulk rheometry and AFM. These measurements are then compared to each other to evaluate for differences seen in local and bulk mechanics of the hydrogel, in addition to various cellular analyses, such as immunostaining, to assess how the mechanics of the microenvironment influence cellular function and fate. Overall, this work demonstrates the potential to apply nano-level measurement, such as AFM, to evaluate the microenvironment response in more complex, engineered reductionist 3D cell culture microenvironments, and presents an opportunity to further correlate and compare the results of nano-level measurements with bulk-level measurements to modify and enhance existing mathematical models for nanoscale mechanical measurements.

Optimizing Kartogenin Delivery By Modifying PLGA-Peg Nanoparticle Size and Surface Charge

Reilly Avila, Abigail Mortimer, Bethany Almeida

Osteoarthritis is a degenerative joint disease affecting over 500 million people worldwide, resulting in irreversible degradation of articular cartilage, further leading to pain, inflammation, and loss of mobility[1]. Kartogenin (KGN) is a synthetic molecule known to enhance chondrogenesis and chondroprotection, resulting in cartilage regeneration[2]. However, its small size and hydrophobicity present limitations for delivery within aqueous cellular and tissue environments. We propose to improve the delivery of KGN through encapsulation in poly(lactic-co-glycolide)-poly(ethylene-glycol) (PLGA-PEG) nanoparticles (NPs). Previous work has demonstrated the efficacy of KGN-loaded PLGA-PEG NPs against free drug and placebo controls, yet optimization of KGN loading and release is still required[3]. Nanoparticle properties, such as size, composition, are known to affect the release profiles of encapsulated drugs[3]. Thus, we seek to investigate the effects of NP size, composition, and surface charge in order to optimize KGN delivery by changing PLGA molecular weight (15 kDa vs 45 kDa), PEG molecular weight (2 kDa vs 10 kDa), and Molar ratio between the PLGA and PEG (1:1 vs 1:5 vs 1:10). Briefly, a dicyclohexylcarbodiimide catalyzed conjugation reaction of PLGA with a terminal carboxylic acid group and PEG-bis-amine in dichloromethane is used to synthesize the PLGA-PEG copolymer, which is analyzed using proton nuclear magnetic resonance (^{1}H NMR). Following copolymer synthesis, KGN and PLGA-PEG are dissolved within acetonitrile to form nanoparticles using nanoprecipitation into aqueous polyvinyl alcohol. We have successfully fabricated and characterized KGN-loaded PLGA-PEG NPs using 15 kDa PLGA, 10 kDa PEG, and a Molar ratio of 1:7 PLGA to PEG (NP1), as well as NPs using 15 kDa PLGA, 2 kDa PEG, and a Molar ratio of 1:15 PLGA to PEG (NP2). NP1 resulted in a mean hydrodynamic diameter of 214 ± 7.20 nm and a mean surface charge of -7.0 ± 2.1 mV whereas NP2 resulted in a mean hydrodynamic diameter of 142 ± 30 nm and a mean surface charge of -0.50 ± 0.20. Current and future work is focused on continuing to characterize the remaining NP combinations as well as investigating the effects of size and surface charge on KGN encapsulation and release using high-performance liquid chromatography.

1. Wieland, H., 2005. *Nat Rev Drug Discov* 4, 331–344
2. Johnson, K., 2012. *Science*, 336(6082):717-721
3. Almeida, B, 2020. *Ann Biomed Eng*, 48(7):2090-2102

Analyzing the Molecular Behavior of Ionic Liquids through Kamlet-Taft Parameters

Ryan Dahl

The American Innovation and Manufacturing Act of 2020 initiated the phase-out of hydrofluorocarbon (HFC) refrigerants for more environmentally friendly hydrofluoroolefin (HFO) refrigerants. In response, Project E.A.R.T.H (Environmentally Applied Research Towards Hydrofluorocarbons) and our collaborators are actively pursuing methods of recycling HFC refrigerants that will no longer be usable. However, widely used mixtures of HFC refrigerants often form azeotropes which are not separable through conventional separation techniques like distillation. Thus, new separation methods are necessary and ionic liquids are proposed as entrainers in extractive distillation for the separation of HFC refrigerant mixtures. To design these complicated new systems, thermophysical data for ionic liquids and ionic liquid mixtures are necessary but uncommon. Our group aims to elucidate the complex behavior of ionic liquid mixtures with respect to their thermophysical properties in hopes to examine potential correlations between thermophysical properties and Kamlet-Taft parameters. Kamlet-Taft parameters π^*, α, and β represent polarizability, hydrogen bond donating ability, and hydrogen bond accepting ability respectively. We hypothesize that the molecular characteristics described by Kamlet-Taft parameters are related to viscosity, density, and thermal conductivity. In which case, the necessary thermophysical data needed to design HFC separation systems could be derived from Kamlet-Taft parameters. This correlation would aid the effort to integrate ionic liquids as an entrainer in extractive distillation systems for the separation of hydrofluorocarbon refrigerant mixtures.

Fabricating Cost-Effective, Wireless Pressure Sensors Using Piezoresistive Electronic Textiles.

Kalvin Huang, Isha Dave, Esther Whang, Jennifer Weiser, Mili Shah

Electronic textiles, also known as eTextiles, are fabrics that contain electronic capabilities. With their ability to conduct electricity, eTextiles have become widespread in the field of soft robotics over the past few years. ETextiles allow people to embed various electronic components in fabrics to transform them into flexible and wearable electronics. For this study, Velostat® is used to create a wearable pressure sensor matrix. The 6"x6" matrix consists of 64 sensor points created by intersecting 8 rows and 8 columns of copper tape. The copper tape is compatible with the pressure sensor matrix due to its flexibility and good conductivity. When pressure is applied on any sensor point, it will relay a value to an ATmega328 microcontroller that corresponds to the amount of pressure placed. As more pressure is placed, resistivity of the fabric decreases, allowing more voltage to flow through that point and the microcontroller relays a higher analog value. The analog values range from 0 to 1023 corresponding to approximately 0N to 3N of force. 2 HC-05 Bluetooth® modules (one connected to the matrix circuit, one connected to a computer) are used to transfer the values from the microcontroller to a computer. The processing code on the computer creates an image of 64 squares that change color based on a gradient that corresponds with the analog values. The circuitry, consisting of the Bluetooth® module, a solderable protoboard, and a 4.5V battery pack, is placed inside a small acrylic box. 16 ribbon cables, attached to the 8 rows and 8 columns, are connected to 2 multiplexers on the board, which relay voltage to and from the matrix. Faux leather is used to encase the Velostat® and copper tape to provide flexibility and comfort to the wearer.

The purpose of this study is to provide an affordable and simple method of using eTextiles, like Velostat®, to create flexible, wearable sensors. The application of these flexible sensors shows promise in the field of exoskeleton research, where external pressures placed on the body by exoskeletons can be evaluated for fit and comfort. Other fields of application include human-computer interaction, prosthetics, and motion capture.

Swelling in CO_2-Responsive Hydrogels

Finlay Smith, Sarah Sergi, Sophia Kosednar, Melissa Gordon

Upon exposure to external stimuli such as temperature, pH, or light, stimuli-responsive polymers exhibit changes in physical or chemical properties that can be advantageous for many applications. CO_2 is advantageous for use as a trigger as it is abundant, affordable, and non-toxic. Here, we show that CO_2 can be used to trigger pronounced, reversible swelling of crosslinked hydrogels by modifying their volume phase transition temperature (VPTT), which distinguishes the transition from a collapsed to swollen state. In this study, we synthesized hydrogels that display a CO_2-switchable volume phase transition that enables controlled swelling in response to CO_2 addition. Using UV-visible spectroscopy, we characterized the effect of varying co-monomer content on the resulting VPTT shifts in these samples. Once the VPTTs were determined, isothermal swelling studies were conducted at a temperature between the VPTTs before and after CO_2 exposure. The extent of swelling was dependent on the composition of the gels, leading to gels that swelled to more than double their mass prior to CO_2 exposure. Further swelling tests were conducted that demonstrated the reversible nature of the swelling via the introduction and removal of CO_2. Overall, we show that CO_2 can be used as a trigger for pronounced, reversible swelling in hydrogels displaying a volume phase transition, which may have applications in the field of drug delivery.

Assessing the Hydrolytic Degradability of Bio-Based Plastics from Birch Bark

Sophia Harrill, Megan Dodge, Devin M. Rosmarin, Melissa Gordon

The disposal of plastics in the environment harms global ecosystems and presents challenges to human health. Most commonly used plastics have slow degradation rates and are often improperly discarded into the environment, damaging ecosystems and food chains as well as releasing harmful chemicals into the environment. Thus, developing bio-based alternatives to petroleum-derived plastics is a promising approach to lessening the adverse environmental impacts of plastic use. This study focuses on the valorization of betulin, a pentacyclic triterpenol, which can be extracted in high quantities from birch bark. We incorporated betulin into a series of polyester networks with bio-based diacids of varying chain lengths using glycerol as a crosslinker. Dynamic mechanical analysis was used to examine thermomechanical properties and thermogravimetric analysis was used to determine degradative temperatures. We now extend this study to investigate the accelerated hydrolytic degradation of these thermoset polyesters. Results show that using shorter chain diacids as co-monomers lead to more crosslinked networks, which degrade slower in water. We used solid-state kinetic models to determine the rate constants describing the hydrolytic degradation. As a control experiment, negligible degradations of these samples are measured when tested at room temperature in deionized water. Overall, this work seeks to consider the viability of these biobased plastics from monomer selection, to polymer synthesis and properties, to end-of-life.

Degradation of PLA in Soil

Harrison Goehrig, Katsuyuki Wakabayashi

Plastic pollution is currently one of the most pressing environmental issues since most plastics take hundreds of years to decompose. Bioplastics, such as PLA, are often derived from organic sources, and are able to degrade much faster than traditional plastics. However, these alternative plastics perform worse than traditional plastics in some areas due to their weaker physical properties, such as having a lower glass transition temperature. Cellulose nanofibers can be used as an additive filler with PLA to form a stronger composite that has a higher stiffness and tensile strength, comparable to other commodity polymers, but there is still little known on the effect that adding these nanofibers has on the length of degradation of PLA. This study investigates how the addition of cellulose nanofibers changes the rate of degradation of PLA in soil. PLA samples with a range of weight percentages of cellulose nanofibers were compared. Different techniques for forming samples were also utilized to compare solid-state shear pulverization (SSSP), solid-state/melt extrusion (SSME), and twin screw extrusion (TSE), and to see if the method of forming samples has an effect on the length of degradation. Following ASTM D5338 standards, we built a custom setup to simulate controlled composting settings to conduct our tests and obtain our results.

Disulfide Containing Poly(β-Amino Ester) Hydrogels Inhibit Glutathione Reductase in Cancer Applications

Celeste Busch, Anastasiia Aronova, Claire Rowlands, Brittany Givens Rassoolkhani, Thomas D. Dziubla

Glutathione is a compound inside the cytosol that acts as the primary antioxidant in mammalian cells; it can be found in two forms- a reduced form (GSH) and an oxidized (GSSG). In its reduced form, it supports the reduction of reactive oxygen species (ROS), an excess of which results in a condition known as oxidative stress. Uncontrolled oxidative stress can result in damage to DNA, lipids, and proteins, leading to cell death. In the glutathione recycling reaction, GSH reduces hydrogen peroxide (H2O2), a common form of ROS, leaving it as an oxidized GSSG molecule. GSSG is then returned to its active form GSH through the enzyme, glutathione reductase. Under oxidative stress, a cell will have significantly more GSSG than GSH. The balance of the reduced and oxidized form is thus a critical part of establishing a cell's ability to resist injury, as a higher ratio of GSH to GSSG indicates the cells preparedness to defend against ROS. In this work, we develop disulfide containing poly(β-amino ester) hydrogels, whose degradation products are structurally similar to GSSG. We demonstrate the ability of these hydrogels to inhibit glutathione reductase, suppressing GSH recycling. The ability to manipulate the ratio of the two forms creates opportunities for therapies in cancer, wound healing, and regenerative medicine.

Wrinkled Graphene Oxide-Based Films Coatings for Liquid Repellency in Personal Protective Equipment

Maria Louiza Dimtsoudi, Jiaman Wang, Aidan Stone, Zidan Yang, Robert Hurt

Graphene oxide(GO) is a two-dimensional material derived from graphite that contains epoxide, carbonyl, carboxyl, and hydroxyl functional groups. Graphene oxide's novel properties make it a breathable barrier that prevents permeation of toxic chemicals. This has generated interest in its usage as a functional layer in personal protective equipment. The hydrophilic groups present in graphene oxide allow both water and potentially toxic polar organic liquids to wet and spread. Therefore, one aspect of developing GO films for personal protective equipment is modifying its surface to repel liquid droplets.

The goal of this study was to explore methods to modify the surface chemistry and surface structure of GO films to increase liquid contact angles. The test films were created by drop-casting 1ml of GO onto a nitrile glove surface and air drying for 24 hours. The films were roughened by a wrinkling process in which the nitrile substrate is stretched before the deposition of the GO solution. Following the air drying, the films were de-stretched, sprayed with one of five commercially available hydrophobic coatings, and left to air dry for an additional 24 hours. Four degrees of pre-stretching were examined: 0%, 33%, 66%, 100%.

To assess the liquid-repellent properties of the treated coatings, water and diethyl-malonate (a chemical warfare agent simulant) contact angles on the surface of the film were measured. Images of 5μL droplets of the two liquids deposited on the surface of the film were analyzed to determine the contact angle. Data was collected for twenty cases with different commercial coatings and degrees of wrinkling. The results show that surface coating can successfully switch GO films from hydrophilic to hydrophobic and can also increase the contact angle with diethyl malonate. Wrinkling enhanced the liquid repellency in most cases. These findings suggest that wrinkled and coated GO films are promising barrier layers in personal protective applications where both stretching and liquid repellency are desired.

Nanoetched Stainless Steel Architecture Enhances Cell Uptake of Biomacromolecules

Maeve Janecka, Thomas Pho, Julie Champion

Nanotextured surfaces promote enhanced cellular adhesion and tissue integration in biomedical implants. These surfaces' high aspect ratio nanoachitecture serves as an ideal interface between implant materials and host cells which is well-suited for localized therapeutic delivery. Despite their potential, nanotextured surfaces have not been widely applied for localized biomacromolecule delivery. We employed a low-cost, industrially relevant nanoetching process to modify the surface of biocompatible stainless and evaluated the feasibility of employing nanotextured SS316 (NT-SS316L) as a material for intracellular biomacromolecule delivery. We generated a functional coating of model biomacromolecule cargoes including superfolder green fluorescent protein (sfGFP) charge variants, cytochrome c and siRNA for intracellular delivery. Nanotextured surfaces enhanced cellular biomacromolecule uptake and delivered functional proteins and nucleic acids through energy-dependent endocytosis. Collectively, these findings indicate that NT-SS316L holds potential as a surface implant modification for localized drug delivery with many biomedical applications.

Control of Hydrogel Properties By Incorporation of Surfactants: Dynamics Interfacial Properties of Precursor

Sarah Onyembe, Muchu Zhou, Reza Foudazi

Hydrogels are characterized by their three-dimensional hydrophilic polymer networks and remarkable capacity to absorb and retain high amounts of water. Hydrogels have a wide range of applications, from wound dressings and contact lenses to drug delivery systems and tissue engineering. Their ability to retain water and provide a moist environment makes them valuable in biomedical applications, such as promoting wound healing and supporting cell growth for regenerative medicine, and environmental applications, such as water retention in soil. This research comprehensively explores the potential to modify network structure and bulk properties of hydrogels by manipulating the macromer assembly prior to crosslinking by incorporating surfactants. The initial phase of this study is an in-depth investigation of the interfacial properties of surfactants and macromer mixtures. We use sodium dodecyl sulfate (SDS) as surfactant and polyethylene glycol diacrylate (PEGDA) as macromer. The dynamic surface tension of precursor is studied to determine the interaction of SDS and PEGDA, providing valuable insights into the alterations in bulk properties initiated by surfactant incorporation. Hydrogel synthesis is achieved through UV irradiation of aqueous solutions containing PEGDA and SDS. Then, the hydrogels are washed to remove the surfactant and unreacted species. Extensive swelling kinetics tests are conducted to evaluate the effect of surfactant on macromer assembly and subsequently on the swelling kinetics and network architecture. Moreover, drying kinetics studies give insights into the kinetics of water loss from the hydrogel, further elucidating its structural characteristics. The results and insights from this research will have profound implications for advancing hydrogel materials across a diverse spectrum of applications. The potential applications are vast and transformative, from tissue engineering, where precise control over hydrogel properties is needed for biomimetic scaffolds, to the sphere of plant growth, where tailored hydrogel formulations may change irrigation and nutrient delivery systems. This study is essential in exploring the multifaceted world of macromer-surfactant interactions in hydrogel synthesis, opening new horizons for innovation and technological advancement in materials science.

Effects of Wicking Characteristics on the Accuracy of Wearable Nylon Sensors

Ellen Ingle, Chelsea Monty-Bromer, Victoria Stege, Shelby Daniels

SweatID wearable sensors consist of electrically spun nylon fabric that is treated and then functionalized to become sodium selective. The functionalized nylon is then assembled into a wearable fabric sensor (WFS), used with a corresponding electronic device, to monitor sweat sodium concentration during exercise. The monitoring is based on a resistance reading measured from the device through the WFS. It is crucial that these nylon sensors are accurate and precise, so quality control measures are needed at multiple stages. One of these measures consists of monitoring the wicking characteristics of the nylon mat at each stage. This ensures that each stage proceeds uniformly and helps to determine the orientation of the functionalized nylon during the assembly process. A modified drop test was used to determine the horizontal wicking speed and water absorbance capacity (WAC) of the nylon through its three stages in the treatment process: nylon, post-treatment pre-functionalization, and post-treatment post-functionalization. These characteristics help determine if adjustments need to be made within the process or if the nylon mats fail the quality check and a new nylon mat just be used. As the nylon mat is being processed during the treatment stage, one of the materials used during this stage is carbon nanotubes (CNT). An effect of this material is that the nylon mat has a distinctive color variation on either side of the mat due to the dispersion of the CNTs. From the wicking data gathered, further alterations can be done on fully assembled WFS, determining which side of the nylon mat is more sensitive to the resistance reading due to the color variation. The third stage in the process is the functionalization stage, where calixarene is used to functionalize the treated nylon post-CNT. The calixarene coats the mat, causing a significant change to the wicking properties of the nylon. A quality control measure can be implemented through the change in properties to ensure a consistent distribution of calixarene over the treated nylon mat. This change is validated through standard in-lab calibration testing and repeat in-person testing on indoor cycles. The future of these WFSs hopes to continue into the eventual integration into fabric athletic apparel, so as more data is acquired about these wicking characteristics, the better the integration of better fabrics can be achieved and the closer these sensors can reach these expectations.

Measuring Cation Concentrations in Sweat Using a Wearable Fabric Sensor

Lisa Chambers, Chelsea Monty-Bromer, Victoria Stege, Shelby Daniels

Many endurance athletes face challenges in recovering effectively after high-intensity workouts. To optimize recovery, athletes should understand the physiological changes taking place within their bodies. Understanding what athletes are losing in their sweat during exercise can help them determine how best to rehydrate. Personalized hydration strategies can improve an athlete's performance by up to 40%. The work presented here will focus on the development and testing of a novel wearable fabric sensor (WFS) to help athletes monitor biomarkers in sweat, specifically sodium.

The WFS allows for the non-invasive collection of sweat sodium concentration during exercise. While various sweat sensors are available, most rely on conductivity measurements to estimate a value for sweat sodium concentration value. While accurate to total electrolyte concentration, this approach falls short in providing athletes with comprehensive insights into sweat sodium composition and more personalized hydration improvement strategies. The wearable fabric sensor presented here selectively monitors for sodium concentration in sweat, giving athletes the personalized information they require.

This study recruited 10 healthy adults (20-68 yrs) through fliers and Internet notices at Cleveland State University from June 2023 to October 2023. Upon arrival to the laboratory, consent was obtained, and exercise participation clearance was established using a PAR-Q+ form. The subjects then wore the WFS, and sterile gauze adhered to the body using tegaderm and cycled continuously for 60 minutes. The sterile gauze was removed and centrifuged at the end of the testing for analysis using a laboratory-quality sweat sodium analyzer (Precision Hydration, conductivity-based measurement) and a flame photometer (Jenway)

Cations within sweat mainly consist of sodium and potassium, rendering concentrations of other cations negligible. Using flame photometry, we were able to quantify the exact concentrations of sodium and potassium in the sweat samples. The average percent of sodium found in sweat was 83%, assuming that sodium and potassium make-up 100% of the cations in sweat. A WFS provides a non-invasive way to verify that sodium levels in the sweat are normal and can give the athlete valuable information. The accuracy of the flame photometer to the SSA was 3.35 mM, or 6%. The accuracy of both the flame and sweat sodium analyzer (SSA) to the WFS was variable to each batch of experimental WFSs, which ranged from 1-22 mM.

These findings contribute to a better understanding of post-exercise hydration needs and inform athletes about their sweat, leading to better recovery and performance.

Nanoparticle Surface Chemistry Tailoring for Enhanced Biomedical Applications

Chuzube Edeh

This study embarks on a fascinating journey into the dynamic interplay between nanotopography, silica surfaces, and their interactions with both artificial and natural membranes. Through meticulous exploration, it unravels the captivating influence of nanotopography on silica nanoparticles' behavior and their interplay with lipid bilayers.

Two distinctive synthesis methods, the Stöber and surfactant-templated methodologies, set the stage for an exploration of diverse surface morphologies and pore size distributions within the synthesized silica nanoparticles. The Stöber particles grace us with their smooth surfaces, while dendritic MSNs introduce us to rugged topographies, enriching our understanding of the intricate world of nanotopography.

With advanced imaging techniques at its disposal, this research unearths a rich tapestry of topographical variations on nanoparticle surfaces. Cryogenic transmission electron microscopy (cryoTEM) initially led us to believe that lipid coatings were absent on larger-pore MSNs, challenging conventional wisdom. However, further scrutiny unveiled a surprising revelation - lipids are not exclusive to smoother surfaces but grace even the most rugged topographies, expanding the realm of possibilities in the design of lipid-coated nanoparticles.

The intrigue deepens as this study ventures into the realm of cholesterol's influence on lipid coating formation. This additional layer of complexity prompts the formulation of hypotheses, unraveling the mysteries of lipid coatings on surfaces with varying topographies. The interplay between surface architecture and lipid behavior becomes a central theme in this captivating exploration.

Beyond these revelations, the research offers a groundbreaking discovery - the ability to load large protein molecules within porous MSNs, reshaping silica surfaces. This innovation heralds a new era of enhanced lipid coatings, promising advances in drug delivery and imaging applications.

In summation, this study serves as a beacon in the field of nanomedicine, highlighting the pivotal role of nanotopography in nano-biointeractions. Its insights open the door to the optimization of lipid-coated nanoparticles, offering a vast spectrum of potential applications. The complex and versatile findings of this research promise revolutionary advances in targeted drug delivery, imaging, and biomedicine.

Enhancing the Strength of Pluronic F127 Hydrogels for Biomedical Applications: Investigating the Complexities of Alginate Integration

Ding Wang, Hengwei Zhu, Surita Bhatia

Our research focuses on the combination of alginate and Pluronic F127, two hydrogels with distinct properties. The challenge we sought to address in this study is how to enhance the strength of the Pluronic F127 gel for applications like wound healing. To achieve this, we aimed to combine Pluronic F127 with alginate, however, the process of effectively integrating alginate with Pluronic F127 presents complexities that require investigation.

The primary goal of our study was to optimize the protocol for creating blended gels of alginate and Pluronic F127. Specifically, we aimed to determine the most effective concentration levels for Pluronic F127 and the best approach to integrate alginate into the system to strengthen the resulting gel.

To achieve this aim, previous researchers in our group employed a protocol of soaking alginate gels in a sequence of Pluronic F127 solutions with concentrations of 5%, 10%, 20%, 25%, and 30% over a seven-day period. In this work, we modified the process to explore concentrations of 5%, 10%, 15%, 20%, and 30%. This adjustment allowed us to assess the effect of concentration shifts on the gel's stability. We also utilized rheology to analyze the resulting materials and characterize their properties.

Our findings revealed that the sudden concentration increase from 20% to 30% in Pluronic F127 caused some issues with the stability of the hydrogel, leading to the appearance of outlier data points, underscoring the challenges of integrating alginate with Pluronic F127 and the need for further investigation.

In conclusion, our study highlights the complexities of combining alginate and Pluronic F127 to create robust gels. While the addition of alginate shows promise for enhancing material strength, the process remains delicate and prone to variations. Additional research is required to optimize the integration process and ensure the reliability of the resulting materials.

This study is significant for the development of biomaterials, particularly in the field of hydrogels for applications like wound healing. Understanding how to enhance the strength of Pluronic F127 gels through the incorporation of alginate can pave the way for more effective biomedical applications. Moreover, our investigation contributes to the broader understanding of material properties and the challenges involved in combining different types of materials for specific purposes.

The Active Sands of Time: Flow of Granular Microrollers through Funnels

Brianna Sauder, Alex Oh, Jocelyn Lu, Tyler Richardson, Douglas Hardy, Jinghui Gao, James Gilchrist

Granular materials play a vital role in agriculture, industry, medicine, and filtration. The behavior of granular media is very complex. Since they behave as a fluid when moving at a high enough speed, they exhibit distinct velocity profiles in certain flows, and can exhibit turbulent and laminar flows, the same as fluids. This work explores the area of "active granular media" where magnetically responsive microroller particles rotate in response to an oscillating field. Being able to magnetically activate granular particles and initiate motion opens a vast spectrum of possibilities for applications in biomedicine and industry. What is unclear is to what degree can systems of many microrollers behave as granular media. This work will observe the flow of functionalized 44 μm polymer particles in funnels and how they mix and disperse within a fluid with various magnetic fields to further explore their viability for navigating complex systems. The polymer particles have evaporated Fe_2O_3 on half of their surface, creating a north and south pole in the metal, allowing them to be moved magnetically. Submerged in ethanol, the passive granular matter without magnetic manipulation settled at an angle of repose of 47° whereas microrollers can fluidize in the ethanol and flow through the funnel in complex ways, even stagnating away from the orifice or flowing deterministically toward the orifice. In trials with applied magnetic fields that individually rotated microrollers clockwise on the left and counterclockwise on the right side of the opening, all of the microrollers moved through the orifice. These preliminary results suggested there is significant impact on the flow rate, final distribution, and fluidization of the particles in the system. Microroller movement prevailed even in a container with flat surfaces surrounding the orifice. Given the properties of these particles and their ability for collective movement, potential use in biomedical and industrial fields shows promise.

Isothermal and Gradient Foaming of Polycaprolactone in Supercritical Carbon Dioxide

Emma Troiano, Dawn Dokyung Rhee, Erdogan Kiran

Polycaprolactone (PCL) is a polymer of biomedical importance. Scaffolds based on PCL are of great significance in tissue engineering applications. Thus, there is much interest in foaming polycaprolactone foams using benign physical blowing agents such as carbon dioxide. In this poster, we will present the foaming behavior, pore size, pore distributions, and bulk foam density of polycaprolactone exposed to carbon dioxide at 100 bar and varying temperatures. These foaming conditions were determined based on the melting temperature reduction as displayed in the changes in the rigidity of the PCL in carbon dioxide. The changes in rigidity are assessed with High Pressure Torsional Braid Analysis (HP-TBA), a technique that allows for determination of T_m and T_c from the assessment of changes in relative rigidity and mechanical damping as functions of temperature or pressure.

The foamed polymers were prepared in gradient and isothermal foaming cells. The gradient foaming cell exposes a strip of polycaprolactone to a range of temperatures and a uniform set pressure. The temperature is controlled at one end of the cell, allowing the development of a gradient temperature distribution within the cell. In the isothermal cell, the temperature is controlled to ensure a uniform temperature throughout. Either cell was then pressurized with carbon dioxide, using an ISCO pump to maintain the set pressure. After an hour of equilibration under the target pressure, the cells were rapidly depressurized to induce pore nucleation and growth. The foams were then freeze-fractured in liquid nitrogen and then sputter coated with gold and characterized for pore morphology with SEM. Pore size distributions were assessed using ImageJ software. The bulk foam densities were also determined by measuring the mass of the sample in air as well as submerged in water.

The foaming experiments were first carried out in the gradient cell over a temperature gradient of 55-43°C, 45-37°C, 40-33°C, and 35-29°C at 100 bar, and then in isothermal cell at 30, 35, 37 and 40°C, at 100 bar. The results of foaming will be presented in terms of bulk foam density and SEM images and compared with the rigidity reduction path in carbon dioxide. We will also provide the carbon dioxide sorption levels during the saturation stage in carbon dioxide prior to depressurization which were determined using magnetic suspension balance (MSB).

Low-Temperature Water Dynamics Study in Fully Hydrated Nafion® Membrane

Yin-Jen Chang, Bratoljub Milosavljevic

Nafion® membrane has been widely used as a key component in high performance fuel cells due to its high proton conductivity and slow chemical aging characteristics[1]. Although many studies have been published on this subject, the hydrated membrane performance at low temperatures is not fully understood[2]. The corresponding ambiguities reported in the literature may be attributed to the state of water in very narrow channels within the membrane. In this work, in order to better understand the water dynamics in this system, we combine the interpretation of our DSC study results with the data obtained using the $Ru(bpy)_3^{2+}$ luminescence probe.

In the DSC experiment performed using 10 °C/min heating and cooling rates, the enthalpy of melting of water was found to be 253 J/g, which can be rationalized as if 24 wt% of water remained unfrozen when the sample was cooled to -150 °C. The difference in enthalpies of the water freezing and the ice melting peaks indicates that water freezing starts at -38 °C; about 34 wt% of water is frozen at -65 °C and water continues to freeze as the temperature decreases to -150 °C. The thermogram also reveals that the frozen water in Nafion® membrane melts at -22 °C, which is consistent with the previously established fact that the water freezing and melting temperatures are sample size dependent[3].

To gain deeper understanding of this peculiar behavior of water, the $Ru(bpy)_3^{2+}$ luminescence probe, which is sensitive to spatial confinement, was applied[4]. The emission spectrum maximum was measured as a function of temperature; the uninterrupted hypsochromic shift down to -105 °C was observed. Two crossover points were noticed (-30 °C, -71 °C); they can be attributed to some phase transitions of the polymer matrix. The crossover at -105 °C can be attributed to the complete freezing of water since the $Ru(bpy)_3^{2+}$ spectrum hypsochromic shift is negligible below that temperature. The NMR measurements reported in Reference 2 were performed in the temperature range from +25 °C to -70 °C, while our measurements were performed from +25 °C to -150 °C and that is the reason for the discrepancy in the water dynamics interpretation.

References:

[1] V. Rao et al, CRC Press, 2015, pp. 567–6143.

[2] Ren-Hao Cheng et al, PCCP, 2021, **23**, 10899–10908

[3] A. E. Carte, Proc. Phys. Soc. B 1956, **69**, 1028–1037

[4] B. H. Milosavljevic and J. K. Thomas; Macromolecules, 1984, **17**, 2244–2248

Selectively Dissolvable Biopolymer Interface for Textile Recycling

Gabrielle Wood, Cecile Chazot

Although usually only accounting for about 1% of the composition of blended materials, 80% of discarded garments in the US contain spandex fibers. Blended spandex fibers are notoriously difficult to reprocess through traditional mechanical and chemical recycling methods. Here, we propose the design of a dissolvable interface layer of chitosan to facilitate spandex separation from blended fabrics through selective dissolution, a common chemical recycling method. In this study, chitosan, an antibacterial biodegradable biopolymer, is coated onto spandex filaments via dip coating. A lab-scale dip-coating and drying process was developed for this study based on limitations of the physical properties of the spandex filaments and the literature available for liquid coating of a fiber. We investigate the relationship between dip coating parameters and final morphology of the chitosan coating and find that both chitosan molecular weight and concentration greatly affect coating thickness. Using FT-IR spectroscopy, we also uncover indications of strong non-covalent interactions between the chitosan coating and spandex filament, suggesting good interfacial adhesion. This work represents foundational knowledge in the development of alternative blended spandex fibers to facilitate textile recycling.

Acknowledgement – This work was supported by the MIT Materials Initiative for Comprehensive Research Opportunity (MICRO) and by the Northwestern University MRSEC REU program funded by the NSF.

Characterizing New Mexico-Grown Cannabidiol (CBD) Hemp (*Cannabis sativa* L.) for Sustainable Development

Madison Vasquez

Hemp (*Cannabis sativa* L.) offers multiple products: flower, fiber, and grain. Extensive breeding efforts have led to hemp varieties with different morphologies, physiologies, and chemical compositions optimized for specific purposes. In the state of New Mexico, the focus on high-cannabinoid varieties for flower production has prompted hemp producers and processors to assess their strategies for recovering value from the fibrous residues from cannabidiol (CBD) crops. Lignocellulosic hemp biomass has potential as a source of bioenergy; a precursor for chemicals to produce paints, plastics, and sealants; a building material in bio-composites; and as a paper or pulp additive. Choosing the right product stream hinges on the physiochemical traits of the lignocellulosic hemp biomass and the available processing equipment.

Currently, the domestic focus on cannabinoids and the absence of local fiber processing infrastructure (decortication facilities) result in the underutilization of high-CBD hemp crops' stalks and stems, constituting up to 70% of the total crop dry weight. These residues are often discarded or repurposed into low-value products such as animal bedding or compost. A first step in the development of biorefinery concepts is the characterization of the available materials to enable comparison to other lignocellulosic resources.

Here, hemp biomass from pilot field trials was first milled, sieved, and then measured for moisture and ash contents. Serial Soxhlet extractions were performed, first with water to remove soluble inorganic materials such as soil residues, then with ethanol to obtain soluble organic compounds including terpenoid compounds, chlorophyll, and plant waxes. The extractives-free hemp biomass was then subjected to acid hydrolysis, autoclaving, ultraviolet-visible (UV-Vis) spectroscopy, and high-performance liquid chromatography (HPLC) to determine the lignin and lignin and structural sugars contents. The higher heating value (HHV) of the hemp biomass was determined using bomb calorimetry and was expected to range between 17.5 and 18.8 MJ/kg.

Investigating the Impact of Dope Solution Composition on PVDF Membranes

Michelle Kryl, Isabel Escobar, David Lu, Rachel Kaplan

This research aims to achieve the fabrication of a successful polyvinylidene fluoride (PVDF) membrane through the incorporation of an additive and support layer. The study involves comprehensive characteristic analysis of these membranes such as permeability, rejection and contact angle measurements. Due to the hydrophobicity of PVDF-based membranes, they are prone to fouling and inefficiently absorb organics [1]. The change in composition of the dope solution and addition of polyvinylpyrrolidone (PVP) will allow for surface modification of these membranes. To enhance the performance of the PVDF membrane this research explores the influence of different compositions of PVDF in the dope solution in addition to variations of PVP. 17% PVDF in the dope solution was found to be the most consistent when analyzing permeability and rejection data. To increase the permeability of the 17% PVDF membrane, 0.5% PVP was added to the dope solution. PVP functioned as a stabilizer and thickening agent due to its hydrophilic nature, therefore decreasing the contact angle of the 17% PVDF membrane. The evaporation time was explored and was found to have a small impact on the permeability of the membrane, decreasing it slightly and increasing the rejection. The successful development of a PVDF single layer membrane is paramount as it serves as a foundational milestone for initiating the fabrication of composite membranes with polysulfone (PSf). The inclusion of PVP in the dope solution may have played a role in the effective mixing of the two dope solutions when casting the composite. Due to the experimentation being done on a lab scale, collaboration with the Georgia Institute of Technology and their slot die coating machine will be helpful in creating large scale membranes for commercial water filtration.

Increasing Oil Fouling Resistance of PVDF Microfiltration Membranes Via Surface Modification Using Polydopamine Combined with Hyperbranched Polyol

Hunter Chunn, Moustafa Zagho, Mohammad Hassan, Sergei Nazarenko

In the face of climate change, reclaiming freshwater from oil-in-water emulsions produced by industrial processes has become an area of increasing interest. Membrane-based filtration techniques offer an efficient solution for filtering these emulsions. Hydrophobic poly(vinylidene) fluoride (PVDF) membranes are widely used for their superior chemical, mechanical, and thermal properties in filtration applications compared to hydrophilic membranes. However, the inherent hydrophobicity of these polymers often causes membrane fouling: the adsorption and deposition of oil-droplets and surfactants onto the surface and within the pores of membranes. This work aims to modify the surface of PVDF membranes using polydopamine (PDA) with reduced pore blockage, which is accomplished in the presence of hyperbranched-polyol BoltornTM H40 (HBP, H40) creating an oil-fouling-resistant surface. It is hypothesized that polymerizing dopamine in the presence of hyperbranched-polyol controls PDA aggregate size while creating an effective fouling-resistant surface with reduced pore blockage. Using scanning electron microscopy (SEM), it was observed that modifying PVDF membrane surfaces with PDA in the presence of H40 resulted in larger surface pores and higher surface porosity, in contrast to membrane surfaces bearing the PDA layer alone. The PDA and PDA/H40 modifications of the surface of PVDF films resulted in a significant decrease in the water contact angle of the film surface, indicating improved hydrophilicity. The deposition of polydopamine in the presence of H40 demonstrated a remarkable boost in the separation of oil from water in tight oil-in-water emulsion while displaying high resistance to fouling over multiple runs versus unmodified and PDA modified PVDF membranes.

Solid-State Shear Pulverization of Post-Industrial Uhmwpe: Particle Morphology and Molecular Structure Modifications Toward Conventional Mechanical Recycling

Nathan Herrold, Katsuyuki Wakabayashi, Haley Scopelliti, Riggs Johnson

Ultra-high molecular weight polyethylene (UHMWPE) is one of the most prominent high-performance thermoplastics for biomedical, leisure, and coating applications. Large-scale recycling of UHMWPE is extremely difficult due to the high melt viscosity of the material as well as its exceptional chemical resistance and impact strength. There is a need for a commercially scalable methodology that can process the waste feedstock for mechanical recycling while sustaining the outstanding physical properties of the material. Solid-state shear pulverization (SSSP) is a continuous, twin-screw extruder-based processing technique in which the low-temperature application of shear and compressive forces impart changes in structure at different length scales to overcome the challenges of difficult-to-recycle polymers. This paper investigates the use of SSSP in mechanically recycling post-industrial scrap UHMWPE (rUHMWPE) material from a local ski and snowboard manufacturer. The SSSP-processed particles are flat, micron-scale flakes with enhanced surface area, which can sinter very quickly when compression molded. The molded rUHMWPE samples in turn exhibit enhanced ductility and toughness compared to the as-received scrap material, based on the tunable mechanochemical modification of the ethylene chains. Polymer characterization is performed on both rUHMWPE samples along with a control UHMWPE in order to determine the effects the SSSP process has on the molecular nature of the molecule. The results show that post-industrial recycling of engineering and high-performance plastics can achieve a highly sustainable and value-added life cycle when executed with a proper processing technique. The rUHMWPE could be easily compression-molded in a conventional way as the molecular architecture of rUHMWPE has been modified to flow and interdiffuse effectively in its sintering mechanism. However, there were no disruptions to the robust ethylene crystal structure. These results have positive implications for sustainability, which can be extended to other types of high melt-viscosity or otherwise difficult-to-recycle engineering and high-performance thermoplastic materials. As society continues to use more advanced thermoplastic materials in real-world applications, the industry has a responsibility to find a more robust, systematic, and universal recycling methodology for these polymeric systems. While this paper focused on the recycling of neat systems, the SSSP technology can be applied to other polymer sustainability solutions. It has the potential to improve the compatibilized blending of mixed polymer streams and compounding of bio-based fillers. Different pathways to sustainability need to be curated to make a significant difference in this polymer-laden society.

Comparative Analysis of Kevlar, PBO and Derived Carbon Fibers

William Beck, Christopher Altamuro, Michael Chauby, Joseph F. Stanzione III, James Newell

Carbon fiber (CF) reinforced carbon/carbon composites (CCCs) are in increasing demand in the automotive, aerospace, and defense industries due to their robust mechanical and thermal properties. Widespread application of these advanced materials is limited because of excessive production costs which are driven up due to long, expensive densification processes that are needed to achieve these properties. After the initial pyrolysis, CCCs are exceedingly porous due to the cracks that form as a result of the rigid CFs that inhibit in-plane shrinkage of the matrix during pyrolysis. This porous structure requires densification to increase strength and prevent thermo-mechanical degradation. Cocarbonization of polymer fiber reinforced polymer composites has the potential to offer CCCs with an optimized pore morphology. Currently the CF market is dominated by PAN- and Pitch-based CFs because of their strength and economic viability. These fibers are not suited to test this hypothesis because they require a lengthy oxidative stabilization step in air which cannot occur in a polymer matrix with nearly zero permeability. Poly(p-phenylene terephthalamide) (Kevlar) and poly(p-phenylene-2,6-benzobisoxazole) (PBO) are high-performance polymer fibers that can be converted to CFs via rapid carbonization processes without oxidative stabilization. While these precursor fibers initially cost more than PAN- and pitch-based CF, the pore structure generated from simultaneous carbonization of fiber and matrix has the potential to offset those costs via facile densification. This study investigates Kevlar and PBO fibers that were treated under different thermal schedules up to 1200 °C. Raman spectroscopy was used to compare the generated carbon structures. For both Kevlar and PBO, 1000 °C was found to produce the most ordered carbon fibers. X-ray diffraction (XRD) was used to quantify the carbon nanostructure which revealed PBO possessed superior crystalline order. Mechanical testing is consistent with XRD results displaying carbonized PBO achieves higher tensile modulus and compressive strength, however carbonized Kevlar unexpectedly out performed carbonized PBO in tensile strength. Comparing these fibers under SEM imaging revealed that the lower tensile strength of carbonized PBO is due to the unique microstructure that forms. This microstructure is shown to be influenced by processing conditions during pyrolysis. The results of this study demonstrate PBO's unique characteristics that distinguish itself as the primary candidate to explore this hypothesis going forward.

Tuning the Spatial Arrangement of Sol-Gel Based Polymer Electrolyte Membranes

Zhao Ren Tan, Thivani Senathiraja, Chris Cornelius

Polymer electrolyte membranes (PEMs) are an integral part of many of the next generation's clean and sustainable energy sources. The spatial arrangement of PEMs plays a key role in deciding the final properties of these membranes. Altering the morphology of PEMs via functional inorganic nanoparticles has been of great interest among researchers due to their synergistic interactions between the organic and inorganic phases in the PEM matrix. In this study, five different molar ratios of inorganic nanoparticles were synthesized using tetraethylorthosilicate (TEOS): titanium isopropoxide (TIP) via sol-gel chemistry in THF. The inorganic phase was introduced into a Penta block copolymer (PBC (1.0), commercially known as Nexar, with an ion exchange capacity (IEC) of 1.0 to form 2 wt% homogenous nanocomposite PBC membranes. To elucidate the interrelationships between the interactions of the sol-gel network with the PBC ionomer and their microstructure, TEOS-TIP was added to the PBC 1.0 domain in two different forms, solid and liquid. It was postulated that the addition of TEOS-TIP in its powder form could probe the domain spacing in the PBC (1.0) ionomer and push them farther away compared to the addition of TEOS-TIP in its liquid state. FTIR shows the successful integration of TEOS-TIP in the PBC (1.0) matrix. The changes in the intensity of the peak at 910 cm^{-1} corresponded to the molar ratio of TEOS-TIP in the system. TGA showed that the decomposition temperature for the PEM decreased when the ratio of sol-gel was increased. However, PBC (1.0) liquid nanocomposites exhibited a greater reduction in thermal stability at high TIP concentrations compared to the PBC (1.0) powder nanocomposite membranes. It was also observed at high ratios of TIP, the proton conductivity of PBC (1.0) is greatly reduced when exchanged with powder form compared to the liquid form. These changes in the properties at a high TIP ratio suggest the disruption in the domain spacing of the PBC (1.0) matrix due to changes in the interaction strength of TIP with the sulfonate groups in the PBC (1.0). Hence this can be a potential route to systematically alter the phase separation of the ionomers and design the membranes for fuel cells and gas separation applications. Future studies will incorporate SAXS to investigate the changes in the microstructure of these PEM nanocomposites, as well as TEM studies to explore the changes in domain size and spacing.

Fabrication of Reinforced Polycaprolactone and Cellulose Nanocrystal Composite Membranes for Cardiovascular Tissue Engineering Applications

Jared Noel, Taylor Norris, Joseph Batta-Mpouma, Jin-Woo Kim

Cardiovascular disease is the leading cause of death worldwide each year. Modern medical treatments for myocardial infarctions primarily consist of stents or blood thinners and are aimed at restoring oxygen flow to the heart. However, these treatments do not address the scar tissue that forms as a result of cardiac events, thus leaving patients at a greater risk for repeat events. Cell scaffolding is a prominent technology in tissue engineering that utilizes biomimetic materials to facilitate native tissue regrowth. Polycaprolactone (PCL) is a synthetic biopolymer that has successfully been used in bone and musculoskeletal tissue engineering in the past. However, PCL is limited for cardiovascular cell scaffolds because of its low mechanical strength and poor cell affinity. Cellulose nanocrystals (CNCs), naturally occurring subunits of cellulose, have a high tensile strength and tunable surface chemistry rich in hydroxyl groups. These unique properties make CNCs ideal candidates as additive materials for PCL-based cell scaffolds. In this research, the preparation of uniformly mixed PCL and CNC composite materials at different ratios was utilized to make membranes. Mechanical properties, including tensile strength, Young's modulus, and elongation strain, were characterized. It was found that an increasing concentration of CNCs in the composites led to increasing mechanical strength values of the membranes, demonstrating the practicality of PCL/CNC composite materials for tissue engineering applications in stress-intensive environments like the cardiovascular system.

Revisiting Experimental Techniques and Theoretical Models for Estimating the Solubility Parameter of Rubbery and Glassy Polymer Membranes

Harold Ly, Matthew Webb, Lucas Condes, William Box, Sepideh Razavi, Michele Galizia

The Hildebrand solubility parameter, δ, is used to predict and correlate miscibility among substances, including low molecular weight compounds and polymers, which affects practical applications such as coatings, drug delivery, material formulation, and membrane separations. Estimation and correlation of the Hildebrand solubility parameter (δ) of polymers and small molecules is a common practice in membrane material science and is accomplished by both experimental and numerical routes. Intrinsic viscosity, as well as swelling measurements, are commonly used to study polymer interactions with solvents and determine polymer δ values. These experimental routes, however, require viscosity data in numerous solvents at various concentrations, and therefore they are time and material consuming, which may be problematic when working with non-commercial, expensive polymers.

Dynamic light scattering offers a quicker solution while consuming less material, by correlating δ to the hydrodynamic diameter of polymers in various solvents. Rubbery and glassy polymers, including a microporous polymers, PIM-1, and a high fractional free volume (FFV) polymer, poly(1-trimethylsilyl-1-propyne) (PTMSP), are among the samples included in this study with great relevance to membrane science. Secondly, in an attempt to enhance the accuracy of numerical estimate of polymer solubility parameters via the group contribution method, we provide updated group contribution parameters, along with their uncertainty, according to the technique recently reported by Smith et al. (J. Membr. Sci. 636 (2021) 119525). These updated group contribution parameters result in a mean absolute relative error of 9.0% in predicting the solubility parameter on a test set of 40 polymers, which is on par with the average 10% error reported previously. We also show, using machine learning techniques, that augmenting the group contribution model with extra parameters or non-linear relationships does not improve its accuracy. Results of the updated group contribution technique and dynamic light scattering measurements were compared to experimental viscometry on four test polymers, and the difference between the three techniques is discussed.

An Exploratory Study on the Development of Sargassum Algae-Based Biodegradable Polymer Composites Via Selective Laser Sintering

Stephanie Garcia, Omar Movil

As a direct effect of climate change and agricultural runoff seeping into the Amazon River, tons of brown macroalgae known as Sargassum, have been accumulating on shores throughout the South Coast of North America and the Caribbean region since 2011. This phenomenon has had severe repercussions for marine ecosystems, human health, and the economies of coastal communities in some U.S. territories and states. Despite its negative impacts, Sargassum holds potential as a resource for various industries, such as cosmetics, fertilizers, and construction materials. However, its potential as a biomaterial for additive manufacturing (AM), a transformative field in the production of goods, remains largely unexplored. AM is expected to play a crucial role in America's future economy and national security. Hence, the primary objective of this research project is to investigate the use of Sargassum/biopolymer composite micro-powders for selective laser sintering (SLS) 3D printing, a highly promising technology in the AM sector. If successful, this research could diversify the range of materials compatible with SLS, which is currently dominated by polyamide 12 (nylon PA12).

To achieve the project goal, the team has established three main objectives: (1) fabricate Sargassum/biopolymer composite micro-powders having algal biomass contents $\geq$ 30wt%, and excellent flowability for SLS, (2) fabricate Sargassum/biopolymer composite specimens via SLS using the obtained powders, and (3) study the effects of Sargassum content and key SLS printing parameters on the microstructure and mechanical properties of the specimens.

Regarding the first objective, Sargassum/biopolymer composite micro-powders are fabricated via wet granulation as follows: Firstly, a suitable amount of dried Sargassum is pulverized via ball milling and then incorporated into an aqueous biopolymer/binder dispersion using a planetary mixer. The resulting brown paste is then dried in an oven overnight to obtain a cake. This dried cake is cut into pieces, which are then fed into an oscillating granulator to obtain a fine powder. Lastly, the fine powders are characterized via optical microscopy, and angle of repose.

To fabricate the 3D printed specimens, the powders are fed into a SLS machine from Sintratec®, where the printing process takes place under controlled conditions. In addition, the team plans to evaluate the effect of some key parameters on the mechanical properties and microstructure of the specimens. The printing parameters to be changed are the laser scan speed, layer thickness, and hatching spacing. In the case of the specimens' properties, these will be evaluated via tensile and flexural tests.

Improving Mechanical and Water Stability of Graphene Oxide and Polymer Composite Films for Protective Barrier Applications

Rebecca Martin-Welp, Aidan Stone, Zidan Yang, Jiaman Wang, Aicha Sama, Indrek Kulaots, Robert Hurt

Graphene Oxide (GO) nanosheet films are of interest as barriers to prevent exposure to toxic liquids and vapors. However, pure GO films are not mechanically stable and not stable to water immersion and need to be toughened while retaining their barrier properties. This project focuses on designing GO/polymer film composites and GO-embedded polymer sandwich structured films. A set of water-soluble polymer binders and cross-linking agents were screened, both individually and with GO for mechanical and water stability. These polymer binders included synthetic polymers, elastomeric coatings, and liquid rubber. The composite film screening criteria included the durability and strength of the films, through tensile, bend, and water immersion stability tests. Films showing promise were further tested for vapor permeability. Nacre films containing cross linked methylcellulose or polyallylamine with glutaraldehyde were found to have significantly increased mechanical and water stability compared to unmodified GO films, passing all three screening tests. Six sandwich-structure films also passed all three screening tests. To affirm these modified films still operated as barriers, water vapor transmission and permeability, as well as hexane permeability, of the promising films were tested. With these successes, robust GO and polymer film composites affirm the potential for GO nano sheets to operate as highly effective protective barriers, which could operate as barriers to UV light, antibacterial coatings, and protection from chemical and environmental toxins. Further testing is also in process to investigate both the stretchability and breathability of these films.

Prediction of Polymer Solubility Using Quality Datasets

Sofia Moreno Briceno, Blair Brettmann

Polymers are large molecules of many repeating units and their high strength-to-weight ratio makes them versatile and appealing to many industries, including food and drink packaging, medical equipment, furniture, construction, and travel. Although plastic is crucial to these industries and more, the majority of plastic generated becomes waste at the end of life. According to the US Environmental Protection Agency, only 8.7 percent of plastic generated in the US was recycled in 2018–but while higher rates of mechanical recycling is ideal, it is much less feasible than the recycling of other materials. Every time plastic is recycled, the polymer chain becomes shorter, losing strength and durability. To resolve this, chemical recycling can depolymerize the plastic to its monomer form for repolymerization, attaining the same quality as a virgin polymer. For chemical recycling, the polymer must first be dissolved in a solvent. However, since polymers are large and complex molecules, predicting which solvents can be used is challenging. Many factors such as polymer and solvent functional groups, polarity, boiling/melting point, and molecular weight impact solubility.

To better predict polymer solubility without experimentally testing a polymer in every possible solvent, machine learning can be implemented. However, machine learning models are only as good as the quality of data they are trained on. In this work, I compare a high-quality dataset collected at Georgia Tech to a lower quality one consisting of literature data to discover mismatches in the solubility classifications. The objective is to identify likely inaccurate classifications within the low-quality dataset in comparison to the high-quality experimental dataset and determine if adding certain data or correcting mismatches improves the performance of machine learning models trained on the low-quality dataset. The experimental data contains 176 unique polymer-solvent combinations across room, cold, and hot temperatures, while the literature dataset does not specify temperature. I analyzed whether mismatches occur more frequently for polymer/solvent pairs listed in the literature dataset as insoluble or soluble and whether more mismatches were at hot or cold temperatures. I found that there were more instances, about 6% of the data points, where the literature data labeled the polymer as soluble where it was found to be completely insoluble in the same solvent across all temperature ranges by the experimental data. This analysis is being used to prioritize the experiments to perform to improve the data set and thus improve the quality of the machine learning predictions.

Ellipsometry of Ultra Thin Polymer Films

Andres Cotto, Ryan Toomey, Pavan Kumar Tilaru

The usage of ellipsometry to characterize thin film properties such as refractive profile, film thickness and more has been an extremely potent tool for scientific research. Ellipsometry leverages the polarization of light being changed as a function of light-matter interactions to acquire a signal that can be decoded into material properties. As incident light moves through the thin film, the axes of polarizations deviate into an elliptical shape, the difference of which can be calculated into an ellipsometric ratio ρ through data collection of the amplitude, Ψ, and angle difference, Δ. However, the lower parameters to which ellipsometry can accurately characterize a thin film are not fully understood. Previous studies have reported anomalous increases in material density at lower thicknesses without change to chemical structure, resulting in large increases in refractive index as measured by the ellipsometer.

Here we report on our ongoing study investigating the the refractive profiles of characterized thin films as a function of their thickness. To diversify the collected data, the combinations of thin films and substrates are expanded to include three different substrates as the basis for the ellipsometry. We utilize spin-coating of varied concentrations of thin-film solutions to deposit films with a range of thicknesses onto optical glass substrates. Importantly, the usage of high-speed spin coating allows depositions of uniform thickness. The coated substrates are then analyzed using rotating compensator ellipsometry to characterize their refractive profiles. The aim of this research is to develop straightforward analytical models to predict the refractive index of different thin films as a function of their thickness. Next steps of the research include the expansion of the thin film formulation set to include chemically unique thin films in the research.

Effects of Copolymer Chemistry on the Formation of Multiphase Complex Coacervates

Amy Lim, Isaac Ramírez Marrero, Sarah Perry

Complex coacervation is an associative phase separation caused by the interactions of two oppositely charged macromolecules. The resulting liquid-liquid phase separation creates a polyelectrolyte-rich coacervate phase and a polyelectrolyte-dilute supernatant phase. In multiphase complex coacervation, differences in the phase behavior of the polyelectrolytes used may lead to further separation into multiple sub-phases. Research on the topic of coacervation is quickly growing and becoming increasingly applicable to our general lives, but research on multiphase complex coacervation remains limited.

While the formation of multiphase complex coacervation has been previously reported, methods for the formation of multiphase complex coacervates are not yet standardized due to their novelty. In response, I am researching and analyzing the specific conditions of the formation of multiphase complex coacervates. Specifically, I want to understand how differences in polymer chemistry and phase behavior allow for their formation, through copolymers of varying charge density and hydrophobicity. Parameters such as salt and polyelectrolyte concentration are extensively studied. Samples are then investigated via turbidity and microscopy analysis. Multiphase complex coacervation allows for pursuit of advanced laboratory experiments, such as those of self-organized compartmentalization and catalytic conversion of chemical reactions, in microscale, while still preserving important charged interactions and phase separation effects that are critical in macroscale. Namely, this research project aims to develop a precise and consistent experimental method to obtain multiphase complex coacervates, to further investigate the numerous advanced laboratory applications of compartmentalization in drug delivery, vaccine mechanisms, and the modeling of various intracellular structures.

Development of Biodegradable Polymers Poly(β-amino ester) and Poly(lactic-co-glycolic acid) for the Controlled Release of Bupivacaine

Marissa Nicholson, Pranto Paul, Claire Rowlands, Brittany Givens Rassoolkhani, Nikita Gupta, J. Zach Hilt

This project explores the potential of a poly(β-amino ester) (PBAE) encapsulated poly(lactic-co-glycolic acid) (PLGA) system to serve as an extended drug delivery method for local anesthetic-bupivacaine. This research addressed the need to provide a longer-lasting pain management regimen for patients and to also develop an alternative to opioids, commonly used medicine for pain management, and the risks that come with prescribing them. This project investigates a new system for controlled release of bupivacaine for 48 hours. To make the PBAE hydrogel, PBAE macromer was first synthesized by reacting diethylene glycol diacrylate (DEGDA), polyethylene glycol 400 diacrylate (PEG400DA), and isobutylamine. Next, a solution of ethanol and the photoinitiator 2,2-dimethoxy-2-phenylacetophenone (DMPA) was added and mixed with the macromer. This mixture was then injected into two parallel glass plates and exposed to a UV flood source to form the PBAE hydrogel. The PBAE hydrogels were characterized by conducting a swelling study and degradation study. PBAE hydrogel was cut into circular chunks, soaked in a bupivacaine-in-ethanol solution. Then to make microparticles, bupivacaine loaded hydrogels were grounded and passed through a sieve. Then the microparticles were encapsulated in PLGA, and in vitro drug release study was conducted in PBS buffer. It was found that PBAE encapsulated in PLGA system could release bupivacaine for 48 hours and with less initial burst release compared to the PBAE or PLGA alone. The PBAE-PLGA based drug release system shows potential and could be further improved for the desired outcome.

Molecular Functionality and Nanoparticle Influence on Surface Tribopolymerization

Ashton Wolford, Lindsay Martin, Reilly LaGrand, Meagan B. Elinski

Tribology has been increasingly relied on due to emerging technologies requiring more efficient lubrication and surface protection qualities. Processes in mechanochemistry rise to mitigate these issues, as research continues to look for tribochemical reactions that allow for antiwear film formation. This work is aimed at developing a deeper understanding of surface chemistry, and the different impacts that various molecules have on tribopolymerization and surface wear. Molecules comparing relative positions of alkene and alcohol functionalities, cis-2-nonen-1-ol and 8-nonen-1-ol, are added to a polyalphaolefin oil (PAO4) and used as lubricants in sliding processes subjected to steel on steel interactions. The use of nanoparticles, specifically nanodiamonds, are also implemented in order to observe impacts on tribopolymerization as well as wear, which is analyzed using optical microscopy. An in depth understanding of tribopolymerization resulting from different surface chemistries and molecular structures could result in advancing surface lubrication schemes.

Polymer Chemistry Effects on the Mechanics of Polyelectrolyte Complexes

Yaozu Chen, Isaac Ramírez Marrero, Sarah Perry

Polyelectrolyte complexation is a phase separation phenomenon driven by the electrostatic interactions between two oppositely charged polyelectrolytes. Such materials are widely used in products like shampoo and are produced by organisms like sandcastle worms as bioadhesives. Processing polyelectrolyte complexes has the benefit of requiring only water as a solvent, avoiding the use of high temperatures or organic solvents. Additionally, their properties can be tuned by altering factors like the length, charge density and hydrophobicity of the polyelectrolyte. Furthermore, it is well understood that salt concentration strongly affects the mechanics of the polyelectrolyte complex. Adding salt typically causes swelling of the material, resulting in a softening of the complexes, in addition to disrupting the electrostatic interactions between the polyelectrolytes. However, the effect of the charged functional group of the polyelectrolytes on the mechanics of the materials is not well-understood. As such, we are interested in studying the effect of the charged functional group on the mechanics and its relationship with salt concentration. In particular, we are examining the linear viscoelasticity of polyelectrolyte complexes with carboxylate, sulfonate, phosphonate, and amine groups with different degrees of alkylation on polyelectrolyte complexes at different salt concentrations. These rheological studies give us insight into properties like viscosity and solid-to-liquid transitions in our material. Furthermore, we apply time-salt superposition to combine data from individual experiments and produce a master curve for each of the systems to allow for a broader examination of how the mechanical properties can be tuned with salt concentration – from glassy complexes at low salt to viscoelastic liquids at higher salt concentration. By understanding how salt concentration and polymer chemistry affect the linear viscoelasticity of polyelectrolyte complexes, we can design materials with desirable properties for applications ranging from adhesives to encapsulants, all while avoiding harsh polymer processing methods using high temperature and organic solvents.

Radical Polymerization of Methacrylate-Based Covalent Adaptable Network with Chain Transfer

Taras Nagornyy, Jason Lennon, Christopher J. Kloxin

Covalent Adaptable Networks (CANs) are a promising category of cross-linked polymers capable of reversible bond exchange in response to external stimuli, such as heat. This research investigates the impact of composition and light intensity on photoinitiated radical polymerization in a vinyl ester polymer film with dynamic bond exchange properties and examines its stress-relaxation behavior as a function of temperature. Surprisingly, stress increased during relaxation tests and introducing a catalyst did not mitigate this effect; however, a distinct reduction in stress was observed when the temperature was lowered. These findings have significant implications on CAN materials, emphasizing the importance of temperature to control over the amount of the desired stress relaxing reaction versus the amount of undesired side reactions. These studies provide insights into tailoring polymer formulations for enhanced performance in practical applications. Understanding the complexities of stress-relaxation behavior in polymer materials is crucial for precise control and optimization of CAN materials.

Fabrication and Characterization of Novel Sargassum-Based Polymer Composite Filaments for 3D Printing

Sebastian Toro Bernal, Zuanichi Figueroa, Abraham Polanco, Jeziel Rodriguez, Omar Movil

As a direct effect of ocean warming and nutrient enrichment, tones of a brown macroalgae known as Sargassum, have been accumulating on shores throughout the Caribbean region since 2011. These unprecedented annual events have been detrimental to marine ecosystems, human health, and economy of coastal communities. In the search for new applications of Sargassum biomass, the present work aimed to explore the use of this seaweed as a raw material for the fabrication of renewable powder that was incorporated into polylactic acid (PLA) composite filaments to 3D print different specimens. The mechanical properties of the resulting structures were evaluated via tensile tests, while studies of biodegradability were performed using burial tests.

The materials characterization results indicate that the Sargassum processed via ball milling exhibited fine particles with a broad particle size distribution (microns). TEM analysis is required to observe the nanosized particles. This work also demonstrated that it is possible to fabricate filaments with Sargassum powder contents up to 30 wt%. However, the required extrusion temperatures and the brittleness of the filaments increase with the Sargassum content. Also, 3D printing of these composite materials required nozzles with sizes ≥1 mm to avoid clogging.

In terms of the mechanical properties, the elastic modulus and yield strength of the 3D printed specimens exhibited a declining trend as the Sargassum content into the PLA polymer matrix increased from 0 to 20wt%. These results are supported by the SEM analysis that shows that the number of defects and surface roughness increases with the biomass content. These defects are usually associated with poor mechanical behaviors. At higher filler contents (≥ 25 wt%), the variability in the elastic modulus was significant, making it difficult to draw precise conclusions. This variability is supported by the inhomogeneities, and defects observed in the microstructure of specimens with high Sargassum contents. The yield strength results followed a similar trend.

Burial test results indicate that the weight loss (wt%) of the fabricated specimens increases ~4.5 times as the biomass content increased from 0 to 30wt% (after 120 days).

Hydrophobic Coating for Cellulose: Boronic Acid Activated Siloxane Polymer

Matthew Bradley

Cellulose, a widely abundant building block of the organic world, is present in many cotton containing textiles and paper products. Fluorine containing hydrophobic substances or surfaces, such as the popular Gore-Tex fabric, is expensive and releases forever chemicals into the atmosphere when disposed of. As a cheaper and more environmentally safe alternative, Boronic Acid activated Siloxane Polymer is an effective way to create a hydrophobic coating for cellulose. The purpose of this project is twofold: first, to investigate the product's (BorSilOx) efficacy on various types of cotton, and second, to investigate the reaction processes as a means to obtain the final polymer product. Building off initial tests with pure cellulose paper, 4 types of cotton were tested and contact angle measurements were obtained for each. In addition, two reactions were investigated, hydrosilylation and dehydrogenation, the latter being a much more cost efficient option to obtain a similar polymer product.

Impact of Polymer Composition and Salt Concentration of Poly(ethylene oxide)/Poly (methyl methacrylate)/Salt Blend Polymer Electrolyte

Paola Méndez, Rui Sun, Yossef Elabd

Poly (ethylene oxide) (PEO) based solid polymer electrolytes (SPEs) have been widely investigated due to their high lithium salt solubility. However, low mechanical strength and high crystallinity limits the performance of PEO based SPEs. Blending PEO with poly(methyl methacrylate) (PMMA), a high glass transition temperature polymer, is an effective method to improve mechanical stability and reduce crystallinity of SPEs. In this study, PEO/PMMA/lithium salt (lithium bis (trifluromethanesulfonyl)imide (LiTFSI)) blend SPEs were prepared by solution casting at various polymer composition and salt concentrations. The impact of salt concentration and polymer composition on the thermal and electrochemical properties of the resulting SPEs was evaluated. Results show that increasing salt concentration leads to reduced degree of crystallinity and higher ionic conductivity. SPEs with varying polymer composition exhibit similar ionic conductivities at the same salt concentration.

Dynamically Switchable Polymer Architectures: Roles of Molecular Weight and Persistence Length for Improved Cyclization of Linear Polymers.

Olusotemidayo Oyedele, Thomas Gartner

The ability to dynamically switch polymer architecture offers an opportunity to tailor materials with responsive, tunable properties for a wide range of applications from advanced drug delivery systems to adaptive structural materials that can adjust to varying environmental conditions. In this work, we aim to determine the polymer design characteristics that affect the dynamics of conformational changes between linear and cyclic topologies. We performed molecular dynamics simulations on isolated coarse-grained polymer chains of varying length and chain stiffness. We quantified the average 'cyclization time'—i.e., the time necessary for the polymer chain ends to touch, which would be the first step in the cyclization process of a linear chain with reactive end groups. We found an optimal range of the ratio of polymer chain length to persistence length to facilitate rapid chain-end convergence, enabling the efficient formation of cyclic polymer structures. Our results provide quantitative insights into the role of molecular weight and persistence length in achieving dynamically switchable chain architectures. The polymer design criteria identified here open potential avenues for designing polymers with dynamic/switchable properties, potentially enabling the efficient transformation of linear polymers into cyclic analogues through controlled cyclization processes.

Computational Studies to Design Block Polymers for Thermal Conductivity Applications

Kaveri Srivastava, Tristan Myers, Sojung Park, Arthi Jayaraman

Block copolymers are macromolecules (polymers) are composed of two or more chemically distinct polymer segments/monomers. Linear diblock copolymer melts' phase behavior is well established from past theory, simulations, and experiments. The four main morphologies accessed by linear diblock copolymers include lamellar, cylindrical, gyroid, and spherical domains; these morphologies can be achieved through control of the interactions of the two or more chemically distinct segments or monomers as well as composition. In this poster, I will share my contributions to a larger scale project being conducted in Prof. Arthi Jayaraman's lab at University of Delaware. The project goal is linking design of block copolymers synthesized in our experimental collaborator Prof. Brent Sumerlin's lab at the University of Florida to observed morphologies. These predictions come from both theory (especially Self-Consistent Field Theory or SCFT) by Dr. Sojung Park and complementary simulations from my work and fellow graduate student Tristan Myers. By understanding how the interactions between the monomer chemistries and composition of the monomer chemistries in the polymer impact morphology (i.e., domain shapes and chain conformations within each domain) we can identify those designs and conditions (e.g., temperature) that yield structural arrangements that are known to increase thermal conductivity.

Crystallization Behavior and Melt Memory Effects in PEO-*B*-PCL Copolymer for Tailoring Drug Delivery

Aidan Stewart, Emily Rotola, Trevor Jonny, Ryan Van Horn

Poly(ethylene-oxide)-block-poly(ε-caprolactone) (PEO-*b*-PCL) is a copolymer that has been researched for a variety of medical applications including drug delivery. Having an understanding of the crystallization behavior of PEO-*b*-PCL is paramount to optimizing its performance and tailoring it to specific needs. In the case of drug delivery, this can manifest when different drugs need to be released at different rates. One drug may need a constant release over a long period of time, or another may need to be released in bursts. This study focuses on understanding which part of the copolymer will crystallize first during a cooling process while casted in toluene, and the effects that melt memory has on the crystallization process. The experimental approach used a hot stage to melt and subsequently cool samples of PEO-*b*-PCL casted in toluene. Samples varied in molecular weight with both 5k10k and 5k5k polymer being used. Fourier-transform infrared spectroscopy was utilized to monitor peaks that correspond to PEO and PCL. Additionally, varying the melting temperature allowed changes due to melt memory effects to be measured. The results show a distinct difference between the crystallization behavior of the two molecular weight variants. In the 5k10k samples, PCL crystallization occurred first and also occurred sooner in trials with a lower melting temperature. The 5k5k samples showed no distinct difference in the temperature in which PEO or PCL crystallization, which suggests coincident crystallization behavior. The differences between the behavior of samples with different molecular weights indicate that PEO-*b*-PCL can be fine-tuned to the needs of specific applications. Being able to control the crystallization behavior, influenced by both melt memory effects and molecular weight, is promising for the future of optimizing the usage of PEO-*b*-PCL in drug delivery.

Cinnamate Functionalized Liquid Crystal Polymers for Elastomer Photoalignment

Katherine Cooney, Matthew L. Smith

Liquid crystal elastomers (LCEs) are a type of material that combines the elastic properties of rubber with the anisotropic properties of liquid crystals. This material exhibits significant reversible shape changes when exposed to light or heat due to transition in the alignment of the liquid crystal units. LCEs could be useful in many types of applications including artificial muscles, robotics, and medical devices. One challenge with this type of material is expanding the processing options available for producing alignment. The goal for this project was to use cinnamate derivatives that were responsive to UV irradiation to create alignment in LCEs. Linear polymers were created using an aza-Micheal addition reaction. Chain extenders with a hydroxyl functionality were used to produce the polymers. Cinnamoyl chloride was then reacted with the hydroxyl groups to produce cinnamate functionalized liquid crystal polymers. FTIR was used to characterize the products. 300 nm UV light was used to photocrosslink the LCEs. UV-vis spectroscopy indicated that photo crosslinking between cinnamate pendants occurred. Future work will examine the extent of crosslinking and investigate processing steps that may lead to photoalignment using polarized UV light.

Utilizing a Novel Acrylate Chemistry for Application in Controllable Polymer Networks

Olivia Damgaard, Claire Niemet, Christopher Bowman

In this study, a novel chemical reaction which reveals latent acrylates in situ was applied to a multistage network of polymer films. A thiol-ene polymerization forms a low modulus material with epoxides covalently incorporated. A small molecule study confirmed that under heat, the electrophilic carbon in the epoxide ring is susceptible for attack by the nucleophilic thioether group, which produces an acrylate in situ. Upon photoinitiation, these acrylates will react to form a higher modulus material. It was noted in this study that it is effective to use two distinct light stimuli to initiate the polymerization reactions separately, so TPO was used to initiate the thiol-ene reaction using 405nm light and DMPA was used to initiate the acrylate polymerization using 365nm light

Several different monomers can be used to form this multistage network. Vinyl ethers heavily favor the polymerized initial stage of this system than allyl ethers do. Allyl glycidyl ether was used to add epoxides to the system. When a higher ratio of epoxide groups to thiol-ene functional groups was used, there were larger differences in modulus and Tg. However, when there were too many epoxide groups in the mixture of monomers, the gel point would not be reached to form a film because there were not enough crosslinking monomers.

X-Ray Vision: Seeing the Impact of Monomer and Initiator Chemistry on Conversion in X-Ray Polymerization Systems

Finnis Ginder, Dr. Sage M. Schissel, Dr. Julie L. P. Jessop

X-rays are a form of high-energy electromagnetic and ionizing radiation that can be used to break molecular bonds and create radicals and ions. Thus, X-ray irradiation can be used to initiate certain chemical reactions. This research focuses on using X-ray irradiation to initiate polymerization in a variety of (meth)acrylates and to activate photoacid generators for epoxide polymerization. (Meth)acrylate systems are prevalent in commercial formulations, in which X-ray polymerization can provide a distinct advantage for these polymerization systems since the ionizing nature means that no initiator molecules are required for polymerization to occur. In addition, X-ray polymerization occurs at ambient temperature and does not require solvents to control the reaction temperature. Thus, this method of polymerization is attractive since it does not require additional chemicals that could become a disposal hazard or leach from the final product. The effect of (meth)acrylate chemistry was explored by measuring monomer conversion as a function of irradiation time using Raman spectroscopy. Methacrylates polymerized more slowly than acrylates; however, all monomers reached similar final conversions. The effect of photoacid generator chemistry, as well as compatibility with X-ray irradiation, was also tested. Epoxide samples were prepared using one of three different photoacid generators, and conversion was measured at three different irradiation doses. Higher doses have resulted in higher initial conversion of the epoxide monomer. In addition, photoacid generators with a hexafluoroantimonate results in noticeably lower conversions than those that contained a hexafluorophosphate or pentafluorophenylborate groups, possibly due to the relative strength of the acid produced. The conversion of these epoxide samples continues to be monitored after the initial polymerization in order to observe any dark-cure effects; however, preliminary results show that dark cure effects after X-ray initiation are very limited compared to other forms of initiation. Understanding how monomer and initiator chemistry impact conversion will allow for successful implementation of X-ray polymerization in future industrial applications.

Development of Poly(arylene ether sulfone) Based Mxene Composite Material with Enhanced Mechanical Properties

Anna Huszar

MXenes are highly favorable 2D structures with a layered thickness of ~1 nm with high electric conductivity rates and thermal properties. Thin films of MXenes are flexible which allows these materials to be good candidates for nanofillers in polymer composites. Modeling the composites can predictively illustrate the effect of different loading concentrations and the interaction of nanofiller with the polymer matrix on mechanical properties. In this research, the effect MXene concentration on the thermal and mechanical properties UDEL P-3500 composite membranes were studied. Composite modeling was used to correlate MXene quantity on UDEL mechanical changes with two loadings of MXene were investigated: UM1 indicating 1 w/v% MXene loading and UM2 2w/v% MXene loading. Visible color changes were observed with increasing MXene concentration. ATR-FTIR analysis showed a mixture of the MXene and UDEL peaks rather than forming new functional groups. The pure MXene powder exhibited stretching vibrations at 1600 cm^{-1} for the C=O, and 3600 cm^{-1} while both MXene composites and the MXene powder exhibited peaks around 600 cm^{-1}, indicating the presence of C-Br bonds due to halogenated etching. Raman Spectroscopy exhibited two distinct sharp peaks related to MXene powder that are expected: A1g(Ti, O, C) around 200 cm^{-1} and A1g(C) around 720 cm^{-1}. TGA analysis showed the addition of MXenes into the UDEL matrix increased the UDEL thermal deflection temperature from its expected temperature of 174 °C to closer to the expected degradation of MXene flakes. SEM imaging clearly shows the preservation of the layered $Ti_3C_2Br_2$ MXene structure after inclusion into the UDEL matrix indicates no damage to layers upon incorporation into the polymer. Modeling studies showed that the mechanical properties of MXene/UDEL composites can be significantly improved by aligning the MXene inclusion, in which maximum MXene loading was found to be 19.99 vol% for the random case and 45.03 vol% for the aligned case.

Wrinkling of Poly(4-Vinylpyridine) Thin Films after Ion Bombardment

Grace Pettis

Ion bombardment (IB) is an efficient method of manipulating & cleaning surfaces of materials. When the beam of ions hits the substrate, a few possibilities that depend on the substrate and bombardment conditions can occur; e.g. ion implantation, breaking of molecular bonds, and sputtering off of surface atoms resulting in surface erosion. In the specific conditions that Poly(4-Vinylpyridine) (4-VP) thin film on a silicon wafer is bombarded at 2 keV with argon ions and subsequently exposed to humidity, periodic wrinkling of the film develops over time and a decrease in surface wettability is observed. Without bombardment, 4-VP degrades in water and is relatively unaffected by the level of humidification that is sufficient for wrinkling. The potential to wrinkle is credited to this affinity for water. Polystyrene (structurally similar) was tested under identical conditions and no changes were observed. In 4-VP, bombardment results in a constant decrease in contact angle with water regardless of humidity conditions, implying bombardment is uniquely associated with a decrease in contact angle and consequently changes in the surface energies that govern it. In humidified samples, increasing absolute humidity increases the rate at which the amplitude of the wrinkles develop, and increasing film thickness proportionately increases their wavelength. This evidence and potential water signals on Raman spectra of wrinkled samples suggest that under these conditions the bombarded polymer slowly absorbs water in the air, consequently causing a uniform bi-axial stress that drives isotropic wrinkling.

Impact of Molecular Weight Ratio and Solubility on Dye Release from PEO-b-PCL Copolymer Films

Madison Horvath, Lily Snyder, Sabrina Gonzalez, Ryan Van Horn

Poly(ethylene-oxide)-*b*-poly(ε-caprolactone) (PEO-*b*-PCL) is an amphiphilic diblock copolymer that is biodegradable and FDA approved. PCL is highly biocompatible and hydrophobic; however, it becomes an effective drug delivery system when attached to hydrophilic PEO. Due to its biocompatibility and unique properties, PEO-*b*-PCL is currently being researched in terms of its sustained drug delivery applications for hydrophobic drugs. Using a solvent casting method, polymer films were created using both PEO(5,000 g/mol)-*b*-PCL(16,300 g/mol) (5k-16.3k) and PEO(5,000 g/mol)-*b*-PCL(10,000 g/mol) (5k-10k). Methyl red (MR), methyl orange (MO) and methylene blue (MB) dyes were used as drug surrogates. Films of 3% dye (wt dye:wt polymer) were incubated in DI water to study dye release over time with minimal PCL degradation at neutral pH. Dye release trends were determined through UV-Vis Spectroscopy. The components impacting release rates of PEO-*b*-PCL copolymer films include varying molecular weight ratios, type of dye used, and the subtle differences in the pH of the environment. In general, initial studies showed that MB released the fastest from films, and 5k-10k films released dye faster than 5k-16.3k films. Analyzing these trends with both previous and current data allows for a better understanding of the future applications of PEO-*b*-PCL in drug delivery.

Polymerization of Three-Arm Poly (e-Caprolactone) Via Reactive Extrusion

Ryan Stearns, Badal Lodaya, Ramani Narayan

Production of Three-Arm Poly (e-Caprolactone) by Extrusion Polymerization

Ryan Stearns

Under direction of Dr. Ramani Narayan, Chemical Engineering

Our goal is to Produce Polycaprolactone (PCL) in a continuous process utilizing a twin-screw extruder to shift away from batch-based production and improve the processing characteristics of PCL. PCL is currently made in a batch process, which is not ideal for production needs. PCL is biodegradable and has been drawing attention for potential in single use packaging, biodegradable blown films, shoe stiffeners, and other applications. Current polymers suffer poor processing characteristics such as low melt strength and poor mechanical properties in some cases. Our research goal is to synthesize Poly (e-Caprolactone) via reactive extrusion. The reason for this is that by transitioning to a single-step continuous process, time and energy will be saved at industrial level. As a biodegradable material, PCL shows promise in an eco-friendlier setting by reducing pollutants for companies, potentially saving businesses money that would normally be used towards pollution permits. In addition, reactive extrusion is a scalable, continuous process, and can be adapted to many different process models. The extrusion will take place in a Leistritz 27mm Co-Rotating Twin-Screw Extruder, where we will pelletize the strand and use three methods of identification of PCL. A differential scanning calorimeter (DSC) will be used to determine the temperature of glass transition and melt. A thermal gravimetric analyzer will be used to determine the degradation temperature of the polymer. Intrinsic Viscosity experiments will help determine molecular weight, along with Mark-Houwink constants.

Conductive Polymers As Flexible Sensors

Cameron Tardy, Kalvin Huang, Esther Whang, Mili Shah, Jennifer Weiser

In recent years, a novel class of polymers with intrinsic electrical conductivity, relying on conjugated bonds with delocalized π-electrons along their chains, has emerged[1]. These organic polymers possess a high band gap that enables electron flow within the material, though they require oxidation or reduction to introduce charge centers in the bandgap. A dopant is used during polymerization to achieve conductivity[2].

This study investigates flexible sensors based on polypyrrole doped with different oxidants, ammonium persulfate (APS) and iron chloride ($FeCl_3$), with varying polymerization parameters. The doped polypyrrole is polymerized onto 7cm x 7cm cotton fabric sheets to create flexible eTextiles. Optimal results are obtained using a 1:1 APS : pyrrole mass ratio with a 110-minute polymerization time. The optimized eTextile gave the best range of resistivities from rest to extended fabric, and then rest to compressed fabric. While $FeCl_3$-doped eTextile demonstrates lower resistance, making them better conductors, it lacks the extensive range of resistivity values compared to APS doped eTextiles.

To showcase the properties and sensitivity of the APS-doped polypyrrole eTextile, a fabric origami accordion is constructed using a larger fabric sheet with the same polymerization parameters. This accordion, connected to a Lilypad Arduino™ with a speaker, produces different tones when compressed, reflecting changes in resistivity. The DIY textile sensor's potential applications are demonstrated through this eTextile Origami Accordion, highlighting its use in motion detection and opening doors for future applications, such as motion capture.

References:

1. MacDiarmid, Alan G. 'The Nobel Prize in Chemistry, 2000: Conductive polymers.'

2. Umoren, S. A., Solomon, M. M., & Saji, V. S. (2022). Conducting polymers. Polymeric Materials in Corrosion Inhibition, 443–466. https://doi.org/10.1016/b978-0-12-823854-7.00002-3

Nature-Inspired Conjugated Organic Polymer: Self-Healing and Stretchability Via Ureidopyrimidine Functionalization

Wala Abdelhalim, Konstantinos E. Kakosimos, Mohammed Al-Hashimi

In the realm of organic electronics, the development of self-healing materials has gained paramount importance, finding diverse applications including electronic skin, sensors, solar cells, supercapacitors, coatings, and adhesives. This study presents a novel class of self-healing polymers derived from conjugated monomers (DPP) featuring Ureidopyrimidine sidechain functionality. These sidechains exhibit a remarkable capacity for intramolecular hydrogen bond formation, imparting self-healing properties, flexibility, and stretchability to solid-state polymeric films. The synergistic combination of self-healing attributes with the inherent advantages of conjugated polymers holds substantial promise for various applications. Ureidopyrimidinone, characterized by its reversible bonding properties and compatibility with conjugated polymers, emerges as a particularly auspicious self-healing group. This research delves deep into the realm of material chemistry and engineering, embarking on a journey to redefine the boundaries of polymer properties. Through meticulous manipulation of these materials, facilitated by Ureidopyrimidinone functionalities, we are dedicated to pushing the limits of what is attainable. Our ultimate goal is to engineer flawless polymers that set new standards for performance in organic field-effect transistors (OFETs). This ambitious endeavor represents a pioneering effort at the intersection of material chemistry and engineering, where we challenge convention and strive to create the perfect materials for the next generation of electronic devices, with a specific emphasis on their utility in electronic skin.

Enhanced Benzothiadiazole-Based Polymer Materials: Introducing Self-Healing and Stretchability through Nature-Inspired Hydrogen Bonding

Amanda Cruz, Konstantinos E. Kakosimos, Mohammed Al-Hashimi

Inspired by the fundamental role of hydrogen bonding in natural processes, such as the formation of DNA's double helix and the unique properties of water, this research explores the incorporation of urea side chains in a conjugated polymer to harness their benefits in developing stretchable and self-healing polymers suited for applications in organic field-effect transistors (OFETs). A conjugated polymer intended for use in OFETs was synthesized, employing a Stille reaction involving two distinct benzothiadiazole (BT) monomers: 10% incorporated with a C16 chain and 90% featuring a urea functional group. These were polymerized in a 1:1 ratio with an IDTT monomer. OFETs, known for their low cost, easy fabrication, and flexibility, are crucial for the development of large-area electronic devices like electronic skin. Drawing inspiration from nature, where hydrogen bonding plays a pivotal role in processes like the double-helix formation in DNA and the properties of water, the conjugated polymer utilizes urea side chains to induce hydrogen bonding. This promotes molecular self-assembly into a well-ordered, planar structure in the solid state, facilitating efficient π-π stacking interactions between the aromatic rings of the polymer, thereby optimizing the rate of electron transfer. Notably, this nature-inspired hydrogen bonding not only enhances tensile strength and maintains electrical performance but also introduces self-healing capabilities to the polymer. The dynamic, reversible bonds allow for autonomous repair, realigning polymer chains, and enhancing material resilience. The unique attributes of this polymer make it particularly well-suited for enhancing the efficiency and longevity of OFET applications.

Micellization of Pee-PEO Block Copolymers in Organic Solvents for Use As Perovskite Nanoreactors

Duncan Hill, Supriya Gupta, Marc A. Hillmyer, Timothy P. Lodge

Block copolymers are made up of chemically distinct polymeric segments connected through a covalent bond. The chemical dissimilarity between segments leads to a variety of self-assembled nanostructures both in bulk and solution. Block copolymer micelles have been studied extensively due to their use in the fields of drug delivery and emulsification. The objective of the present study is to control the formation of poly(ethyl ethylene)-*b*-poly(ethyl oxide) (PEE-PEO) block copolymer micelles in organic solvents such as decane and squalane. The intent is to use PEE-PEO micelles as nanoreactors to grow perovskite nanocrystals, a material used to improve optical-electrical properties in solar cells and energy storage devices. As the micelle shape, size and distribution can greatly affect the stability and effectiveness of the perovskite nanocrystals, the micelle size and its distribution have been optimized by varying polymer concentration, temperature, solvent quality, and the method of preparation. Multi-angle dynamic light scattering measurements were carried out to determine the average hydrodynamic radius of the micelles and the micelle size distribution. Micelles were also characterized using small-angle X-ray scattering to evaluate the shape. Overall, the best micelles for the application were found to be prepared through direct dissolution of the polymer in decane at room temperature, surprisingly without any heating or annealing steps.

Sustainable Polylactic Acid / Cellulose Nanofiber Composites: Mechanical and Thermal Properties

Ian Kineon

Plastics have been an integral part of society for decades.Today's sustainability goal call for these plastics to be more biobased and biodegradable. Polylactic acid (PLA) is a major contender in this field since it replicates many of the properties of its non-biodegradable counterparts. However, commercial PLA loses its stiffness and strength at higher temperatures. Even for simple applications like packaging, neat PLA cannot withstand 65oC without deforming. Thus we developed a composite of PLA with cellulose nanofibers (CNF) to improve its physical performance . The methods of combining PLA and CNF under investigation are solid-state shear pulverization (SSSP), solid-state/melt extrusion (SSME), twin screw extrusion (TSE), and melt mixing (MM). SSSP and SSME yield higher, more consistent strength at higher temperatures when compared to TSE and MM. We also tested the PLA and CNF composites with a range of neat PLA to 10% CNF to determine the optimal amount of CNF. We measured the heat deflection temperature (HDT), performed a creep test, and used X-ray diffraction (XRD) to determine the optimal amount of CNF as well as the optimal process to manufacture the composites.

Structure-Property Elucidation in Diketoenamine Vitrimers from Vinyl Polymers

Jason Hillman

Covalent adaptable networks (CANs) offer the (re)processability of thermoplastics while maintaining the thermomechanical robustness of thermosets. Specifically, associatively crosslinked networks, known as vitrimers, provide predictable flow behavior and a constant crosslink density at all temperatures due to the degenerative exchange mechanism of covalent crosslinking. In this study, we present the novel fabrication of diketoenamine vitrimers from vinyl pre-polymers through Reversible-Deactivation Radical Polymerization (RDRP) of innovative triketone-containing methacrylate monomers in combination with various commercial *n*-alkyl methacrylates. The triketone methacrylates exhibit rapid and well-controlled homopolymerization and copolymerization with statistical incorporation, resulting in narrow dispersities (Đ). We showcase the adjustability of the glass transition temperature (T_g) values, ranging between 17°C and 116°C, by facilely tuning the pendant length of the triketone monomer as well as the identity or feed ratio of the accompanying comonomer. This tunability translates to predictable T_g values and expected viscoelastic behavior within the resulting vitrimer networks. Moreover, the networks generated via catalyst-free diamine condensation exhibit remarkable creep resistance at elevated temperatures, along with both chemical and mechanical recyclability. Finally, we report the seminal synthesis of a vitrimer network from an ultra-high molecular weight (UHMW; $>10^6$ g·mol^{-1}) pre-polymer, which leads to a marked reduction of creep in comparison to its analogous low molecular weight networked pre-polymer. Our results establish the expansion of the attractive triketone motif into vitrimer networks from chain-growth polymers with facile tunability and topological control.

Dynamics of Chemical and Physical Crosslinking of Methacrylated Silk Fibroin Hydrogels for Applications in 3D Printing

Hannah Bagnis, Elizabeth Aikman, Marisa O. Pacheco, Whitney Stoppel

Silk fibroin (SF) derived from Bombyx mori silk is a useful material in the biomedical field. Its biocompatible and tunable characteristics allow for applications in drug delivery, tissue engineering, and regenerative medicine (1). SF on its own is inert, therefore the addition of photocrosslinkable active sites via methacrylation (SilkMA) allows for interactions with Lithium phenyl-2,4,6-trimethylbenzoylphosphinate (LAP), a photoinitiator. Using ultraviolet (UV) light (405 nm), SilkMA can be crosslinked to form a hydrogel (2). SF hydrogels created via chemical crosslinking can be tuned to increase their load bearing capacity to mimic the mechanical properties of tendons or ligaments (3). However, a significant challenge lies in maintaining these properties for extended periods, as silk hydrogels tend to lose elasticity due to the formation of beta sheet crystals (4). This physical crosslinking is irreversible as the hydrogel has reached its thermal equilibrium. Therefore, to mitigate the formation of these physical crosslinks, which are formed by hydrogen bonding leading to beta sheet formation and a more crystalline polymer network. We investigate variables that contribute to the chemical crosslinking of SF hydrogels to hopefully interfere with beta sheet formation. Namely, we hypothesize that decreasing SF molecular weight and increasing chemical crosslinking density will slow crystalline growth. Beta sheet formation is characterized by FTIR spectra and optical transparency, while mechanical properties are assessed with shear rheology, thermal gravimetric analysis, and differential scanning calorimetry. Thus, these results explore the factors of chemical crosslinking that play a role in reducing or increasing physical crosslinking in SF hydrogels.

References

1. Rockwood DN, Preda RC, Yücel T, Wang X, Lovett ML, Kaplan DL. Materials fabrication from Bombyx mori silk fibroin. Nature Protocols. 2011;6(10):1612-31. doi: 10.1038/nprot.2011.379.
2. Kim SH, Yeon YK, Lee JM, Chao JR, Lee YJ, Seo YB, Sultan MT, Lee OJ, Lee JS, Yoon SI, Hong IS, Khang G, Lee SJ, Yoo JJ, Park CH. Precisely printable and biocompatible silk fibroin bioink for digital light processing 3D printing. Nat Commun. 2018;9(1):1620. Epub 2018/04/26. doi: 10.1038/s41467-018-03759-y. PubMed PMID: 29693652; PMCID: PMC5915392.
3. Zheng H, Zuo B. Functional silk fibroin hydrogels: preparation, properties and applications. Journal of Materials Chemistry B. 2021;9(5):1238-58. doi: 10.1039/D0TB02099K.
4. Su D, Yao M, Liu J, Zhong Y, Chen X, Shao Z. Enhancing Mechanical Properties of Silk Fibroin Hydrogel through Restricting the Growth of β-Sheet Domains. ACS Applied Materials & Interfaces. 2017;9(20):17489-98. doi: 10.1021/acsami.7b04623.

Chemical Recycling of Polyesters and Urethanes Via "Imidazolysis"

Lauren Sullivan, Mousumi Rani Bepari, Kathryn OHarra, Jason Bara

Polyesters (especially PET) and polyurethanes (PUs) are of particular interest for chemical recycling (i.e., depolymerization) as they are formed by condensation reactions (as opposed to free-radical polymerization). While post-consumer PET has been the subject of many studies in this area, PU materials are more challenging because they are often found as highly formulated thermosets (e.g., foams, rubbers). Common methods for chemical recycling of PET and PU have included hydrolysis, alcoholysis, ammonolysis, aminolysis, etc. Recently, we have developed methodologies for "imidazolysis" of PET and PU materials using imidazole (and related compounds). Imidazolysis offers unique opportunities with respect to processing PET and PU waste, recovery of products, and versatility of the imidazole-carbonyl products for further chemical transformations. This presentation will discuss our efforts in this area.

Understanding the Effects of Secondary Structure on Mechanical Properties of Protein-Based Copolymers through Protein Screening

Claire Su, Wui Yarn Daphne Chan

One of the most significant environmental issues today is plastic pollution, as plastic takes up to 450 years to degrade with more than 6 billion tonnes of it discarded in the environment since the 1950s. With up to 99% of plastics derived from non-renewable fossil fuels, it is time for an alternative. Protein-based copolymers are a strong candidate as plastic feedstock replacements since proteins are renewable and have functional groups that can facilitate cross-linking. Through understanding how secondary structures of protein-based copolymers affect mechanical properties, this will allow for the design of a plastic's replacement with biodegradable material that can compare to plastic's mechanical qualities. Three proteins with vastly different secondary structures (whey, zein, and gelatin) were copolymerized with comonomers forming rubbery segments and the influence of secondary structures was analyzed with tensile testing and FTIR spectroscopy. It can be concluded that copolymerization can be utilized in various proteins due to the presence of primary amines as copolymers with zein and gelatin were successfully prepared. A combination of proteins result in weaker mechanical properties compared to their pure counterparts. Gelatin was observed to have an increase β-sheet to α-helix ratio in heated samples, but there was no indication of an improvement in mechanical properties. Although β-sheet to α-helix ratio cannot be used to predict mechanical properties across proteins, this ratio correlates with specific properties in gelatin, where it has the potential to be utilized within the same protein type. Future work includes further microstructural characterization through electron microscopy or small angle x-ray scattering.

Characterization of Carbon Dioxide-Reacted Polypeptide Complex Coacervates

Manushi Samarathunga, Emily Charleson, Raymond Tu, Elizabeth Biddinger

In this work, we examine the formation of complex coacervates using oppositely charged polyelectrolytes, poly(L-lysine) and poly(L-glutamic acid), and, subsequently, explore the ability of these complex coacervates for enhanced carbon dioxide sequestration. We create a 'core-and-shell' coacervate droplet using a 2:1 ratio of poly(L-lysine) to poly(L-glutamic acid). We find that this ratio results in atypically small coacervates creating a higher surface area to volume ratio. This greater ratio allows for more possible reaction sites. The proposed mechanism is that the poly(l-lysine) on the surface of the complex coacervates will react with the soluble carbon dioxide to form carbamate. Samples were sparged with carbon dioxide for varying amounts of time, and the change in their size, charge, and chemistry was explored through Dynamic Light Scattering (DLS), Zeta potential, and Attenuated Total Reflectance Fourier Transform Infrared Spectroscopy (ATR-FTIR). DLS measurements for samples that were sparged for 1 hour show a slight increase in size, followed by a slight decrease in size. Zeta potential measurements show two significant charge peaks for samples that were sparged for 1 hour, suggesting the reaction resulted in a secondary charged species. ATR-FTIR is being explored to quantify the evolution of the carbamates in the complex coacervate droplets.

Extracting Rare Earth Elements Using Food-Grade Acids

Teresa Galvan

Rare earth elements (REE) are 17 metallic elements found in the middle of the Periodic Table, that are known for their conductive and magnetic properties. Studies have shown that REE can be extracted from coal fly ash, which is the fine particle ash created from the combustion of coal. This extraction is usually done using concentrated inorganic acids. Though this process works, using strong acids creates environmental sustainability concerns. This study represents the early stages of investigation of the efficiency of extracting REE from coal fly ash using food-grade organic acids and supercritical CO2.

Charge Selective Separation of Dyes and Flavonoids

Saloni Patel, Barbara Knutson, Stephen Rankin

The conductivity of mesoporous titania provides a unique opportunity to examine charge-based binding and release of solutes in high surface area thin films and membranes. The main objective of this project is to develop separation strategies for biomolecules using titania-coated membranes and supports with applied voltage. Mesoporous titania-coated thin films (approximately 100 nm thick with 7 nm diameter pores) were synthesized on the surface of a fluorine doped tin oxide (FTO) coated glass slides. Using an electrochemical workstation, a working electrode was attached to the surface of the titania-coated thin film to apply a positive/negative charge. The effect of applied voltage (magnitude and sign of voltage) on the adsorption and desorption of methylene blue (a model dye with positive charge), acid orange II (a model dye with negative charge, and quercetin (a flavonoid that strongly binds to titania) was examined. The positively-charged methylene blue exhibited adsorption when a negative voltage was applied and desorption when a positive voltage was applied. Quercetin bound to mesoporous titania, most likely due to chelation interactions, but did not demonstrate charge-based binding or release. Additionally, being able to modulate biofouling with applied voltage is also important. Biofouling was examined using BSA (bovine serum albumin) as a model protein. The outcomes of this work provide insight into the design of membranes that exploit new charge-based mechanisms for bio-separation.

Analyzing Phase Changes of Rare Earth Element-Chelate Complexes for Supercritical Fluid Extraction

Anacristina Muniz, Sergio Martinez-Monteagudo, Catherine Brewer

Supercritical carbon dioxide extraction has been investigated as a greener extraction method for many applications. Use of carbon dioxide reduces or eliminates the need for organic solvents and the associated environmental concerns. The properties of supercritical fluid, such as low viscosity and surface tension, have been shown to be conducive for extractions. This research focuses on the extraction of rare earth elements (REE) out of coal fly ash using food-grade organic acids as chelating agents for the target metals and supercritical carbon dioxide as the solvent. The chelators include acetic, citric, and lactic acids. For process evaluation and design, more information is needed about the solubility of the REE-chelator complexes in carbon dioxide at different supercritical temperatures and pressures. Some complexes are more desirable than others, so it is useful to be able to separate by solubility if the pressure and temperature can be tuned.

The purpose of using phase analyzer is to collect a wide range of data about phase diagrams without needing large numbers of experiments. The instrument is called the SFT Phase Monitor II and is designed to be primarily used with carbon dioxide as the solvent. The sample is placed in a capillary tube, lowered into the phase analyzer, and the sample chamber pressurized with carbon dioxide to supercritical conditions. The sample chamber has a viewing port and camera to observe dissolution and crystallization, and software for continuous data collection. The first trials will be with naphthalene to check the instrument function against known phase diagrams and to create an internal lab standard for future checks. Then, tests on the three solid acids will follow to determine the range of temperatures and pressures for their solubility (outside of complexes). Lastly, REE- chelator complexes will be synthesized to test each specific solubility by element and chelator.

Rare Earth and Critical Minerals Extraction Using Biochar

Erika Weisman, Oishi Sanyal, Deniz Talan

This research focuses on using biochar for the extraction of rare earth minerals (REEs) and critical minerals from acid mine drainage. These elements have no current domestic supply chain and yet an opportunity exists to selectively extract them from legacy mine sites, i.e., acid mine drainage. Biochar, like activated carbon, provides high surface area for adsorption and unlike fossil fuel derived activated carbon, biochar could provide a sustainable route to REE and critical minerals recovery. My research had been focused on reproducing the previously obtained results and deriving the kinetic constants of biochar adsorption for La^{3+} uptake. I am currently looking at the effects that REE adsorption and microwave regeneration has on the structure of biochar. In the future, I plan to look at the kinetic data and create a fixed bed system for adsorption and regeneration.

Microfiltration and Ultrafiltration for Viral Vector Purification

Catherine Cox, Mara Leach, Dibakar Bhattacharyya, Malgorzata (Gosia) Chwatko

Membrane filtration is an emerging field in the realm of bioprocessing. One application is the purification of viral vectors-a crucial production step in vaccines and gene therapy. Past methods of purification such as ultracentrifugation and chromatography are time intensive, expensive, and inefficient.[1] Membrane purification, on the other hand, offers a flexible, scalable, and efficient way to separate viral vectors from contaminants.[2] This research focused on the purification optimization of adeno associated-viruses (AAV's) from cell debris and proteins through microfiltration and ultrafiltration membranes. 20 nm silica particles were used to model AAVs, bovine serum albumin (BSA) was used to model HEK293 protein, and HEK293 lysed cells were used to model cell debris in viral vector supernatant. It was found that high hydrophilicity, large volumetric porosity, and large pore size of microfiltration membranes resulted in higher flux and recovery of 20 nm silica particles. Ultrafiltration was also successful in rejecting BSA and studying the effects of surface fouling.

References

[1]Florea, M., et al., "High-efficiency purification of divergent AAV serotypes using AAVX affinity chromatography," Molecular therapy Moethods & Clinical Development, 2023, vol.28, pp. 146-157.

[2]Zydney, A., "New developments in membranes for bioprocessing – A review," Journal of Membrane Science, 2021, vol. 620, pp.1-16.

Acknowledgements

This work would not have been possible without the NSF EPSCoR REU Supplement grant, and the Engineered Bioactive Interfaces and Devices REU program at University of Kentucky.

Development of a Microfiltration Process for the Purification of Precipitated Monoclonal Antibodies and Crystallized Proteins

Lawrence Azzariti, Mirko Minervini, Ali Behboudi, Andrew Zydney

Monoclonal antibodies are one of the most important drug classes because of their ability to target specific pathogens or cell receptors. However, they are also one of the most expensive classes of drug on the market, in part because of the high cost of the downstream purification process. In this work, we examined the use of continuous tangential flow microfiltration for the purification of precipitated / crystallized proteins. Experiments were performed with suspensions of two model proteins: immunoglobulin G (IgG), which was selectively precipitated using $ZnCl_2$ and polyethylene glycol, and human serum albumin (HSA), which was crystallized using $CeCl_3$. Microfiltration was performed through polyethersulfone hollow-fiber membrane modules, with the critical flux evaluated using a flux-stepping procedure to determine the maximum flux before the onset of fouling. The size and morphology of the precipitated IgG particles, evaluated by optical microscopy, were found to be a function of the precipitation conditions, such as the addition of specific salts, buffers, and pH. These changes in morphology had a significant impact on the fouling behavior of the membrane and could thus be leveraged to optimize the precipitate dewatering and washing performance. This work also provided the first demonstration that it was possible to dewater crystallized protein by tangential flow microfiltration, with the yield of protein crystallization and the membrane performance being a function of the concentrations of $CeCl_3$ and polyethylene glycol used for the crystallization. These results demonstrate the potential of developing a low-cost platform for purification of high-value therapeutics proteins using precipitation / crystallization in combination with tangential flow microfiltration.

Analysis of Manufacturing Methods of Ion Exchange Membranes for Desalination

Anna Harris, Inara Oliviera, Deniz Ipekci, Geoffrey Geise, Jeffrey McCutcheon

An emerging approach to membrane based desalination is to remove ions from water via electrodialysis. Membranes for this application need to be highly selective, and thin membranes can offer advantages by reducing mass transfer resistance. A recently reported additive manufacturing approach can enable access to ultrathin and highly selective polyamide-based membranes and could be adapted to prepare conductive ultrathin ion exchange membranes (IEMs) for electrodialysis. These materials could allow for more efficient ion transport and increased membrane selectivity. It is theorized that these polyamide-based IEMs can be manufactured and optimized using an electrospray technology to deposit monomers directly onto a supporting membrane. Once such experimental membranes are created, ion transport properties must be characterized.

Molecular additive manufacturing may enable the production of highly tunable, selective, low resistance ion exchange materials, and linear polyamide IEM analogs can be used as control materials to inform transport property characterization and analysis. Linear polyamide membranes composed of *m*-phenylenediamine (MPD), isophthaloyl chloride (IPC), and terephthaloyl chloride (TPC) can be produced via stirred interfacial polymerization. The addition of a sulfonated diamine comonomer is necessary because strong ion exchange groups are required to facilitate charge transport in electrodialysis applications. Sulfonated copolymers have been studied for their potential use in electric field-driven membrane processes, but the polymerization reaction used to prepare sulfonated polyamide copolymers is complicated by competition between the charged (sulfonated) and uncharged diamine comonomers. Yield calculations and FTIR analysis suggest that incorporation of sulfonate comonomers is indeed challenged by these factors. Ionic conductivity (determined using electrochemical impedance spectroscopy, EIS, and a direct current method) was measured to characterize the materials considered in this study and to provide insight into the properties, stability, and structure of the electrospray membranes. By analyzing the ionic conductivity properties of membranes produced by both linear polymerization and additive manufacturing, this presentation considers the viability of electrospray technology as a means to make ultrathin polyamide-based IEMs.

Understanding Biomolecules Diffusion for Continuous Vaccine Manufacturing

Barbara Stewart, Seth Kriz, Idris Tohidian, Caryn Heldt

Increasing the availability of vaccines by reducing manufacturing costs is a key strategy to reduce global disease burden. Batch processing is a common method in vaccine manufacturing, but it could be improved by moving to continuous manufacturing, which has been shown to reduce capital and operating costs. One method to continuously purify viral vaccines is by using an aqueous two-phase system (ATPS), which is readily operated in continuous mode and is shown to recover up to 80% of infectious viruses. This system must be scaled up to an industrial setting, and to do so, the behavior of viral mass transport, which includes its diffusion, is key to understanding the system. For this study, we quantified porcine parvovirus and bovine serum albumin dynamics by measuring their diffusion in various compositions of polyethylene glycol – sodium citrate ATPS to measure their diffusion coefficients. The biomolecules were tagged with a fluorescent dye, and the rate of diffusion in the bulk phases of ATPS was observed in a microchannel. Image analysis using a Python code determined the diffusion coefficient of each molecule. These values will be critical to determining optimal mixing to achieve maximum recoveries in industrial vaccine manufacturing.

Can Multiple Solubilities Co-Exist in a Single Crystal? Hazards and Opportunities

Claire Schleper, Joseph Kratz, Mitchell Paolello, Fredrik Nordstrom, Gerard Capellades

Crystallization is a separation process commonly found in the production of bulk chemicals, food additives, and pharmaceuticals. In the pharmaceutical industry, crystal solubility dictates a drug's bioavailability, and thus its ability to have an effect on the patient. In a recent publication, we found that 79% of challenging scenarios on impurity rejection in pharmaceutical crystallizations involved impurities incorporated inside the drug's crystal lattice [1]. Challenging impurities pose an issue due to effects on drug stability and toxicity to the body, which has been the subject of recent drug recalls [2].

This study investigates how impurities are distributed through crystal lattices, and their effect on drug solubility and dissolution using systems of dyes incorporated to a common pharmaceutical (acetaminophen, the active ingredient in Tylenol). Our results illustrate that anisotropic impurity incorporation into specific crystal facets can lead to anisotropic dissolution behaviors, where regions containing the impurity quickly dissolve upon exposure to a solvent. This may be due to localized solubility gradients within a single crystal, owing to an uneven distribution of impurity and thus an uneven disturbance of the lattice stability by the incorporated impurities.

These results have various implications, some concerning and others which present opportunities for drug purification and materials design. A potentially concerning implication in the case of pharmaceuticals is the faster crystal dissolution rate of the crystal regions containing the impurity, leading to a fast release and absorption of a potentially toxic impurity by the patient. On the other hand, the ability to tune dissolution kinetics in a crystal by controlling its purity can provide new avenues using safe additives as lattice modifiers, furthering the field of crystal engineering. The possibility of controlled-release medications utilizing a rapid onset could be made possible by enhanced dissolution from a purposeful additive, followed by the extended dissolution of the remaining API for longer lasting residual effect.

[1] Nordstrom, F. L.; Sirota, E.; Hartmanshenn, C.; Kwok, T. T.; Paolello, M.; Li, H.; Abeyta, V.; Bramante, T.; Madrigal, E.; Behre, T.; Capellades, G. Prevalence of Impurity Retention Mechanisms in Pharmaceutical Crystallizations. Organic Process Research & Development 2023, 27 (4), 723-741. DOI: 10.1021/acs.oprd.3c00009.

[2] US Food and Drug Administration. Guidance for Industry: M7(R1) Assessment and Control of DNA Reactive (Mutagenic) Impurities in Pharmaceuticals to Limit Potential Carcinogenic Risk, 2018. URL: https://www.fda.gov/media/85885/download.

Gas Transport Phenomena in PDMS/Matrimid Thin Film Composite Membrane

Xiang Li, Hammed Balogun, Ryan P. Lively

Membrane-based separation plays a vital role in various industrial sectors, facilitating the effective partitioning of mixture components at low energy and capital costs. The efficacy of this separation is largely determined by the membrane materials and their interactions with the mixture components. Polydimethylsiloxane (PDMS) membrane, a silicon-based organic polymer, is notable for its gas separation capabilities due to its high permeability and appreciable gas selectivity. Beyond its gas separation attributes, PDMS offers additional advantages including thermal stability, mechanical flexibility, and potential for hybridization with other materials, making it a versatile choice for numerous separation applications. In practical scenarios, PDMS membranes are often designed as thin film composites (TFCs). This configuration comprises a thin PDMS selective layer cast on a porous support, which ensures membrane integrity under high feed pressures while maintaining a desirable flux. Most transport models in reverse osmosis, for membranes designed as TFCs, assume negligible resistance within the support. This study helps validate such assumptions by evaluating the pore sizes of Matrimid supports and their contributions to gas flux. PDMS was spin-coated on Matrimid supports, followed by gas permeation experiments to measure the gas permeances of N2, He, and CO2 through the membranes. The acquired data were utilized in a permeation resistance model, developed to evaluate the surface porosity and pore sizes of both the PDMS layer and the Matrimid support. Understanding such transport resistance within the support aids in designing support materials with optimal pore sizes for defect-free, high-flux membranes. This insight is pivotal for optimizing PDMS membranes for enhanced separation efficiency and providing a foundation for further explorations into innovative membrane designs, particularly in CO2 capture applications.

PDD-VA Copolymers for the Separation of Azeotropic Refrigerant Mixtures

Sarah Dixon, Abby Harders, Brock Hines, Mark B. Shiflett

From central air conditioning in homes to safely storing perishable vaccines, refrigeration technology is vital to the modern human lifestyle. However, the use of refrigerants has a significant environmental impact on our world today. After chlorofluorocarbons (CFCs) were phased out due to their impact on the ozone depletion, hydrofluorocarbons (HFCs) were implemented across the world to replace them. However, it was found that some HFCs have high global warming potential (GWP). Recent legislation has been enacted to regulate the phase-down of these compounds. The 2020 American Innovation and Manufacturing (AIM) act has committed the United States to reducing 85% of HFC usage by 2035. Many of these HFCs are used in azeotropic mixtures, so finding innovative technology able to separate the mixtures is crucial to be able to reclaim and reuse the individual components. This study focuses on the separation of refrigerants using a copolymer of 75 mol% perfluoro(2,2-dimethyl-1,3-dioxole) (PDD) and 25 mol% vinyl acetate (VA). Common refrigerants such as HFCs, hydrochlorofluorocarbons (HCFCs), and hydrofluoroolefins (HFOs) were studied. The permeability, solubility, and diffusivity of HFC-32 (CH2F2), HFC-125 (CHF2CF3), HFC-134a (CF3CH2F), HFO-1234yf (CH2=CFCF3), and HCFC-22 (CHClF2) were measured in polymer films at varying temperatures of 308.15, 323.15, 348.15, and 373.15 K. Arrhenius relationships were used to obtain the activation energy of permeation for each refrigerant. Arrhenius solubility and diffusivity relationships were then applied to HFC-32 and HFC-125 to obtain the activation energy of diffusion and heat of sorption. The overall results indicate that selectivity of the gases decreases as the temperature of the membrane increases. This is explained due to the higher activation energy of diffusion for larger molecules. A high selectivity of 35 can be obtained for HFC-32/HFC-125 while maintaining a high permeability at 308.15 K. Other high selectivities were achieved with refrigerant mixtures of HFC-32/HFC-125, HFC-32/HFO-1234yf, HFC-32/HFC-134a, and HCFC-22/HFC-125. These promising results show the potential of this copolymer having future commercial use in the separation of refrigerant mixtures.

A Review of the Use of Ionic Liquids for Separation of Hydrofluorocarbon Mixtures

Dorothy M. Haggard, Kalin R. Baca, Mark B. Shiflett

Hydrofluorocarbon (HFC) refrigerants can significantly contribute to global warming and climate change, with some refrigerants being thousands of times more potent than carbon dioxide. HFCs replaced chlorofluorocarbons (CFCs) as refrigerants, as CFCs were lined to the depletion of the Earth's ozone layer. HFCs are now being phased out in favor of hydrofluoroolefins (HFOs), which have greatly reduced global warming potentials. Many HFCs are sold commercially as azeotropic mixtures that provide the necessary application properties as well as ensure leaked refrigerant maintains composition. However, azeotropes are difficult to separate using distillation and recycle the components. Ionic liquids (ILs) are being studied as potential entrainers for use in an extractive distillation process to separate azeotropic HFC mixtures. Separated refrigerants can then be recycled into HFO mixtures or repurposed into fluorinated feedstocks for other processes. ILs are commonly referred to as salts that have a melting point below 100°C, composed of an organic cation and an organic/inorganic anion. ILs are promising materials for separation due to their high tunability and low volatility. The NSF has sponsored research that has led to the design of specific ILs for the separation of HFC mixtures. Developing this new separation technology requires extensive data on the interactions between various HFC and IL combinations. This poster will review the extensive research done on IL/HFC systems for separation. Key areas of work include physical property measurements, experimental techniques, interaction simulations, and modeling of various system combinations. In these areas, studied IL/HFC systems will be identified. The direction of future work will also be discussed, and an overview will be given on the ongoing design of an extractive distillation process.

Evaluation of Ion-Exchange Resins and Electrospun Membranes for Ac-225 Purification

Luke Andrew Venturina, Maura Sepesy, Tuli Banik, Megan Sibley, Christine Duval

Targeted alpha therapy (TAT) is a cancer treatment strategy taking advantage of radioactive decay products to attack tumors. Though promising, progression to clinical scales is hampered by scarcity of suitable radionuclides, especially actinium-225. Production of Ac-225 involves thorium irradiation with high-energy protons that also produce undesired byproducts like lanthanum, neodymium, promethium, and barium. Current standard purification methods include a series of acid-based extraction and cation-exchange chromatography using columns packed with ligand-functionalized resins, but diffusion transport limits performance to low flow rates and high pressures. Membrane chromatography is an alternative method, using the same ligand-metal adsorption chemistry as resin-based ion-exchange. A key advantage to membranes is relatively fast convection - meaning mass transport does not limit process performance. Membrane chromatography methods are still largely in development, but show potential to increase the scale of radionuclide production. Candidate materials for membranes must be evaluated for their physical (permeability, mechanical strength) and chemical (adsorption, solvent and reaction compatibility) properties for implementation at any scale.

In this work, blends of poly(vinylbenzyl chloride) and polystyrene in electrospun membranes are evaluated for stability in chemically relevant solvents, porosity, and pure-water flux. With these initial characterizations, commercial resins are evaluated for their adsorption performance to benchmark sorption in novel DGA membrane adsorbers. Performance at 4M and 6M nitric acid acid solutions provide valuable insight to design conditions for future process elution and retention steps. Equilibrium of adsorbed- and solution-phase metals is analyzed by ICP-OES and fit to a Langmuir isotherm model to determine maximum surface adsorption and association constant parameters. Novel membrane adsorbers show lower lanthanum capacity than standard resins, but slightly higher association. Work remains to explore ligand functionalization chemistry of membranes and to evaluate their performance regarding multi-metal and dynamic adsorption, post-functionalization permeability. With all the relevant chemistries characterized, a foundation may be laid to develop further materials and processes to bring Ac-225 to clinically practical scales.

Chemical Modification of Polymer-Based Membranes for Carbon Capture

Zoe Reddecliff, Joshua Moon, Ryan Johnson, Nathaniel Rudman, Karim El Hajj Sleiman

The threat of climate change is strengthening as a result of the continued release of greenhouse gases into the atmosphere. Our group is cultivating polymer-based membrane technology for gas separation which aims to reduce carbon dioxide levels in the atmosphere when applied to industrial gas byproducts or direct air capture. Gas separation by distillation is widely used in industrial systems but it is energy intensive and costly. Polymer membranes are relatively cheap and have the potential to reduce energy usage of separations.

Our group is exploring how altering the chemical structure of polymer membranes affects permeability and selectivity of carbon dioxide by functionalizing a PEG based membrane with an active ester side chain using "click chemistry." Click chemistry allows various amine groups to be easily added to the base membrane. Our hypothesis is that the addition of amine ligands to the polymer's crosslinked scaffolding would create favorable interactions with carbon dioxide, thus increasing selectivity. Gas permeabilities and solubilities will be tested on samples substituted with amine ligands with different pKa values such as pyridine, imidazole, and piperidine. Gas diffusion coefficients will then be determined using the solution diffusion model. We hypothesize that amines with lower pKa values may have weaker interactions with CO_2 that could facilitate faster diffusion jumps and lead to higher permeability.

Effect of Solvent on CO_2 Sorption Capacity in Supported Ionic Liquid Membranes

Taylor Adams, Jarod Harris, Sarang Ismail, Mona Bavarian

Anthropogenic emissions have turned into one of the greatest issues of our time, warming up our planet. With rapidly increasing demand for energy, the dependence on fossil fuels has exacerbated greenhouse gas emission levels and contributed to global warming. To that end, carbon capture and storage technologies are currently utilized to reduce carbon emissions. The monoethanolamine (MEA) system, for example, works to absorb CO_2 from flue gases; however, this system has a high operational cost and energy demand. Consequently, membrane technologies have gained significant attention as a potential solution for CO_2 separation. Polymeric membranes containing ionic liquids (ILs) have emerged as promising candidates for efficient gas separation. This is primarily attributed to the unique properties of ILs, including their favorable CO_2 absorption characteristics and tunable chemistries. However, the high viscosity of ILs due to CO_2 absorption presents challenges in their practical use, especially in absorption columns. To overcome this limitation, the concept of supported ionic liquid membranes (SILMs) has gained attention.

In this study, we employed 1-ethyl-3-methylimidazolium bis (trifluoromethyl sulfonyl) imide ([Emim][TF2N]) as the ionic liquid and poly (vinylidene fluoride) (PVDF) as the host polymer to fabricate the SILMs. We conducted CO_2 sorption measurements using a quartz crystal microbalance (QCM) gas sorption system. To cast and fabricate SILMs, different solvents were investigated. Dimethylacetamide (DMAc) was shown to be the best solvent choice over dimethylformamide (DMF), dimethyl sulfoxide (DMSO), or triethyl phosphate (TEP). This was attributed to the variation in PVDF crystallinity and the formation of distinct phases when cast from different solvents. Our polarized microcopy data revealed a direct correlation between macroscale structural variations and gas sorption capacities in SILMs.

Effects of Pore-Forming Additives on Adsorbent Loading in Structured Contactors for CO_2 Capture

Shreya Ghosh, Hannah Holmes, Ryan P. Lively

The global mean temperature continues to increase above pre-industrial levels due to human activities, and it will reach critical levels if carbon dioxide emissions continue at current rates. CO_2 capture technologies play a significant role in limiting the increase in global mean temperature by reducing both emissions and existing atmospheric CO_2. Some widely used CO_2 capture technologies include wet scrubbing systems, membrane separators, and solid adsorbents. Wet scrubbing, a commonly utilized industrial method, uses liquid absorbents to remove contaminants from a gas stream. Solid absorbents could be a better option for efficient CO_2 capture. In comparison to liquid absorbents, solid adsorbents have lower regeneration energy, lower capital costs, and more modularity. Adsorbents are typically implemented in structured contactors, which are highly porous polymeric structures. Compared to packed beds, structured contactors have a higher rate of mass transfer, lower pressure drop, easier and more efficient cooling/heating during adsorption/desorption cycles. A higher adsorbent loading is desirable for increasing the contactor CO_2 productivity and reducing energy requirements.

For structured contactors fabricated using phase inversion techniques, high porosity can be achieved by adding pore-forming agents. These additives are water-soluble polymers that are removed during quenching and post-processing steps, leaving behind open pores. However, it is unclear if and how these pore-forming additives affect the sorbent loading in the final fibers. Here, we study the effect of pore-forming additives on the loading of solid adsorbents in structured contactors by fabricating and characterizing fiber sorbents with varying loadings of the pore-forming agents. The sorbents studied are 13X and UiO-66, and the pore-forming agent is polyvinylpyrrolidone (PVP). Solutions with polymer, sorbent, and PVP were prepared with varying amounts of PVP. The solution was then extruded into fibers using phase inversion techniques. Typical fiber post-processing procedures were followed, including extensive solvent exchange steps. The fibers were then characterized using thermogravimetric analysis (TGA) to determine the fiber composition and final sorbent loading. Scanning electron microscopy (SEM) was used to qualitatively evaluate and compare the fiber porosity. Ultimately, these results provide guidelines on the amount of pore-forming additives that should be added to optimize sorbent loading and porosity, which will enhance the performance of structured contactors in CO_2 capture applications.

The Effect of Polymer Properties on Carbon Dioxide Absorbance Capacity of Supported Ionic Liquid Membranes

Jarod Harris, Mona Bavarian, Sarang Ismail, Taylor Adams

With today's energy demands, our dependence on fossil fuels has grown significantly. This leads to devastating consequences for our environment and ecosystem, resulting in global warming. To address the generated carbon footprint, carbon capture techniques, such as absorption using amine systems, are implemented. This technology has limitations in terms of operational and environmental efficiency, toxicity, and operational costs. One promising candidate is to utilize a greener alternative, such as Ionic Liquids (ILs). As green solvents with low vapor pressure, high thermal stability, tunable structure, low flammability, and high affinity for CO2, ILs are a great solution to replace conventional carbon capture technologies. Several recent studies have demonstrated enhanced CO2 solubility for confined ionic liquids within polymeric domains, suggesting a potential for stabilizing ILs within macromolecular domains and taking advantage of their non-ideal properties. With the potential for ILs to be incorporated into polymeric domains, understanding the underlying phenomena that govern their non-ideal properties is a critical task. In this study, the effect of the molecular weight (MW) of polymer on the CO2 adsorption capacity of supported ionic liquid membranes (SILM) was investigated. Quasi-solid-state films were cast from mixtures containing 1-Ethyl-3-methylimidazolium bis(trifluoromethylsulfonyl) imide and Polyvinylidene Fluoride (PVDF) with different Molecular Weight (MW) of 180K and 534K, and the solubilities of CO2 in these films were measured using a microgravimetry method with a Quartz Crystal Microbalance. The adsorption isotherms were obtained at four different temperatures. Our result indicates that the MW of PVDF has a significant impact on the CO2 solubility in SILMs, highlighting the importance of the intermolecular interaction and swelling properties of polymer on CO2 sorption within the mixture. According to our findings, the 1:1 IL-polymer mixture prepared with 180K polymer presents the greatest promise for fabricating SILMs.

Incorporation of Gutter Layers in Polyamide Thin Film Composite Membranes By Electrospray 3D Printing

Martina Jagielski

Incorporation of Gutter Layers in Polyamide Thin Film Composite Membranes By Electrospray 3D Printing

Martina Jagielski[1,2], Mert Can Hacifazlioglu[1,2], Jeffrey McCutcheon[1,2*]

[1]Department of Chemical & Biomolecular Engineering, Center for Environmental Sciences and Engineering, University of Connecticut, Storrs, CT, United States, 06269

[2]Connecticut Center for Applied Separations Technology, University of Connecticut, Storrs, CT,

United States, 06269

*jeffrey.mccutcheon@uconn.edu

ABSTRACT

Polyamide membranes are very prevalent in the membrane desalination field and are applied to nanofiltration (NF) and reverse osmosis (RO) membranes. The successful commercialization and application of these membranes is due to Interfacial Polymerization (IP), which is a reaction between a diamine and an acid chloride. This process forms a thin and densely cross-linked layer on a porous substrate. These membranes have competitive selectivities and much higher water permeances compared to their predecessors (Loeb-Sourirajan membranes). However, the IP process cannot control thickness and chemistries directly, and relies heavily on the support membrane for the characteristics and limitations of the membrane. This support dependence is one of the biggest issues for water transport, since supports, especially those of lower porosity, disrupt the diffusive flow of water and restrict membranes' permeance. Historically, gutter layers have been utilized in gas separation to address this issue of support dependence, but they are unprecedented in desalination. A gutter layer is a thin layer between the support substrate and the selective layer which acts as a barrier between the two to lessen water's disruptive flow. It creates a shorter and less resistant transport path for water and stops pore blockage, which would solve the issue of support dependence. IP is not a compatible process for forming gutter layers but electrospray 3D printing offers more possibilities. This method utilizes an electric field to atomize solvent droplets that are sprayed on the substrate and react directly at the surface to form membrane structures. This method provides the possibility to make gutter layers with varying and controllable thickness using different monomers. In this work, we focus on the creation of tunable gutter layers to see whether they can increase the permeances of RO and NF membranes. Improvement in membrane performance via gutter layers would advance future membrane-based desalination technologies, allowing for routes such as customizable membranes and broadening current niche purposes in the market.

Moisture-Swing Polymer@Textile Composites for CO_2 Capture

Maren B. Thompson, Matthew D. Green, Marlene A. Velazco Medel

Quaternary ammonium-containing sorbents have significant potential in direct air capture (DAC) of CO_2. In DAC sorbents, the adsorption of CO_2 can be achieved at lower concentrations, which is advantageous in mitigating global warming caused by the excess of CO_2 already in the atmosphere. DAC could reduce emissions from nonstationary sources like transportation and agriculture where carbon capture technology would otherwise be difficult to implement.

This carbon capture in ambient air can occur through different mechanisms, for example, via a moisture-swing mechanism where the sorbent absorbs CO_2 at lower moisture levels and desorbs it at higher moisture levels.

The project investigated the modification of textiles into polymer@textile composite membranes with improved mechanical properties for long-term stability and DAC performance. This was done by immobilizing polymeric sorbents with quaternary ammonium groups into the textiles.

Membrane synthesis via free-radical polymerization was performed at various monomer concentrations with the textile submerged in the polymerization solution. Crosslinking agents were included to encourage modification of the textile. Modification was successful with a 15-20% degree of modification (DM%) quantified by mass difference and a clear increase in the stiffness of the material.

The stability was measured after washing the membranes and a conventional ion exchange process was performed to add $-HCO_3$ counterions, which are involved in the moisture-swing mechanism.

The CO_2 capture performance was tested in a closed system where the modified polymer@textiles composites showed moisture-swing activity that makes them a promising candidate for DAC.

Additionally, this work acts as a starting point to eventually use discarded textiles as a material in DAC membranes. Textile waste is a major source of pollution in the environment and new methods of textile reuse, especially for blended textiles which are otherwise difficult to separate and recycle, could help alleviate this problem. Green chemistry techniques such as using water as a solvent in membrane synthesis might also be investigated based on the findings of this work.

Direct Air Capture of CO_2 on Activated Carbon Via Temperature Swing Adsorption

Julian Lam, Amanda Simson, Benjamin Davis

Current climate models require the implementation of large-scale carbon dioxide (CO_2) removal to avoid the detrimental effects of global warming. Direct air capture (DAC) has been demonstrated as a promising method to sequester CO_2 from the atmosphere due to its scalability and low land usage compared to other methods of carbon removal. The relatively high cost of conventional CO_2 sorbents has inspired recent research investigating the performance of cheaper sorbents. Activated carbon is a highly porous carbon material typically produced from the pyrolysis of organic material in an inert environment at high temperatures; its high surface area and low cost make it an ideal sorbent for CO_2 capture. The production of activated carbon can also be a method of carbon capture by pyrolyzing biowaste or other sustainable organic materials using renewable energy sources, and thereby keeping their carbon in solid form out of the atmosphere.

In this work, the performance of an activated carbon DAC device was quantified to assess the potential for biochar as a cheap and abundant CO_2 sorbent. Ambient air was fed through a packed adsorption bed to produce a stream of air with lower concentration of CO_2 (<400 ppm). The bed was regenerated using a temperature swing adsorption (TSA) process, in which the sorbent is heated and the captured CO_2 is released, producing a stream with a high concentration of CO_2, which could be sequestered. Initial data show ambient loadings of 73.6 μg CO_2/g sorbent on unactivated commercial biochar; we hope to demonstrate capacities on activated carbon approaching those found in literature (2.28 mg CO_2/g sorbent) [Zhang et al., 'Direct air capture of CO2 by KOH-activated bamboo biochar,' *Journal of the Energy Institute*, 2022].

Synthesis of Composite Lithium Orthosilicate Based Materials for High Temperature CO_2 Capture

Connor Fawcett, Michael Smith, Charles Coe, Charles Barski

Lithium orthosilicate (LOS) is a highly effective, reversible sorbent for CO_2 capture at high temperatures (600-650°C), with a theoretical working capacity of 36.7%. Suggested applications are reforming operations to produce hydrogen, water-gas shift operations, blast-furnace gas upgrading, and IGCC energy production. The material undergoes the following reaction during CO_2 sorption:

$$Li_4SiO_4\ (s) + CO_2\ (g) \leftrightarrow Li_2CO_3\ (s) + Li_2SiO_3\ (s) \quad (1)$$

Bulk LOS phases form a shell of lithium carbonate and lithium metasilicate, which the CO_2 needs to diffuse through for further sorption with the unreacted core. This has generally been modelled with a double-exponential curve fit, but we have designed a model based on physical properties of the materials. In our lab, we prepared a nanostructured LOS composite material synthesized with a surfactant that shows rapid uptake kinetics and cyclic stability up to 80 cycles. We hypothesize the surfactant results in grain boundaries that facilitate diffusion of reacting species.

Our goals are to reduce production costs as well as synthesis duration. It would also be ideal to slightly reduce the optimal temperature for carbon capture in the material for use in reactions with a water gas shift, to allow for an increase in H_2 production. Studies indicate that the inclusion of alkali dopants can lower the temperature for carbon capture in the material, making it more suited for inclusion in water gas shift processes. Other studies indicate that mesoporous silica sources such as SBA-15 and KIT-6 can be converted into LOS. This is typically done by a solid-state conversion of silica with a lithium salt such as lithium nitrate. We hypothesize that through pyrolysis rather than calcination, the carbon-carbon bonds can be preserved in the silica sources which will lead to measurable porosity in the LOS product. This would therefore increase the diffusivity of CO_2 in the material, as well as provide a cheaper and faster production method.

Selective Lithium Extraction from Geothermal Brines

Alex Hawkins

The development of future lithium-ion battery technologies will create a demand for cleaner and more versatile lithium sources. Currently, lithium is primarily obtained through ore mining in a limited number of countries, a process that has been associated with significant ecological impacts in the regions where mining occurs. In the United States, there is a growing interest in obtaining lithium domestically from environmentally sustainable sources. One promising option is to extract lithium from geothermal brines found in California's Salton Sea. However, the complex chemical composition of these geothermal brine solutions will require the development of a highly selective separation process to make lithium sourcing from brines an economically viable process.

Researchers have been investigating battery materials as potential ion sieves to selectively capture lithium from geothermal brines. One such material is iron (III) phosphate (FP), which has demonstrated the ability to capture and release lithium through redox cycling mechanisms. When reduced, FP has a strong preference for capturing lithium over other cations found in geothermal brines. Subsequently, a separate oxidation process can be employed to release lithium in a more concentrated and cleaner form suitable for further downstream processes.

The goal of this research is to determine the optimal reaction conditions to maximize the selectivity of lithium capture during the reduction of solid-phase FP. Using these optimized conditions, a laboratory-scale plug flow reactor was used to demonstrate lithium capture and assess its viability for larger-scale production processes. The findings obtained from this study can provide insights for the development of other separation techniques utilizing redox-coupled adsorption processes.

Solvent-Resistant Thin Film Composite Membranes By Polymerizing Partially Fluorinated Monomers during Spin Coating

Tyler Oddo, Zane Parkerson, Matthew Vasuta, G. Kane Jennings

Membrane technologies can offer an order of magnitude greater energy efficiency than thermally driven separations such as distillation. The fabrication of robust, solvent-stable active layers on polymeric membranes is essential for the insertion of this technology into industry. To generate robust films, we introduce a method known as spin coating ring opening metathesis polymerization (scROMP) of 5-(perfluoro-n-alkyl)norbornenes (NBFn) with perfluoroalkyl side chain lengths (n) of 4, 6, 8, and 10 on a polyacrylonitrile (PAN) support. The scROMP method combines the polymerization and deposition of the membrane selective layer into a rapid and sustainable 2 min process. The polymer films grown by this innovative new method were then characterized and their effectiveness as pervaporation membranes for the dehydration of a 90 wt% aqueous ethanol solution was analyzed. pNBFn membranes exhibited greater solvent stability than their polynorbornene (pNB; n = 0) counterpart while retaining excellent thermal stability, evidenced by reduced swelling and membrane selectivities independent of operating temperature. Furthermore, pNBF8 has shown the greatest promise for ethanol dehydration, obtaining a selectivity of 177 and a flux of 180 g/m^2h while showing the ability to maintain a high level of performance after >40 h of continuous operation for ethanol/water separations. This study successfully demonstrates the ability of the scROMP technique to screen specialty polymers efficiently and effectively for membrane applications and sets the stage for additional research into the use of pNBF8 membranes for other chemical separations.

Mapping Nucleation and Growth Kinetics of Pharmaceutical Solutes Using Data-Rich Kinetic Screening

Joshua M. Zaharof, Ibrahim Joel, Kennedy Tomlinson, Ethan Shumaker, Kevin P. Girard, Alpana Thorat, Gerard Capellades

Separation via crystallization is one of the most important steps in producing lifesaving medications in the pharmaceutical industry. Crystallization is one of the most popular separation processes in pharmaceutical manufacturing, yet there is a lack of fundamental understanding of how to design these processes with limited amounts of time and material. Crystallization kinetics play a critical role in gaining the required understanding needed for crystallization process development. However, there is no standard method to screen for kinetics, which depend highly on variables like scale, impurities, or techniques for measuring crystal growth. Each new compound requires a specifically and uniquely designed process, which becomes time and material intensive.

In a recent publication, we described a set of automated methods for the rapid screening of nucleation and growth kinetics, requiring one week and material at the 1-10 g scale. These methods were found to be consistent across multiple operators, instruments, and sites. By implementing that framework over multiple different systems, we've now created a database including nucleation and growth constants for common pharmaceuticals, all collected in consistent conditions of scale, mixing, and temperature, and decoupling the effect of supersaturation using population balances.

The database allows for the prediction of crystal behaviors based on the similarity to other previously studied systems when subjected to consistent methods. Process development scientists will be able to screen an entirely unknown compound as fast as one week and begin process development earlier. The presented methods will serve as the standardized workflow for crystallization process development, which will aid scientists in the design of their processes based on leveraging the behavior of already well-characterized systems.

Hemp-Based Biochar As a Heavy Metal Adsorbent

David Opiela, Hema Ramsurn

Heavy metal impurities are toxic even at low concentrations (parts per billion) and can cause health issues as well as contamination if not removed from crude oil. Metals including arsenic, barium, cadmium, chromium, copper, lead, tin, vanadium, zinc, and mercury are present as stable organometallic compounds in crude oil. Although several methods exist to remove metals from crude oil, only a few have been used industrially due to various limitations. For instance, the solvent extraction process reduces the yield of the end product substantially, whereas distillation only concentrates the metals in the heavier fractions (or in petroleum coke). Therefore, there is still a need for processes that can lower crude oil's metal contents. We propose testing hemp-derived bio-adsorbents for this process as a cheaper and greener alternative. Post-extracted hemp is an abundant waste biomaterial found in Oklahoma and growers and extractors need to pay to dispose of this material. It is proposed to use hydrothermal carbonization (at 300 ^{o}C with subcritical water as solvent) as a green technology to thermally convert this hemp material into biochar, a precursor to bio-adsorbents, and then in the future to chemically modify the biochar to trap sulfur and heavy metal impurities from crude oil. As baseline studies, the focus was on chromium removal, using both activated and non-activated hemp biochar which were tested over a range of different metal concentrations (10-500ppm). It was surprising to see that the results were not as expected since our lab had worked on activated cellulose biochar as bioadsorbent with better results. It is suspected that our source material had a high ash content due to either the soil of Oklahoma or the fertilizers added during the hemp growth. It is proposed to analyze the hemp material and its biochar for ash content and then reflux the biochar prepared to remove the ash content before retesting the refluxed biochar for Cr removal.

Purification of Lactic Acid with Ultrafiltration Membranes

Alexandra Bradbury, Malgorzata (Gosia) Chwatko, Emily Ingram, Tyler Barzee

In recent years, there has been a drive to commercialize degradable polymers such as poly(lactic acid) to help tackle the current plastic crisis. For the technology to be successful, it is imperative to find cost-effective methods of extracting lactic acid from different waste streams such as agricultural waste, food waste, or reaction byproducts. One promising technology is pressure-driven membrane filtration where molecules are separated by their size. In this work, we aimed to study how different commercially available membranes can extract lactic acid from different streams. To accomplish this goal, we evaluated the performance of different commercial nanofiltration membranes at filtering the lactic acid solutions. First, we identified the best method to determine lactic acid concentration in different solutions. We evaluated two methods: UV-Vis Spectroscopy based on and high-performance liquid chromatography (HPLC). We determined that the HPLC method, while more accurate, took a significantly larger amount of time. Having identified the method, we proceeded to test membranes made up of different polymers and of different pore sizes. We evaluated the resistance of the membranes to lactic acid and tested their performance in both pure water and in lactic acid solutions. We found that poly(ether sulfone) membranes exhibited significant degradation under short contact (<30min) in lactic acid containing solution. Further studies will evaluate the purity of the samples and ways to simplify the separation process.

Employing Gutter Layers on Printed Gas Separation Membranes to Improve Membrane Permeance

Arjun DeJesus, Noah Ferguson, Srivatsa Bettahalli, Edward Wazer, Jeffrey McCutcheon

Upgrading biogas into biomethane using membranes is of increasing interest for small to medium sized anaerobic digesters. While a number of commercial upgrading membranes exist, we note that the thin film composite (TFC) membrane is not commonly employed in this application. Many membranes for gas separations are made using the non-solvent induced phase separation process and result in integrated asymmetric membranes. TFCs offer better control of chemistry and selective layer thickness in order to hit performance targets. To make a good TFC selective layer requires both a method of processing a thin film in combination with a smooth, homogeneous support to achieve the thinnest possible membrane for maximum permeance while preserving selectivity. In this work, we explore the use of an emergent electrospray printed membrane approach to making gas separation membranes while using a gutter layer of a highly permeable polymer to provides a smooth surface for the selective layer and improved distribution of penetrating gases into the support. We discuss methods for making these layers and how we improved compatibility with the selective layer (Pebax). We chose polydimethylsiloxane (PDMS) as a traditional gutter layer material, but we noted that its hydrophobicity prevented good Pebax film formation due to its hydrophobicity. To remedy this, we treat the PDMS gutter layer with plasma to introduce chain defects and increase hydrophilicity. This allows compatibility between the gutter layer and the polar solvent used to print the Pebax and results in a more homogeneous and selective layer.

Humidified Gas Permeation through Modular Polymer Membranes for Enhanced CO2 Separation.

Karim El Hajj Sleiman, Ryan Johnson, Zoe Reddecliff, Nathaniel Rudman, Joshua Moon

Over the last several decades, polymer-based gas separation membranes have emerged as a competitive, energy-efficient technology for a wide range of applications including carbon capture. The effect of humidity on gas separation membrane performance remains an underexplored topic that can significantly influence separation process efficiency and viability. Despite the many advancements in soft materials for membranes, many promising materials developed in the lab fail to exhibit high selectivities under industrially-relevant conditions, which often involve the detrimental effects of membrane swelling and plasticization resulting from water vapor. In facilitated transport membranes in particular, there is a need to better understand how humidity impacts gas transport mechanisms.

To achieve this, our research involves incorporating functional ligands into PEGDA membranes to tailor the crosslinked polymer functionality and impart high selectivity for carbon capture through facilitated transport. Introducing vapor changes how transport occurs within the membrane due to swelling and plasticization and reactions in the presence of CO_2 and active ligands (amines). Data collection involves the design of a custom temperature-controlled air bath system for humidified permeation measurements with a high-pressure gas feed. Using multicomponent gas and vapor permeation, we are exploring how varying the levels of humidity affects the permeability and selectivity of carbon dioxide for our polymer membranes. As water uptake increases, we expect to see an increase in fractional free volume and an increase in gas permeability due to enhanced segmental mobility. At the same time, we hypothesize that higher CO2 selectivities may result due to the formation of bicarbonate in the membrane.

Thermoresponsive Membranes from Lyotropic Liquid Crystal Templates

Ann-Sha Headley, Mostafa Tabatabaei, Younes Saadat, Reza Foudazi

We study the fabrication of thermoresponsive membranes from the polymerization of lyotropic liquid crystals (LLCs) composed of poloxamers, diacrylated Poloxamers, oil, and water in this experimental work. After formation of Lyotropic Liquid Crystals, the oil phase (containing monomers) and Poloxamer can be crosslinked to make mechanically robust membranes with pore sizes in the range of 1-10 nm, thereby conducting nanofiltration/ultrafiltration of samples. The overall purpose of this study is leveraging the stimulus-responsive behavior of the membranes since they show great potential in a variety of fields, such as pharmaceuticals, oil processing, protein purification and chemical detection. Poloxamers are non-toxic, amphiphilic (hydrophilic head and hydrophobic tail), non-ionic surfactants and their structure consists of poly(propylene oxide) which are hydrophobic and poly(ethylene oxide) blocks which are hydrophilic. Lyotropic liquid crystals are a result of the self-assembly of these amphiphilic molecules, water, and oil in a ternary system, where depending on the composition, different crystal structures, such as lamellar, cubic, and hexagonal phases, are produced. The study finds that, as shown by significant liquid crystalline transitions, a 50/50 mixture of diacrylated and pristine Poloxamers is the most effective ratio for creating thermoresponsive membranes with a normal hexagonal (H1) lyotropic liquid crystal structure. Additionally, the combination of diacrylated and pristine Poloxamers improves the structural stability of the mesophase of the lyotropic liquid crystal, resulting in a thicker gel whose viscosity decreases at lower temperatures and increases at higher temperatures. The change in viscosity based on temperature demonstrates the thermal responsive behavior of the membranes. When testing permeability of the membranes, a water flux of 34.25 L/m^2.h is calculated and molecular weight cut off test are completed to observe rejection capabilities of the membranes. The potential of the LLC templating method in creating customized thermoresponsive membranes, which are reproducible for protein separation applications is highlighted by this study.

Separation of HFC-32 and HFC-125 Using Imidazolium Based Ionic Liquid + Polyvinyl Acetate Ionic Liquid Polymer Membranes

Hannah Uhl, Abby Harders, Irene Xu, Kalin Baca, Tessie May, Gabby Zaher, Mark B. Shiflett

Hydrofluorocarbons (HFCs) are greenhouse gases commonly used in refrigeration systems. Current legislation has called for the phase-out and regulation of HFCs due to many of them having high global warming potentials. Since many refrigerants are azeotropic HFC mixtures, innovative and effective methods for separating these compounds are crucial. In the past, ionic liquids and polymers have been used separately for this application. The incorporation of ionic liquid into polymer membranes creates a great possibility for more efficient separations due to both entities having separation potential. Pure polymer membranes are able to separate gas mixtures by differences in the solubility and diffusivity of the gases. In this work, the permeability, selectivity, and solubility of difluoromethane (HFC-32) and pentafluoroethane (HFC-125) are investigated in ionic liquid polymer membranes of polyvinyl acetate (PVAc) with three ionic liquids: 1-hexyl-3-methylimidazolium ([C_6C_1im][Cl]), 1-hexyl-3-methylimidazolium bis(trifluoromethylsulfonyl)imide ([C_6C_1im][Tf_2N]), and 1-hexyl-3-methylimidazolium tris(pentafluoroethyl) trifluorophosphate ([C_6C_1im][FAP]). The composite membranes were created with 0, 1, 5, 10, and 20 wt% ionic liquid. The membranes were evaluated for their glass transition temperature, density, molecular composition, and mechanical stability to investigate the effect of increasing ionic liquid concentration. Results of this study indicate that incorporating ionic liquids into polymer membranes can create solubility tunability for the separation of hydrofluorocarbons.

Reducing Physical Aging of Microporous Polymer Membranes through Porous Polymer Network Blending

Aditi Gali, Lucas Condes, Matthew Webb, William Box, Cara M. Doherty, Leoncio Garrido, Tran Le, Jing Deng, Laura Matesanz-Niño, Angel E. Lozano, Cristina Álvarez, Alberto Striolo, Anita J. Hill, Michele Galizia

The Sixth Assessment Report by the Intergovernmental Panel on Climate Change (IPCC) highlighted that global average temperature increase relative to pre-industrial levels will likely exceed 1.5°C by 2050. To mitigate the effects of global warming, efficient and practical carbon capture technologies are urgently needed. Membrane techniques for CO_2 separation from gaseous streams, based on microporous polymers, would greatly enhance the efficiency when compared to conventional, thermal-based separations. However, several issues arise when utilizing membrane materials, one being their rapid physical aging and instability under long-term industrial operations and harsh conditions. Hence, in this study, we evaluated the effect of using relatively inexpensive, under $50 per gram, hyper-crosslinked amorphous, microporous structures, known as porous polymer networks (PPNs), composed of triptycene and isatin. These networks were blended with a model microporous polymer, poly(1-trimethylsilyl-1-propyne) (PTMSP) to properly understand the potential of PPN to decrease the transient free volume collapse of glassy polymer membrane materials. PPN, in particular, was chosen as it is compatible with organic polymers, has high surface areas, and tunable surface properties. Samples of neat PTMSP, 5% wt. triptycene-isatin PPN incorporated into PTMSP (PTMSP-5PPN), and 20% wt. triptycene-isatin PPN incorporated into PTMSP (PTMSP-20PPN) were fabricated. Membranes were characterized using ATR-FTIR and SEM to verify the material morphology and compatibility between PPN and PTMSP. Positron Annihilation Lifetime Spectroscopy (PALS) measurements were additionally performed on aged samples of neat PTMSP and PTMSP-5PPN to measure the pore size distribution changes. Furthermore, physical aging was tracked using N_2 permeability measurements of 20 μm neat PTMSP and PTMSP-20PPN films at 35°C for one month. Permeability and gas uptake of 80 μm samples were measured as well to analyze the effect of PPN on gas permeability and selectivity. Dilatometry measurements of neat PTMSP, PTMSP-5PPN, and PTMSP-20PPN were used to directly measure volumetric contraction of physical aging and make comparisons with permeability measurements. The potential for PPN to hinder the permeability decline due to physical aging is evaluated, and different mechanisms of action for the observed differences are assessed in terms of either a hindered excess free volume collapse by PPN or rearrangement of PTMSP's free volume distribution and interconnectivity due to incorporation of PPN. This study is one of the first addressing, at a fundamental level, the molecular mechanism by which permeability decline due to physical aging can be reduced, opening the doors to a more rational design of membrane systems for large scale, energy-efficient separations.

Exploring Physical Properties of Polyvinyl Acetate + Ionic Liquid Composite Membranes for the Separation of HFC-32 and HFC-125

Irene Xu, Hannah Uhl, Tessie May, Kalin Baca, Abby Harders, Mark B. Shiflett

Refrigerants play essential roles in our everyday lives, from providing comfort to food preservation to industrial processes. However, HFC refrigerant mixtures are currently being phased out due to their high global warming potential. Since most refrigerants mixtures are azeotropic, finding a way to separate these mixtures is necessary to repurpose refrigerants in both a cost effective and sustainable fashion. Two possible methods for the separation of refrigerant mixtures are through membranes and ionic liquids. This poster will give an overview on several physical properties of ionic liquid and polyvinyl acetate composite membranes, which aim to combine the separation abilities of both materials in a single membrane. The membranes were created by mixing three ionic liquids, 1-hexyl-3-methylimidazolium ([C_6C_1im][Cl]), 1-hexyl-3-methylimidazolium, bis(trifluoromethylsulfonyl)imide ([C_6C_1im][Tf_2N]), and 1-hexyl-3-methylimidazolium tris(pentafluoroethyl) trifluorophosphate ([C_6C_1im][FAP]) with polyvinyl acetate in four different weight precents. The densities of each membrane were found using a densitometer, and the membranes were evaluated using a High-Pressure View Cell to find their melting and deterioration points at a certain pressure. Fourier-transform infrared spectroscopy was also used to find the percent transmittance. Results from this study indicate that trends occurring with percent transmittance and density in a specific combination of ionic liquid and polyvinyl acetate may be able to reveal more about the intermolecular interactions between the ionic liquids and polyvinyl acetate.

Packed Bed Fluidization Conditions throughout a 4-Step Pressure Swing Adsorption Cycle

Dylan Ernst

One of key problems that arise when designing large, packed beds is the existence of high velocities that may lead into its mechanical disintegration. When gas is fed through a packing material at such a high velocity, the bed can loosen until the particle-fluid mixture expands and behaves as if it were a fluid. When fluidization occurs, the adsorbent grinds against itself and the wall of the column resulting in poor separation. Additionally, dusting, lifting, churning, and coning of beds have all been attributed to fluidization. In this effort, the conditions for fluidization phenomena in a bed of BAX 1100 and the effect of the amount of weight compressing a bed of BAX 1100 were determined. For this purpose, a single bed PSA unit specifically designed to evaluate pressure drop and fluidization conditions was constructed from top to bottom of 5" of 6 mm glass beads, 65" of BAX 1100, and 5" of 3A Zeolite.

To determine the conditions of fluidization, a 4-step pressure swing adsorption cycle was created. First, the bed undergoes a pressurization (Press) step, wherein the bed is pressurized from the top of the column from atmospheric pressure (14.7 psia) to 70 psia. Next, the bed undergoes a feed step, wherein the bed is fed from the bottom of the column at a constant flow rate at 70 psia while simultaneously flowing gas out of the top of the column. Then the bed undergoes a co-current depressurization (CoD) step wherein the bed is depressurized in the same direction of the feed from 70 psia down to 40 psia out of the top of the column. Finally, the bed undergoes a counter-current depressurization (CnD) step wherein the bed is counter-currently depressurized back to atmospheric pressure through the bottom of the column. These steps were individually tested in the cycle to determine for the Press, CoD, and CnD steps the shortest time the step pressure can switch between the targeted values or for the Feed step at what flowrate the step can be run at without fluidization conditions occurring. Keeping the bed height of 75", runs were tested with 2", 5", and 10" of glass beads while adjusting the height of BAX 1100 accordingly. In the end, Feed fluidization velocity did not appear to be strongly dependent on the weight of glass beads while the CoD fluidization step time did.

Corner Capillary Rise As a Mechanism for Slurry Dewatering

Sara D. Arzate-Lujan, Bibiana Morales-Alvidrez, Risha Shetye, Enrique Ortiz-Nadal, Vignesh Thammanna Gurumurthy, Enrique A. Lopez-Guajardo, PhD., Stephen Garoff, Robert Tilton

Slurries of tailings produced by the mining industry represent an environmental threat not only for their inadequate management, contamination of soils and bioaccumulation of heavy metals, but also for the intensive use of water resources. Dewatering of these slurries is an important factor in reducing the environmental footprint of the mining activity. Partial dewatering is typically carried out by adding a thickening agent, resulting in a concentrated slurry. However, water recovery from concentrated slurries is often energy-intensive. This work explores the use of capillary-driven flows in sharp corners as a passive process for dewatering slurries. That is, capillary flows in the angle formed at the point of intersection between two solid surfaces (corner flow) to reduce the local concentration of the slurry particles within the liquid as it rises in a vertical column. Laboratory tests were conducted to test the premise of the proposed process, using kaolin and dolomite slurries at different concentrations as model systems. The process was analyzed by image processing to determine relative concentration profiles during the slurry rise. Results showed that there is a direct influence of the slurry concentration on the system capability to dewater while affecting the dynamics of liquid rise. At high concentrations, no appreciable particle moved into the corner flow and only water was extracted. Likewise, for both systems it is observed that, as the concentration of the slurry increases the particles tend to agglomerate near the corner region of the system. Therefore, a local decrease was observed in concentration near the contact line (outermost region of the system). Finally, size exclusion near the corner for kaolin was observed due to interparticle interactions and house-of-cards structure. This was tested by adding salt to disrupt structuring in the slurry, which increased the uniformity of the particle distribution in the rising slurry. This research is the first, to the best of our knowledge, to explore corner flow with fine particle systems and provides insights for potential new engineering solutions to enhance slurry dewatering processes.

Degradation and Separation of Perfluorooctanoic Acid (PFOA) from Drinking Water Sources

Maria Rincon Perez, Paola Perez, Oishi Sanyal, Jianli Hu

Perfluorooctanoic acid (PFOA), a perfluorinated alkylated substances (PFAS) has been identified by the US EPA as detrimental to human health. Its consumption is related to thyroid disease, and increased kidney and testicular cancer. PFAS is estimated to be present in the blood of almost all residents of the United States (USGS Report, 2022). PFOA is ubiquitous across many consumer and household items including non-stick and stain-resistant products and in fact, West Virginia waterbodies record 101-349 ng/L PFOA. This project intends to use membrane technology to remove PFOA from the water as well as microwave technology to degrade it and combine these technologies in the future to obtain a one-step separation process. To address the mitigation of ultra-low PFOA levels (ppb to ppt) in drinking water systems, a pre-concentration step of adsorbing the PFOA on zeolite is currently being explored. The PFOA-loaded zeolite is then exposed to microwave irradiation to degrade it into smaller chain fluoro compounds. Successful completion of these steps will be followed by embedding the zeolite on a membrane surface to accomplish the one-step degradation and separation process.

Electrochemical Characterization of Activated Carbon Cloth for Capacitive Deionization

Lily Callen, Lauren Valentino

The global water demand continues to increase as population growth, urbanization, and industrialization intensify. The resulting pressure on limited water resources and environmental concerns about current treatment methods have inspired research and development of alternative separation approaches including electrochemical separations, such as capacitive deionization (CDI). CDI is a low cost and low energy electrochemical separation technology that works by applying a potential of ~1.2 V to a system to remove ions from water. Activated carbon cloth (ACC) is a promising electrode material for CDI because of its electrical conductivity, porosity, and large surface area. However, previous studies show that oxidation of ACC is leading to decreased performance over a long period of time. This work aims to connect CDI performance with quantitative electrochemical properties. Various CDI experiments were performed by cycling an electrochemical cell more than 1,000 times. Using a three-electrode electrochemical cell, cyclic voltammetry was performed to calculate the specific capacitance of pristine ACC in comparison to the post-CDI ACC. Additionally, scanning electron microscopy (SEM) and energy dispersive X-ray spectroscopy (EDS) were performed, and the results are consistent with carbon oxidation. This work shows that cyclic voltammetry is a valuable tool for investigating and understanding the electrochemical properties of ACC electrodes and their connection to CDI performance. Oxidation was not observed following 1,000 CDI cycles with an inorganic salt solution (sodium chloride). However, due to successive voltage cycling and prolonged voltage application in the presence of an organic salt solution (sodium butyrate), a decrease in specific capacitance and an increase in oxygen content were observed. More tests need to be performed to fully understand the conditions that lead to oxidation and the extent of its damage.

Competitive Vs. Synergistic Surfactant Adsorption

Diana Barrios Perez

The adsorption behavior of both nonionic and charged surfactants onto minimally charged nanoplastic surfaces holds potential for advancing the comprehension of antibiotics' removal onto the nanoplastic present in post-cleaned water systems. The aim of this ongoing research is to measure the adsorption of hexadecyltrimethylammonium bromide (CTAB) and varying nonionic ethoxylated surfactants such as tetra- (C12E4) and hexa-methylene glycol monododecyl ether (C12E6) onto polystyrene. Employing the quartz crystal microbalance with dissipation (QCM-D) apparatus, the adsorption behavior of these surfactants onto polystyrene quartz crystal was evaluated to create isotherms as the concentration of the surfactants exceed the critical micelle concentration (CMC). CTAB, a cationic surfactant, caused desorption onto the polystyrene quartz crystal, facilitated by the electrostatic repulsion driven by the slight negative charge the polystyrene surface. This makes it a good cleaning solution. Conversely, the C12E6 isotherm illustrates adsorption onto the crystals surface. Subsequent investigations will encompass more surfactants to facilitate a further understanding of the underlying relationships.

Gas Permeation and Liquid Filtration Cells for Understanding Transport Mechanisms through Covalent Organic Framework (COF) Membranes

Dhruv Singh Tomar

Title: Gas Permeation and Liquid Filtration Cells for Understanding Transport Mechanisms through Covalent Organic Framework (COF) Membranes

In the ever-evolving landscape of membrane technology, understanding the transport mechanisms of novel Covalent Organic Framework (COF) membranes holds great promise. COF membranes have the potential to address the complex challenge of separating the mixtures (e.g., xylene isomers having closely matched boiling points) that are otherwise challenging with other energy-intensive separation techniques (e.g., distillation), making them invaluable for a wide range of separation processes, environmental sustainability, and resource management.

This research, initiated in collaboration with a multidisciplinary team, focuses on elucidating the transport mechanisms through COF membranes. It is important to note that this project is in its early stage, and the researcher has recently joined the team and an important milestone has already been reached: the construction of a gas permeation cell following the ASTM2476-20 standard. This cell is now being employed to explore the behavior of gas permeance of COF membranes under various pressure conditions and gas mixtures.

The gas permeation cell's adherence to ASTM standards ensures the rigor and accuracy of our experiments as we understand the fundamental COF membrane transport properties. Through systematic testing of different pressure conditions and gas mixtures, we aim to unravel the underlying mechanisms governing gas permeance through COF membranes. These insights are vital for the development of more efficient gas separation processes (e.g., xylene isomer separation) and applications.

Looking to the future, our research is poised to expand into liquid filtration cells. For this, we will utilize TpHz COF membranes that are readily available, thanks to the work conducted in Dr. Kailong Jin's lab. These liquid filtration cells will enable us to study the separation of dyes and salts, a task of critical importance in various industrial and environmental applications.

Overall, this abstract offers a glimpse into the early stages of our research journey, underlining the importance of gas permeation cells and the promise of forthcoming liquid filtration studies. As we proceed, we aim to unlock the full potential of COF membranes, contributing to innovative solutions in separation science and technology.

Simulations to Predict Temperature and Plasticization Effects for Gas Separation in 3D Canal Ladder Polymers

Brandon Tapia, Jing Ying Yeo, Francesco Maria Benedetti, Zachary P. Smith

Approximately 10-15% of the world's energy consumption is used for chemical separations via traditional distillation and absorption methods. Gas separation research focuses on membrane-based technologies to reduce these energy requirements by up to 90%. Industrial membranes currently fall well-below the Robeson upper bound due to high selectivity's required for industrial useful separations, yielding low permeabilities. Polymers of intrinsic microporosity (PIMs) have been used to create membranes that have pushed the limits of the Robeson upper bound due to their rigid and contorted structure that leads to poor packing—leading to high permeability but selectivity's below industrial separation requirements. Previously, our group reported PIMs made of fused norbornyl benzocyclobutene repeat units made via catalytic arene-norbornene annulation (CANAL) with 3D contortions that exhibit strong permselectivity. Specifically, CANAL-Me-Me_2F has a selectivity two times higher and a permeability 100 times higher than cellulose acetate membranes for CO_2/CH_4 separations.

This work presents the first simulated results for system design, material properties, and sorption for CANAL polymers. Simulated CANAL-Me-Me_2F was validated against experimental with bulk and skeletal densities, wide angle x-ray scattering (WAXS), and small gas adsorption isotherms. Nonlocal density functional theory (NLDFT), a conventional but indirect method of obtaining pore size distribution (PSD), is seen to be unable to detect the angstrom-sized pores due to experimental limitations. To understand the crucial importance of size-sieving in gas separation membranes, geometric PSD calculated via these simulations is seen to be a far more physically consistent approach. Sorption of plasticizing gases was accurately simulated by integrating molecular dynamics into iterative grand canonical Monte Carlo (GCMC) simulations. Sorption at temperatures inaccessible to experimental determination, but relevant for industrial separations, was used to examine the unique physical aging behavior of CANALs. These methods allowed for generation of sorption coefficients without the necessity of the sorption-diffusion framework. Dual-mode sorption (DMS) models were used to predict multi-gas sorption selectivity. The physically consistent results suggest that these simulations are a vital companion to experimental results, providing previously inaccessible data, validating assumptions, and suggesting new experimental directions.

Monovalent Ion Transport in Sulfonated Poly(arylene ether sulfone) Membranes

Beatrice Tremblay, Sean Bannon, Geoffrey Geise, Mou Paul, Abhishek Roy

Currently, 1.42 billion people - nearly one-fifth of the world's population - live in regions experiencing high water vulnerability. As the climate emergency and population growth increase the strain on freshwater systems, it is clear that alternative methods of obtaining clean water, such as seawater desalination, are needed to supplement existing sources. Thermal methods for filtering water, e.g., multi-effect distillation, are highly energy intensive and, because of this, cannot realistically meet modern needs. As such, current water purification systems rely on polymer membranes, which enable lower cost, higher efficiency, and lower energy filtration. This said, effective, mechanically and chemically robust, and eco-friendly polymer membranes are needed.

Modern membrane filtration technologies predominantly use polyamide-based thin film composite membranes, which boast impressive separation capabilities, but are susceptible to degradation, most notably via chlorine attack. As such, researchers have looked for alternative chemically stable polymers, e.g. polysulfones. These polymers exhibit several appealing traits: high mechanical and thermal stability, solubility in common organic solvents, and resistance to oxidative attack. Unfortunately, non-functionalized polysulfone membranes are quite hydrophobic leading to low water flux, which hinders performance in water filtration applications. To combat this issue, polysulfones can be sulfonated to increase the hydrophilicity and, in turn, the water flux, which is coupled to the productivity of the membrane in a desalination application.

This presentation reports experimental characterization of a series of sulfonated poly(arylene ether sulfones). The ion separation properties of these materials are of increasing interest due to a shift in focus from simply salt rejection, which has been used for decades to characterize desalination membranes, to the use of membranes for ion fractionation (or ion specific separations). Focusing primarily on transport properties, we discuss the impact of polymer sulfonation and counter-ion form on competitive monovalent ion transport. Specifically, we report sodium (Na^+) and lithium (Li^+) transport data for a series of sulfonated poly(arylene ether sulfone) (BPS) polymers to inform the use of such materials in emerging ion separation applications that could be important for critical materials extraction, recovery, and recycling.

Selective Ion Transport through Copolymer Membranes Functionalized with Imidazole Ligands

Evan Wood, Jonathan Ouimet, William Phillip

Lithium-ion batteries are crucial for the transition to green energy. Recycling lithium-ion batteries is increasingly important to safely dispose of this hazardous waste as well as to supplement the supply of valuable and rare metals such as cobalt and lithium. Nanofiltration membranes can separate dissolved molecules from each other through a combination of electrostatic and steric hindrance in an energy-efficient manner. While small molecules pass through these membranes, larger particles cannot and thus a separation occurs. If membrane ligands that selectively interact with one metal over another could be developed, then these membranes could separate similarly-sized and similarly-charged ions from each other. Functionalized membranes could then replace complex and waste-intensive solvent extraction methods (hydrometallurgy and pyrometallurgy) used in lithium-ion battery recycling processes. This project aims to create metal-selective membranes by using imidazole ligands to selectively partition a target metal into a copolymer membrane–through ligand-metal complexes–while excluding metals with lower affinity for the ligand.

Ligands with an imidazole moiety were investigated due to their well-known complexation with transition metals. To begin, the strength of the interaction between the ligand and metal was measured in solution. Specifically, propargyl 1h-imidazole-1-carboxylate was dissolved in water and mixed with varying levels of copper or cobalt. The absorption change in the ultraviolet-visible light spectra correlated to the imidazole complexing with copper while no spectroscopic effects were observed in this region of light for cobalt-imidazole complexes. The imidazole ligand was attached to the membrane through a copper-catalyzed azide alkyne coupling (CuAAC) click reaction. The success of the reaction was confirmed using Fourier-transform infrared spectroscopy. X-ray fluorescence spectroscopy was used to demonstrate that membranes exposed to a solution with equimolar concentrations of both copper and cobalt, bound copper selectively. Copper diffusion cell experiments demonstrated that the imidazole-functionalized membranes had twice the copper permeability as non-functionalized membranes. Diffusion cell experiments with equimolar mixtures of copper and cobalt demonstrated modest selectivity which we attribute to the large pore size of the current membrane.

Current membrane pore sizes–around 3 nm diameter–were quantified using neutral solute rejections experiments with poly(ethylene) glycol molecules of varying size. This study will continue to alter different aspects of the membrane fabrication process to produce membranes with smaller pores.

AUTHOR INDEX

AIChE
120 Wall Street, FL 23
New York, NY 10005-4020

ISBN 978-1-7138-9322-6

CURRAN ASSOCIATES INC.
proceedings.com